Wine
A Global Business

Second Edition

Edited by

Liz Thach
and
Tim Matz

Wine
A Global Business

Second Edition

Copyright © Miranda Press 2008

No part of this publication may be reproduced, stored in a retrieval system, or transmitted in any form or by any means, electronic, magnetic tape, mechanical, photocopying, recording, or otherwise, without permission in writing from the publisher.
The publisher and the publisher's agents represent that the data provided were formulated with a reasonable standard of care. Except for this representation, the publisher makes no representation or warranties, expressed or implied.

Miranda Press
An imprint of Cognizant Communication Corporation

U.S.A.   18 Peekskill Hollow Road, Putnam Valley, NY 10579

Library of Congress Cataloging-in-Publication Data

Wine : a global business / edited by Liz Thach and Tim Matz. -- 2nd ed.
   p. cm.
Includes bibliographical references and index.
ISBN 978-0-9715870-3-8 (alk. paper)
 1. Wine industry. 2. International trade. 3. Globalization--Economic aspects. I. Thach, Liz, 1961- II. Matz, Tim, 1959-
   HD9370.5.W555 2008
   663'.20068--dc22
                                  2008034459

Printed in the United States of America

Printing: 1 2 3 4 5 6 7 8 9 10 Year: 1 2 3 4 5 6 7 8 9 10

About the Authors

Dr. Liz Thach (pronounced "tosh") is a management and wine business professor at Sonoma State University in the Wine Business Center where she teaches in both the undergraduate and MBA wine programs. In addition, she has 15 years of executive and management level experience at Fortune 500 companies and has done consulting and research projects for more than 20 different wineries. Liz has a doctorate in Human Resource Development from Texas A&M University and has traveled to most wine regions of the world. She has published over 80 articles, the 1st edition of *Wine: A Global Business* and *Wine Marketing & Sales*. A fifth generation Californian, Liz finished her Ph.D. at Texas A&M and now lives on Sonoma Mountain where she tends a small hobby vineyard and makes homemade wine.

Tim Matz is president of Jackson Wine Estates International where he is responsible for the operations, sales, marketing and finance for all of Kendall Jackson Wine Estates International business globally. Prior to KJ, Tim was Vice-President & General Manager for Imports at Beringer-Blass Wine Estates where he managed all of their international wine brands in the US and Canada. Other experience includes heading up Sales and Marketing at Southcorp before moving into the position of President of Americas. Tim started his career in the wine industry in 1985, with Brown-Forman with management and executive positions in Sales, Marketing, Strategic Planning, and International. He has an MBA from University of Kentucky and a BBA from Kent State University.

Contents

Foreword ix
James Halliday

Acknowledgements xi

Chapter 1. The Landscape of New & Old World Wine 1
Denis Gastin and Mack Schwing
  Defining New World Wine
  The Drivers
  The Key Players
  The Old World Responds
  Opportunities and Challenges
  Conclusion

Chapter 2. Wine Business Strategy 17
Liz Thach
  Defining Wine Strategy
  The Wine Business Strategy Development Process
  When, How Long, and Who Should Be Involved in
    Strategy Development
  Prerequisite for Strategy Development Process:
    A Wine Business Concept
  External Analysis Phase
  Internal Analysis Phase
  Strategic Development Phase
  Strategy Implementation and Evaluation Phase
  The Strategy Reassessment Phase
  Conclusion

Chapter 3. The Business of Viticulture 63
*Richard Thomas, Mark Greenspan, David Beckstoffer,
and Liz Thach*
  Vineyard Selection
  Planting the Vineyard
  Preparing the Land

Vineyard Management
Important Issues in Viticultural Business
Vineyard Management at Beckstoffer Vineyards

Chapter 4. The Business of Enology   **95**
Linda Bisson, Roy Thornton, and Peter Gago
    Business Decisions in the Winemaking Process
    The Use of Multiple Labels in Winemaking
    The Role of the Winemaker at Penfolds
    New Technologies in Winemaking
    Current and Future Issues in Enology
    Conclusion

Chapter 5. Wine Supply Chain Management and Quality Control   **123**
Thomas Atkin and Jon Affonso
    What Is a Supply Chain?
    Determining Your Core Competencies
    Commodity Strategy
    Major Supplies in the Wine Industry
    Quality Control Systems Impacting Supplies
    Trends in Supply Chain Management
    Conclusion

Chapter 6. Marketing and Branding Wine   **155**
Linda Nowak, Paul Wagner, and Jean Arnold
    Defining Marketing and Branding
    The Five Ps of Wine Marketing
    The Importance of Branding
    Conditions Favorable for Successful Branding
    Know Thy Customer: Wine Consumer Segmentation
    Wine Brand Positioning
    Being Consistent With Brand Image
    Characteristics of a Good Brand Name
    Building Brand Loyalty
    The Challenges of Taking a Brand International
    A Tale of Two Brands
    Summary

Chapter 7. Wine Distribution   177
Bruce Herman and Gary Long
    Wine Distribution on a Global Basis
    Wine Distribution Outside the United States
    Wine Distribution in the United States
    The Role of the Wholesaler
    The Role of the Broker
    Finding a Distributor and Broker for Your Wine
    Working Successfully With Distributors and Brokers
    A Day in the Life of a Distributor Sales Rep
    Conclusion

Chapter 8. Professional Wine Sales   197
Armen Khachaturian and Elizabeth Rice
    Defining Professional Wine Sales
    Selling Wine in the Three Major Channels of the US Market
    Creating the Annual Sales Plan
    Implementing the Sales Plan
    Evaluation and Revision to the Wine Sales Plan
    Challenges in Selling Wine Internationally
    Exporting Wine From the US to International Destinations
    Successful International Selling Tips
    Conclusion

Chapter 9. Direct Wine Sales and Wine 2.0   219
Janeen Olsen and Josh Hermsmeyer
    Advantages and Challenges of Direct Wine Sales
    Channels for Direct Wine Sales
    Compliance and Regulation Issues With Direct Wine Sales
    The Advent of Wine 2.0
    Conclusion

Chapter 10. Wine Tourism   235
Donald Getz, Jack Carlsen, and Liz Thach
    Defining Wine Tourism
    Benefits of Wine Tourism

Motivations of Wine Tourists
The Experiential Appeal of Wine Regions
The Wine Tourism Development Process
Case Studies in Wine Tourism
Conclusion

Chapter 11. Wine Media and Public Relations   259
Megghen Driscol, Tim Matz, and Tor Kenward
Defining the Role of Wine Media
The Importance of Wine Media
Role of the Wine Journalist
Wine Scores and Rating Systems
Defining Wine Public Relations
Public Relations Role
Relationship Between the Wine Reporter and the Public Relations Professional
Communication Vehicles
The Future of Wine Media and Public Relations

Chapter 12. Global Marketing and Exporting   273
Larry Lockshin and Tony Spawton
Key Trends in the Global Wine Sector
Success Strategies for Global Marketing
Definitions and Examples of Exporting
Examples of Three Australian Exporters
Tactics for Exporting
Export Readiness Checklist
Branding the Export
The Export Value Chain and Logistics
Conclusion: Future Predictions for the Global Wine Market

Chapter 13. Financial Aspects of Wine   313
Robert Eyler and Tony Correia
Financial Foundations
Net Present Value
Financial Decisions in Vineyard Projects
Real Estate Markets and Winery Investments

    Winery Economics and Finance
    Segment Analysis
    Equity Versus Debt Financing, Mergers and Acquisitions
    Conclusion: How Wine Financing Is Different

Chapter 14. Wine Accounting and Tax   333
Jeff Sully, Terry Lease, and Jon P. Dal Poggetto
    Product Costing: Introduction
    Produce Costing: Defining Cost Centers
    Product Costing: Using Cost Centers
    Costing the Vineyard
    Costing the Winery
    Tax Issues in the US Wine Industry

Chapter 15. Managing Human Resources in the Wine Industry   359
Liz Thach and Lillian Bynum
    Wine Labor Issues Through the Centuries
    Historical Impact on Wine Labor Issues Today
    Common Employee Positions in a Vineyard/Winery
    Defining Human Resource Management in the
        Wine Industry
    Future HRM Issues

Chapter 16. The Legalities of Wine   385
Cyril Penn and Wendell Lee
    The Historical Context of US Wine Legalities
    Taxes and Compliance
    Trademark and Label Issues
    Farming and Land Use Issues in the US
    A Global View on Wine Legalities
    Trade Issues
    Shipping Internationally

Chapter 17. Environmental and Social Responsibility Issues   401
Jeff Dlott, Karen Ross, Allison Jordan, and Kari Birdseye
    Defining Sustainability

California Sustainable Winegrowing Program
Overview of International Sustainable Winegrowing
   Efforts
Sustainable Winegrowing Practices: The Next 5 Years

Chapter 18. Looking Toward the Future   **417**
Paul Dolan, Tim Matz, Liz Thach, and Richard Cartiere
From Supplier to Consumer Driven
Trend 1: Rising Interest in New Varietals From Around
   the World
Trend 2: Increased Environmental Concerns
Trend 3: Focus on Health and Responsible Drinking
Trend 4: Thirst for Innovative Products
Trend 5: Desire for Both Value and Premium Wines
Trend 6: Trust in "Word of Mouth"
Trend 7: Increased Online Purchasing
Trend 8: Desire for Experiences Rather Than Possessions
Future Forecasts for the Global Wine Industry

Contributors   435

Bibliography   449

Index   459

List of Figures

1.1	Map of major New World wine countries	4
2.1	Major phases of strategy development process	22
2.2	The wine business strategy star	22
2.3	Example of strategic group map for luxury tannat winery	28
2.4	Generic wine industry value chain	36
2.5	Example of grape grower value chain	37
2.6	Five major strategies adapted to the wine industry	48
2.7	Example of contents of a strategy document	58
3.1	Cyclical nature of wine industry	71
5.1	Supply chain management	125
5.2	Supply chain decision matrix	128
5.3	RFID data flow system	153
6.1	US wine consumer segmentation by shopping behavior	165

7.1	Distribution process in US open and control states	184
7.2	Distribution channels	191
8.1	Three major sales channels for wineries	199
9.1	Direct wine sales channels	222
9.2	Social media components	231
10.1	Wine tourism development process	244
10.2	Lake Okanagan wine region	249
10.3	Blending seminar at Sandaford Winery	254
12.1	Global market shares of the major beverage categories	280
13.1	Average prices, grapes/ton, selected California districts	318
14.1	Vineyard accounting worksheet	343
14.2	Facilities allocation summary	345
14.3	Facilities allocation	346
14.4	Example of bottling information worksheet	352
15.1	The five categories of human resource management	373
16.1	The three-tier wine distribution system in the US	388
17.1	The three Es of sustainability	403
18.1	How consumers select wine in retail stores	425

List of Tables

1.1	New Versus Old World Wine	3
1.2	The major wine players	6
1.3	Top global wine brands	10
1.4	Top 10 world wine-producing countries	11
2.1	Top 10 wine selling companies	18
2.2	Levels of strategy within a wine business	20
2.3	Macroenvironmental analysis questions	31
2.4	Example of Porter's five forces	34
2.5	Critical success factors for grape growers	40
2.6	SWOT considerations for wine businesses	41
2.7	Steps 2 and 3 of SWOT analysis	42
2.8	A company's MVV	45
2.9	The five strategies defined	53
3.1	Famous grape varietals in New World countries	67
4.1	Standard wine price categories	97
4.2	The crush process	99
4.3	The five "Ss" of wine appreciation	108
4.4	Sample bottling costs for standard 750-milliliter bottle	109
5.1	Common winemaking supplies	137

6.1	Defining wine marketing and branding	157
8.1	Checklist of decisions for annual sales plan	202
10.1	Benefits of wine tourism for wineries	237
10.2	Top 10 motivations of wine tourists	240
10.3	Wine tourism development partners	243
10.4	List of wine tourism programs	245
11.1	Three major wine scoring systems	264
12.1	Six major trends driving globalization of the wine market	276
12.2	Top exporters of wine in the world, 2003	277
12.3	Definitions of indirect and direct wine exporting	284
12.4	Major considerations when branding for export	298
12.5	Promotion strategies in wine export markets	306
13.1	Vineyard cost categories	317
13.2	Winery cost categories	326
13.3	Equity versus debt financing	327
13.4	Selected winery and vineyard property acquisitions, 2004–2007	328
13.5	Winery stocks and ticker symbols	330
14.1	Example of basic cost allocation method	338
14.2	Example of cost allocation by block	339
14.3	Example of cost allocation with different block techniques	340
14.4	Example of cost allocation based on block farming costs	341
14.5	Example of per ton calculation	342
14.6	Bulk wine cost summary	348
14.7	Example of capitalized interest	355
14.8	Comparative depreciable lives	356
14.9	LIFO calculation example	356
15.1	High-level job categories in vineyard/winery	365
17.1	The 14 chapter topics in the Code of Sustainability	405

Foreword

The first edition of this outstanding book was published in 2004, the scope of its content breathtaking. It must have been akin to the labors of Hercules for editors Liz Thach and Tim Matz to harness the 41 authors who (in teams of two or three) wrote the 18 chapters of this book. The experience and skills of those authors come from all parts of the New World, and their pre-eminence in their particular fields shines through the chapters they respectively contributed.

Why, then, is it necessary for a second edition to be published so soon after the first? Much of the basic thrust remains unchanged, even if statistics and rankings have altered. Does it really matter that Yellow Tail had barely taken flight in 2004? Even if it does, there will be other such brands emanating from South America and (probably) the Iberian Peninsula in the next decade.

The answer to the question lies in the rapidly shifting sands of the global wine market. In 2004 the remorseless march of Australian exports seemed certain to continue into the foreseeable future. It was possible to construct an argument that it would overtake Spain, Italy, and France as the largest exporter (by value, if not volume) in the second half of the century. A combination of the most severe drought in the last 100 years and the impact of climate change has dramatically changed the dynamics of the Australian industry; which of the two events is most responsible for the predicament and possibly catastrophic fate of grape growers along the Murray River is irrelevant. The fact is this always water-starved continent has moved into uncharted territory, and requires a new whole-of-industry strategy if it is to maintain its credibility as a major player in the export league, and forget any thoughts of toppling the major players of today or the future. Chile and Argentina are now the favorites in those sweepstakes.

The other dramatic change since 2004 has been the emergence of the long-awaited mainland Asian markets, with China at the head, India still to turn dreams into reality. Here events unfold daily, so vast have been the changes in demand and the building of distribution infrastructure. It is ironic that Australia is theoretically best placed to take advantage of the spoils at the very time its supply chain has moved with equal speed from short-term surplus to medium-term (perhaps long-term) shortage. Others, however, will happily move to supply the demand, and should do so swiftly.

In order to capture all of these changes, the editors have not only updated statistics, but added three new chapters on wine tourism, professional selling, and a strategy development process. Today this book is still the only comprehensive book on the global wine industry. It does an excellent job at providing an overview of the essentials of wine business by touching on all of the major aspects from strategy and production to marketing, accounting, and many other topics. It is extremely useful for wine students, as well as newcomers to the industry and anyone currently working in wine who wants to gain a broader perspective on the various functions of a successful wine business.

Most Sincerely,
James Halliday

Acknowledgments

We are very grateful to all of the people who helped make this 2nd edition of *Wine: A Global Business* a reality. It took over 16 months to obtain all of the updates for each chapter, plus recruit writers for three new chapters in this edition. We extend our deep appreciation to every contributor for their efforts.

We also want to extend a very special thanks to James Halliday for taking the time to write such a compelling Foreword. His beautifully crafted words add style and authority to the book, as well as give the foundation for this new edition.

We are pleased to extend a special thanks to our publisher, Robert Miranda of Cognizant Communication Corporation, and to Christine Stagg, his highly responsive and detail-oriented production manager. Without their encouragement and excellent follow-up skills, this book would have been very challenging to complete.

It would be remise not to acknowledge our families as they supported us through many long days and nights typing away at the computer, as well as providing helpful advice, assistance, and reassurance.

Finally a very special thank you and acknowledgement to all of our supporters in the wine industry who provided quotations, as well as the many students and professors around the world who are using this book in classes.

Most Grateful,
Liz Thach and Tim Matz

LEARNING OBJECTIVES:
- Differentiate between New and Old World wine
- Identify the drivers of New World wine
- Describe how the Old World has responded
- Describe opportunities and challenges for the global wine industry

CHAPTER 1

THE LANDSCAPE OF NEW & OLD WORLD WINE

Denis Gastin
Australian Wine Writer

Mack Schwing
Co-founder, WISE Academy LLC

What is New World wine? In a word, New World wine represents *freedom*. It means freedom to imagine new wine styles and innovative varietal combinations, to experiment with new practices in the vineyard and new winemaking methods, and, above all, freedom to interpret what the drinker wants and then to go ahead and do it. In this world, the market is the ultimate arbiter. If the consumer likes your product, you have a business; if not, you are on your own.

This chapter introduces the concept of a New World wine model. It provides some basic definitions and identifies the drivers. It discusses formative trends, presents some important statistics, and discusses the response of Old World Wine players. It concludes with an overview of the challenges and opportunities in the current and future landscape for New World wine.

DEFINING NEW WORLD WINE

When asked what New World wine represented to her, Jancis Robinson, one of the world's most widely published wine commentators (www.jancisrobinson.com) and editor of the *The Oxford Companion to Wine*, as well as co-editor of the *World Atlas of Wine*, described it as follows:

> Without an inherited model such as the centuries of tradition in European wine production, the New World has been able to identify what's important for business survival: successful selling of a product designed for the consumer rather than the producer. So sought-after grape varieties have been planted and then farmed efficiently, often sold to large wine producers with the muscle and sophistication required to actively sell those products into the major retailers, wherever in the world they may be. From my perspective in Britain, the stereotypical contrast is between one of the dominant Australian wine companies that parachutes its sales force into the UK supermarkets' buying offices with all the equipment needed (targeted price points, regular promotions, vast marketing budgets, etc.) to secure long-term co-operation, on the one hand, and, on the other, the typical French vigneron, one of tens of thousands, who makes wine more or less as his father did and waits for potential customers to drive up. (D. Gastrin, personal email interview, August 2003)

Free of the burden of history, liberated from geographic boundaries, and uninhibited by onerous regulations designed principally to preserve tradition, New World wine producers have drawn the most out of what the grape has to offer in the finished wine. They have experimented creatively with new grape varieties—and indeed have commercialized their own signature varieties. They have experimented judiciously with varietal blending, and have found inspired flavors and textures forbidden in the Old World (Table 1.1). They have pioneered new viticultural and winemaking techniques and technologies that have been replicated far and wide, including, more recently, in the traditional heartland of wine.

Most importantly, the consumer is in their sights from the vineyard to the table—not just at the *end* of a heavily regulated process. And the consumer has rewarded this diligence handsomely—initially in their home markets where consumers were enticed away from other beverages to wine, but in the traditional wine markets too, where consumers have been increasingly attracted by the alternatives on offer from the New World producers (Figure 1.1).

The leaders in this wholesale change to the world of wine were Californian, quickly followed by Australians, then more recently New Zealanders, Chileans, Argentineans, and South Africans.

On the back of the successes by the New World pioneers, wine is now being exported to world markets from countries that would never have contemplated it even a decade

Table 1.1. New Versus Old World Wine

Old World	New World
Traditional taste styles & methods	Designed for consumer; innovative tastes
Less fruit & oak; more subtle; minerality	More pronounced fruit & oak
Major countries: France, Italy, Spain, Portugal, Germany, Austria, (Eastern Europe)	Major countries: US, Canada, Australia, New Zealand, South Africa, Chile, Argentina, (Asia)

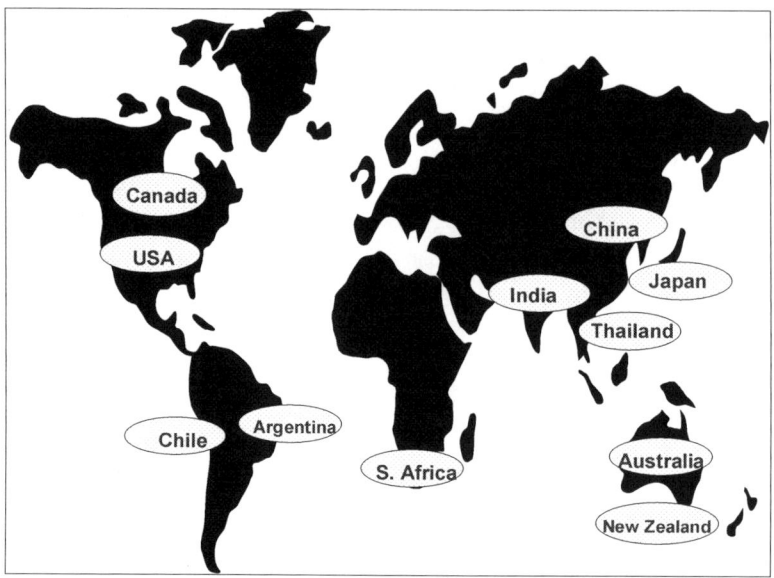

Figure 1.1. Map of major New World wine countries.

ago—like Mexico, Peru, and Uruguay. And in this "emerging New World" there are some dazzling new wine experiments and nascent new players—like China [now, according to the Office International de la Vigne et du Vin (OIV, 2008), the world's sixth largest wine-producing nation], and even India and Thailand.

THE DRIVERS

Historically, the drivers of the New World wine phenomenon were closer to the soil than to high commerce. They were farmers looking for new crops or ways to create value beyond traditional farming. But it was also, fundamentally, influenced by immigration. Settlers and clergy from the "old" countries still wanted to enjoy the pleasures and traditions of wine in their adopted homelands: the only way to do this, in most cases, was to make it themselves. And from this impulse many businesses were spawned.

But no business is a success without a market. Few businesses are "born global," and New World wine businesses are no exception. The domestic market was the core of the original business rationale. In fact, it is really only in the last 30 years that New World wine producers began to make an impression on a global market scale. But what began as a market extension maneuver quickly became an industry phenomenon.

Corporatization of the Wine Sector

The growth of family businesses is one thing, but the corporatization of an entire industry sector is quite another. This has been perhaps the most distinguishing feature of the New World business model: despite their relatively recent origins, 8 of the world's top 14 producers and marketers of wine in 2006 were New World corporations (Table 1.2). Furthermore, one of the European-based entities, Pernod Ricard, has the majority of its wine operations in the New World (Australia).

Corporatization of the industry brought an overlay of technological, business, financial, and organizational skills on Old World tradition—without extinguishing it. This has been reinforced by regulatory frameworks that permit and even encourage change. At the basic level this has introduced:

- Broad acre cropping, adapting practices and techniques previously developed for grains and other crops to a traditionally small-scale agricultural industry.
- Adding mechanization to tradition in picking and processing grapes.
- A constant striving for productivity enhancement (e.g., through clonal selection and innovations in production and storage equipment).
- Focused market research and deliberate market responsiveness.

Table 1.2. The Major Wine Players

			2006 Wine Sales	
Rank	Company	Base	9-L Cases	Wine Revenues in US $ (million)
1	Constellation	USA	100	2,756
2	E.&J. Gallo	USA	73	1,430e
3	Fosters	Australia	41	N/A
4	Pernod Ricard	France	40	1,620e
5	Castel Frères	France	38	1,290
6	The Wine Group	USA	35	1,000
7	Bacardi	USA	25e	650ee
8	Penaflor	Argentina	22	130
9	Concha Y Toro	Chile	21	405
10	Schloss Wachenheim	Germany	21	N/A
11	Grands Chais de France	France	21	N/A
12	Distell	South Africa	19	450e
13	Henkell	Germany	17	609
14	Caviro	Italy	15	N/A

Source: Rabobank International (2007). Reprinted with permission.

- Effective supply-chain linking of production through to retail.
- Modern distribution practices.
- Conscious brand-building cachet.
- Effective exploitation of media as a business resource.
- Open-market capitalization.
- Creating "shareholder value."

Naturally, corporatization has also broadened the definition of "market." Whatever the limits to the market ambitions of farmers, corporations see their market as global. However, in spite of this trend to corporatization, the wine industry remains highly fragmented as only a relatively small proportion of the world's wines is produced by the largest wine companies. Most of the world's wine is still produced and sold by small, and generally family owned and operated, wine companies. This will continue to change as consolidation occurs in the future, but there will always be a large number of small, passionate producers of high-quality wine.

Constructive Industry Regulation

Regulation has also been a positive factor supporting the internationalization of the New World wine industry. In some cases this has amounted to officially conceived and validated quality assurance. In the Australian model, for example, the Australian Wine and Brandy Corporation (AWBC) applies a quality control system to reassure consumers (and the industry itself) that faulty or low-quality wine, or indeed dishonorable exporters, will not be allowed to damage their confidence in "brand Australia." No wine can be exported from Australia other than by a licensed exporter, and, as licenses are renewed annually, any holder behaving in any way that could bring the "brand" into question can quickly be removed.

Furthermore, no wine can be exported without an export approval number issued by the AWBC after the wine meets basic technical standards and is individually assessed and approved by an AWBC sensory assessment panel. The regulations are certainly onerous, but widely applauded by the industry. Although the AWBC is the industry regulatory body created by federal statute, it is jointly administered with and funded by the industry. The effect of this model is that maximum creative freedom is provided for in the growing and production phase but the consumer is totally protected at the end point of wine in the bottle.

The Emergence of New Signature Varieties

Consumer imagination (and, ultimately, consumer loyalty) was stimulated by winemakers doing vastly different things with classic Old World grape varieties (such as Shiraz/Cabernet blends in Australia, Fume Blanc in California) and discovering new styles with less well-known varieties from

Europe (such as Verdelho as a dry white table wine in Australia and Petite Sirah/Durif in California and Australia).

But consumer *loyalty* was most effectively forged with special "signature" varieties that have come to typify particular wine countries or wine regions: the locally evolved varieties, such as Zinfandel in California and Pinotage in South Africa, or the reinterpretation of some of the more obscure varieties, such as Carmenère in Chile, Malbec in Argentina, and Tannat in Uruguay. Even in Japan, a minor wine-producing nation, there is a new respect being won for its very own *vinifera* grape, Koshu, and determined efforts to build consumer loyalty behind the indigenous *amurensis* (wild mountain grape).

A more recent phenomenon has been the willingness of consumers in traditional wine markets to see beyond a generic national picture in New World wine countries and to delve down into the regional detail where particular specialties or specializations can be found. In Australia, for example, AWBC, the industry's governing body, now formally recognizes 103 zones, regions, and subregions across the continent under its Geographic Indications System introduced in 1993 (www.awbc.com.au).

As of April 2007, the US government recognizes 188 American Viticultural Areas (AVAs) or appellations (Alcohol and Tobacco Tax & Trade Bureau, 2007). These are in addition to the approved geographic AVAs, which include all state and counties in the US. The largest AVA in the US is the Ohio River Valley AVA. It covers over 26,000 square miles and includes land within the states of Indiana, Ohio, Kentucky, and West Virginia. The largest AVA in California, the state in which half of the nation's wineries are located, is the Central Coast AVA with 6250 square miles. The smallest AVA is Cole Ranch, in California. It covers a little less than a quarter square mile (www.wineinstitute.com).

The Key Players

Corporatization of the wine industry business model had its origins in California, with the application to wine of the same business logic that had earlier transformed food processing—including the development and application of new technologies, the exploitation of economies of scale, national marketing and promotion strategies, and brand building. The early models included Mondavi, Gallo, and Sebastiani—family businesses that grew and transformed into more conventional corporate entities. This was relatively recent; it began in the 1970s and gathered pace quickly through the 1980s and 1990s.

A similar pattern emerged in Australia in the early 1980s when nonindustry money first appeared, with the takeover of the renowned Penfolds wine business by the Adelaide Steamship Company, later to become Southcorp. BRL Hardy was another example, created when the Hardy family business merged with the large bulk wine operation of the largely grower-owned Berri Renmano and then listed on the Australian Stock Exchange, the first wine company to do so. Another example was the management buy-out of the Orlando wine business and, subsequently, its acquisition by French multinational Pernod Ricard—looking for a preestablished distribution base in Australia for its range of beverages.

Each of these companies, following further expansion and consolidation, and aggressive international market growth, grew quickly to join the ranks of top wine producers globally. In turn, as global wine companies, they have developed global brands. The top 20 brands (Table 1.3) account for almost 7.2% of the global market for bottled still wine. This trend is particularly important for the New World wine companies in the top 20, as they, alone, account for 5.98% of the total market (Shanken, 2007).

Table 1.3. Top Global Wine Brands

Rank	Brand	Company	Origin	Millions of 9-L Case Depletions, 2006	Percentage
1	Franzia	The Wine Group	USA	24	0.95%
2	Gallo/E.&J. Wine Cellars	E.&J. Gallo Winery	USA	23.7	0.94%
3	Carlo Rossi	E.&J. Gallo Winery	USA	12.6	0.50%
4	Tavernello	Eavior Societa Cooperativa arl	Italy	12.6	0.50%
5	Yellow Tail	Casella Wines	Australia	10.5	0.42%
6	Beringer	Foster's Wine Estates	USA	8.9	0.35%
7	Concha y Toro	Vina Concha y Toro SA	Chile	8.6	0.34%
8	Almaden	Constellation Brands	USA	8.4	0.33%
9	Sutter Home	Trinchero Family Estates	USA	8.2	0.33%
10	JP Chenet	Les Grands Chais de France	France	7.5	0.30%
11	Jacob's Creek	Pernod Ricard	Australia	7.4	0.29%
12	Hardy's	Constellation Brands	Australia	7.3	0.29%
13	Woodbridge	Constellation Brands	USA	7.2	0.29%
14	Riunite	Cantine Cooperative Riunite Scrl	Italy	6.4	0.25%
15	Peter Vella	E.&J. Gallo Winery	USA	5.5	0.22%
16	Lindemanns	Foster's Wine Estates	Australia	5.3	0.21%
17	Charles Shaw	Trader Joe's	USA	4.6	0.18%
18	Blossom Hill	Diageo	Italy	4.3	0.17%
19	Inglenook	Constellation Brands	USA	4.2	0.17%
20	Rosemount Estate	Foster's Wine Estates	Australia	4.1	0.16%
Total				181.2	7.20%

Source: Adapted from Shanken (2007).

The early successes of the Californian and Australian wine corporations inspired similar business ventures in other New World locations—most notably in South Africa, following the progressive deregulation of the industry and the removal of the special status that had protected the national industry marketing monopoly (KWV), but also in Chile, Argentina, and New Zealand.

Table 1.4 shows that a significant and growing proportion of world wine trade and worldwide wine production is accounted for by New World wine producers. In 2007 over one third of the wine produced by the Top 10 wine-producing countries is from the New World (cf. 29% in 2001) and 30% of the volume of wine exported by the Top 10 wine-producing countries is now accounted for by New World producers (cf. only 19% in 2001). The table also shows that some New World countries, individually, export

Table 1.4. Top 10 World Wine-Producing Countries

	Wine Production (Provisional 2007)		Wine Exports (2004)		
	World Ranking	000 hl	World Ranking	000 hl	Percentage of Total World Exports
France	1	48,400	1	14,210	18.5
Italy	2	48,000	2	14,197	18.4
Spain	3	34,700	3	14,042	18.2
USA	4	20,000	6	3,874	5.0
Argentina	5	15,000	11	1,553	2.0
China	6	12,000	N/A	20	0.0
Germany	7	10,300	8	2,709	3.5
South Africa	8	9,800	9	2,685	3.5
Australia	9	9,600	4	6,457	8.4
Chile	10	8,200	5	4,740	6.2
New World as percent of Top 10		30.00%		29.97%	

Source: OIV (2004, 2008).

large portions of their production: in Australia's case, just under half, and in Chile's case, over half.

THE OLD WORLD RESPONDS

As in any competitive business situation, the Old World of wine hasn't been watching the success of the New World wines without responding. In order to preserve the traditional ways, regulatory, trade, marketplace, and advertising steps have been initiated to combat the intrusion of the New World ways. These have served to moderate, but not prevent, the spread of the New World influence.

Through diplomatic negotiations and in international forums such as the World Trade Organization, the European Union (EU) has attempted to limit the potential for New World wines to challenge established traditions in Europe. There are now accords in place that restrict the use of certain geographical indicators on wine, the use of certain nomenclature on labels and in marketing, and even the use of certain bottle shapes claimed to be the right of various traditional locations. These efforts have been strongly challenged by most countries outside the EU but have gradually gained currency.

In contrast, some EU winemakers have actually embraced New World labeling standards and viticultural and winemaking innovations. There are now French wines being exported, for example, that feature the grape variety more prominently than the regional appellation. Italy is protecting its market position by strongly promoting the linkages between the food dishes of specific regions and the wines that have been traditionally associated with those foods. Interestingly, Italy is trying to win back some of its traditional heritage by labeling some of its Primitivo wines as Zinfandel. Finally, in an attempt to replicate many New World practices, some European wineries (and many new wineries in Asia and Eastern Europe) have employed New

World winemakers and technical staff from Australia and the US. Today there are around 500 Australian-trained wine industry professionals working in Europe.

OPPORTUNITIES AND CHALLENGES

The past three decades have seen rapid and substantial transformation of the wine industry globally, and the emergence of a whole new phalanx of industry leaders. The future will, undoubtedly, bring further changes. Some of the more significant opportunities and challenges are outlined below.

Production and Consumption Equilibrium

One of the major challenges for the industry in this period has been to achieve growth while maintaining demand and supply equilibrium. The output growth pattern is cyclical but, because of the considerable elapse of time between vineyard development and eventual wine release, each incremental supply response to a demand signal can extend into the medium term.

For white wines the supply response may be as short as 3 or maybe 4 years, but for red wines it could be as much as 8 years. As a result it is difficult to forecast and accurately match production growth and consumer demand. Accordingly, the industry is characterized by periods of over- and undersupply, with the attendant impact on prices and profitability.

Overall, wine consumption has declined over the past three decades, although there has been a modest recovery in recent years, principally because of strong growth in demand in nontraditional wine-consuming countries—most particularly in China and Russia. In the short term, wine consumption (and industry profitability) can also be impacted by one-off events such as the New York City terrorist attacks of 2001 and the marked slowdown in travel and eating out that followed, as well as by the prevailing macroeconomic climate.

Continued Corporatization of the Global Wine Industry

Consolidation trends by major wine companies have been a feature of the industry in the past decade, but the Top 10 producers still only hold a relatively small market share. This is in major contrast to the rest of the beverage industry. The wine industry is still one where passionate newcomers and small producers rule. New winery startups continue and new countries are entering the marketplace.

Success requires long-term commitment and financing. Many large wine businesses are still family owned and face succession challenges. Tax laws, inheritance laws, and general family business issues make this difficult. Also, as family companies get larger they tend to inherit corporate character.

Antialcohol Groups and "Sin Tax" Advocates

In some parts of the world the demand for wine—and, indeed, industry profitability—is impacted by community tolerance or even advocacy for regulatory, legal, and taxation hurdles to limit growth of the wine industry. This is particularly pronounced in the US, where existing controls on interstate trade and campaigns for further restrictions by neo-prohibitionists continue to pose threats to the industry. Despite a ruling by the US Supreme Court in 2005 (GRANHOLM, GOVERNOR OF MICHIGAN, et al v. HEALD et al), many states have been slow to enact legislation that will permit in-state and out-of-state wineries to operate on the same footing. This has often been the result of protracted political intervention by antialcohol and wholesaler groups. The continuing political and legal battles over direct-to-consumer sales will most likely continue in the US for the next decade.

In many other countries, alcohol is often viewed as a luxury or at least a nonessential purchase by many governments, thus making it an easy target for taxation. Prices are artificially driven upward, impacting on demand and, in some cases, industry profits.

Continued Globalization of the Wine Industry

There is a substantial and growing global market in bulk wine, allowing inexpensive wines to be bottled in many locales (J. O'Neill, 2004) The largest wine companies have moved to global operational models where marketing decisions may be centralized and wine production is localized. Global wine corporations, such as Torres, Mondavi, Beringer Blass, Gallo, Kendall Jackson, Constellation/Hardy, and Southcorp, all operate in this way. A relatively recent trend has seen many of the larger global players committing to indigenous wine production (either as new businesses or through acquisition) in a range of countries as a means of more effectively expanding global market share and global presence.

New Wine Producers in Asia

The market for wine in Asia and, indeed, domestic production to meet this demand is growing at a rapid pace, though from a small base. Wine emulating contemporary Western styles is now made using modern winemaking facilities in 12 countries in Asia. There are now almost 800 wineries throughout Asia, many of them making wine at the higher end of quality expectations (Gastin, 2008). More than half of them are in China, in 26 provinces, virutally all of them established since the early 1980s. Over a quarter are in Japan. The remainder are spread sparsely over the continent, from India to Indonesia—including Thailand, Korea, and Vietnam. Fledgling operations can even be found in Taiwan, Sri Lanka, Bhutan, Myanmar, and Cambodia.

As in so many other arenas, China has rapidly emerged as a global wine giant in its own right. It now has the fifth largest viticultural area in the world, with almost 500,000 hectares of vines, of which at least 70,000 hectares are conventional wine grapes, and is now the world's sixth largest wine producer (Office International de la Vigne et du Vin [OIV], 1995–2008). Strong domestic consumption growth is the primary driver but, increasingly, export opportunities are opening up for the better and larger Asian producers.

Conclusion

The foregoing is a chapter in the history of the industry, but time does not stand still. There are already new trends emerging in the industry that will write future chapters. One of these is the likely extinction of the sharp boundaries between "old" and "new" that have been a feature of the past three decades. In the fifth edition of *The World Atlas of Wine* Hugh Johnson wrote: "It was I, I confess, who coined this now much-maligned wine world split. Times have changed. Much of the 'Old World' has become 'New'; a little of the 'New' is deemed to be 'old'" (Johnson & Robinson, 2001, p. 6).

The creation of this new book should add to the body of knowledge of a recent business phenomenon and, in so doing, contribute to the mobility of good business practice for the benefit of those in the world of wine wherever they may be now.

LEARNING OBJECTIVES:
- Define strategy and the levels of strategy within a wine business
- Describe the strategy development process
- Identify the five major external analysis tools for wine businesses
- Identify the three major internal analysis tools for wine businesses
- Explain how to write wine business mission, vision, and value statements
- Describe the five major generic strategies for the wine industry
- List the eight supporting strategic options for wine businesses
- Describe the five points of the wine business strategy implementation star

CHAPTER 2

WINE BUSINESS STRATEGY

Liz Thach
Professor of Management & Wine Business, Sonoma State University

Classic business strategy can be applied to any industry, but in many cases there are special considerations or differences between industries that need to be taken into account when crafting and implementing the strategy. No place is this more true than in the wine industry, where global labels number in the tens of thousands and competition is intense. There are very few other industries that incorporate both agriculture and manufacturing processes, plus must unravel complicated, and in some cases, bizarre, marketing and distribution regulations in order to bring thousands of competing and sometimes confusing labels to the consumer. The only other industry that comes close to producing so many labels is the music industry.

Another unique feature about the wine industry is, though there are recognized global leaders (Table 2.1), the largest pure wine market share held is only around 2% because most large corporations sell other products besides wine. This is quite different from other industries and even beverage companies, where global players like Coca-Cola currently hold 40% of the market in their category (Beverage Digest, 2007). Viewed from this vantage point, the wine industry is incredibly fragmented with thousands of small players around the world and only a handful of large corporations—with a few struggling midsize players in between. Given this situation, the burning question becomes one of strategic competitive advantage: How can I compete successfully in this crowded global market?

Table 2.1. Top 10 Wine Selling Companies by 2006 Total Sales

Rank	Company	2006 Sales (US$ Millions)
1	Heineken	15,618
2	Sabmiller	15,307
3	Diageo	13,186
4	Pernod-Ricard	8,255
5	Constellation[a]	4,603
6	LVMH	3,953
7	Fosters	3,611
8	E.&J. Gallo	2,700
9	Brown-Forman	2,412
10	Kendall-Jackson[b]	471

Source: Korolishin (2007).
[a]Largest wine company.
[b]Sells wine only.

Therefore, the purpose of this chapter is to provide some answers to this question by applying classic business strategy to the wine industry, but with a focus on the unique attributes of the industry. It begins with a definition of strategy and an overview of the strategy development process. From there, each phase of the process is explained, including relevant examples in the wine industry. Various business options and strategic alternatives are investigated, with the end result being a step-by-step process for you to examine in crafting your own wine business strategy.

Defining Wine Strategy

A company's *strategy is the comprehensive plan explaining how the company will achieve its mission, vision, values and goals.* A shorter definition is that it is the company's "game plan for success." When applied to the wine industry, the definition of strategy is no different from other industries in that—whether the company is a winery, distributor, retailer, grape grower, supplier, or a combination of some of these—it needs to determine how to compete in the marketplace and be successful.

What is sometimes confusing about the term "strategy" is that it can be used at different levels within a company. Corporate, or overall business strategy, is defined as above, but larger wine businesses may also have functional, department, and/or brand strategies. Each of these should be linked to the overall corporate strategy. Table 2.2 describes the various levels of strategy within a wine business.

Many small wine business owners question the usefulness of engaging in a strategy development process. In many cases, they have been growing grapes and making wine for years—often times an occupation that has been passed down for generations in the family. The wine is good and it seems to sell, so what is the point of crafting

Table 2.2. Levels of Strategy Within a Wine Business

Level of Strategy	Wine Business Definition
Corporate/Overall Business Strategy	The comprehensive plan (game plan) explaining how the specific wine business will achieve its mission, vision, values and goals.
Division/Business Unit Strategy	The game plan for each division within the wine business—must be linked to corporate strategy. For example, large wine corporations may be divided into premium and fine wine business units, or beer, wine, and spirit divisions.
Department/Functional Strategy	The game plan for each department within the wine business—must be linked to corporate strategy. For example, the marketing department will have a wine marketing strategy.
Brand Strategy	The comprehensive plan for communicating the brand story for your product or service. For example, many wineries have multiple wine brands within their product portfolios. Each brand should have a strategy that links to the corporate strategy. Brand strategy can also be applied to a "branded vineyard," a signature service, or a wine supply, such as a special type of cork.

a wine business strategy? This is a valid question, because strategy can be both formal and informal. Sometimes the strategy in a small family wine business resides within the "head and heart of the owners." It may not be on a piece of paper, but they can clearly explain their formula for success. It may be because of an emphasis on a certain grape varietal, or a unique style of winemaking. This is considered an informal strategy, and works fine as long as the business is run by that owner. However, if it is passed along to a new successor, or begins to grow in size, then a more formal business strategy is usually recommended so that the focus and competitive strength of the business can continue. It is because of this, that medium and large wine businesses generally have a formal and documented strategy process.

Research has shown that businesses with a clear strategy development and implementation process generally outperform those that do not have such a process (Hunger & Wheelen, 2007; Thompson et al., 2005). This is because the business is continually scanning the external environment, analyzing the impact on their internal operations, and proactively adjusting to meet the changing dynamics of the global market. Even a small winery or grape grower that sells locally within one region or country is still operating in a global environment. They cannot afford to ignore the fluctuations of grape/wine supplies on the worldwide scene, changing consumer demands, new regulations, and even their suppliers' situation, such as the introduction of a new type of glass, cork, barrel, yeast, irrigation system, or software package. All of these changes impact strategy, so it behooves even small family wine businesses to consider adopting at least an informal strategy process that they assess at least once a year.

THE WINE BUSINESS STRATEGY DEVELOPMENT PROCESS

The strategy development process can be roughly divided into four phases: 1) External Analysis; 2) Internal Analysis; 3) Strategy Development; and 4) Strategy Implementation and Evaluation. These four phases are illustrated in Figure 2.1. *External analysis* involves gathering data on wine consumers, market conditions, competitor positions, social and political trends, changing regulations, and other important external considerations that may impact the wine business. *Internal analysis* is a review of current operations, competencies, personnel, and other resources. If the company is just beginning and doesn't have current operations, then a review of forecasted operating standards and/or outsourced operations, such as the use of custom crush operations for a new winery, is included.

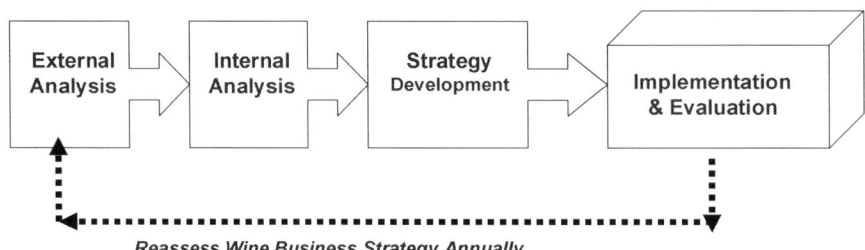

Figure 2.1. Major phases of strategy development process.

Strategy Development involves the actual writing of a strategy document. This includes mission, vision, values, long-term goals, and the specific strategy (game plan) the business will use. *Strategy Implementation and Evaluation* begins with an action plan for each component on the strategy star (Figure 2.2). The action plan includes a timeline, goals, and clear accountability for people responsible for the various processes. Finally, a set of evaluation

Figure 2.2. The wine business strategy star.

tools is developed and installed so that both continual and year-end measurement of progress takes plans, along with needed corrections and revisions to the smooth implementation of the strategy.

Strategy reassessment is a miniversion of the strategy development process and is usually conducted each year as part of the annual executive team meeting. The purpose is to verify that the current strategy is still relevant and to make any needed updates or revisions. In wine businesses that are in emerging or fast-changing areas such as online wine sales or wine software packages, it is recommended that strategy reassessments be conducted more frequently—usually at least once every 6 months.

When, How Long, and Who Should Be Involved in Strategy Development

The four-phase process raises obvious questions about when to use it, how long it takes to complete, who should be involved, and when to engage in a strategy reassessment process. Answers to each of these questions are provided below:

- **When to Use the Strategy Development Process?** In general, a new strategy should be developed anytime a new business is created and/or a new division within a larger business is launched. Also, if the environment has changed drastically and the current business strategy is not working, then a company should engage in a new strategy development process.
- **How Long Does It Take?** In general, the process can take from 2 to 6 months to complete for a new business strategy. The length of time is dependent on the amount of external and internal data that must be gathered and how long it takes to analyze.
- **Who Should Be Involved?** The CEO and top execu-

tive team are always involved in strategy development and reassessment. Often an internal or external strategy facilitator helps to guide the process, as well as to oversee the collection of internal and external data for analyses. In some companies a team of employees who represent a cross section of the business are included as part of the executive strategy team or as a subteam providing feedback to the executives.

- **How Often Should Strategy Assessment Be Conducted?** At a minimum a strategy reassessment process should be implemented on an annual basis. It is a miniversion of the strategy development process and involves gathering new external and internal data and validating that the current strategy is still working. Minor adjustments and tweaks to the current strategy may be made. In general, a strategy reassessment takes 2 to 4 weeks to complete. The information is usually gathered before the annual meeting of the executive team and decisions on revisions made at that meeting.

PREREQUISITE FOR STRATEGY DEVELOPMENT PROCESS: A WINE BUSINESS CONCEPT

Before embarking on a strategy development process, the executive team obviously must have a business concept in mind or an existing business that needs a new competitive strategy to stay in the game. The concept should include a draft mission statement, as well as a clear idea of the target market for the wine business product or service. Four examples are as follows:

- **Grape Growing:** A grape grower intends to grow and sell tempranillo and grenache grapes in the Sierra Foothills appellation of California. His/her target market is US wineries.
- **New Winery:** A group of investors is opening up a winery in Uruguay to produce and sell a luxury-priced

tannat wine targeted at high-end wine shops and on-premise establishments in eight key global cities: London, New York, Los Angeles, Sydney, Tokyo, San Francisco, Frankfurt, and Beijing.
- **Wine Bar & Retail Shop:** Two friends with an extensive knowledge of wine decide to open a wine bar and retail shop in Boise, Idaho. Wine selection is global, but with a focus on wines of the Northwest: Idaho, Oregon, Washington, and British Columbia.
- **Winery Supplier:** An inventor has developed new winery software that monitors and provides data on winery operating systems via wireless technology. It is targeted at global wineries with production of more than 1 million cases.

Each of these is an example of preliminary wine business concepts that can be run through external analysis to determine if it can be developed into a viable business strategy. It is also necessary to have a general idea of the financial investment required to launch the business as well as expected return on investment for the first 5 to 10 years. Some of these business concepts will be used as examples in the rest of this chapter.

EXTERNAL ANALYSIS PHASE

External analysis is one of the most time-consuming phases because it involves gathering data on a variety of variables. Ideally an existing wine business will be doing this on a regular basis as part of standard operations, but a new business will have to gather these data as an initial phase in designing a successful strategy. Although there are many types of external analysis processes, in the wine industry five key analyses should be conducted to determine if the business concept is viable. These are: 1) Market Analysis; 2) Competitor Analysis; 3) Distributor Analysis, 4) Macroenvironment Analysis; and 5) Porter's Five Forces Analysis.

Market analysis is gathering information on what customers are currently buying in the market and reviewing emerging trends pointing towards future sales. For examples, grape growers should obtain and analyze data on the types of grape varieties that have sold well over the past 10 years and review expert opinions on varietal trends for the future as well as popular appellations. In addition they should assess the feasibility of obtaining grape purchase contracts from potential buyers in their target market. Likewise, a new winery should analyze wine sale trends in their target markets over the past 10 years and review expert opinions on future wine sales trends. For the wine bar/retail shop, information on the growth of wine bars in the target markets, as well as current and future sales forecasts of wine sold in retail shops, should be gathered and analyzed. Business concepts that are in emerging areas, such as a new type of software, may not have existing sales trends, but similar products or services can be reviewed to obtain market trend data.

In reviewing market data, special attention should be paid to pricing, discounts, margins, and other financial figures to determine if the business concept is feasible based on preliminary financial estimates. It is also useful to look carefully at changes in market trends to determine if they are gradual or rapidly changing. For example, if the focus is on chardonnay grapes or wine, how much have market demands for this varietal changed over the last 10 years—or is it more a change in pricing, wine style (oaked or unoaked), or appellation?

In order to gather market data, you can do some of it via Internet and trade journal research, as well as interviewing potential customers. However, it is usually recommended to hire a market research firm to gather some of the data, because it is not always easy to find on your own. Furthermore, some of it might be proprietary, and in some cases you will have to purchase it.

After examining all of the data resulting from the market analysis, you ask the question: How viable is this business concept? It is often useful to rate the results of each analysis on a scale of 1 to 5, with 1 being poor and 5 being excellent. In general, if you score a 3 or above on market analysis, move onto competitor analysis.

Competitor Analysis is a review of the major competitors already operating in the arena of the business concept. In addition to identifying the names of major competitors, information on annual revenues or market share, number or geographical breadth of customers, and pricing should be obtained. With publicly held companies, most of this information is available on the Internet. However, because the majority of companies in the wine industry are private, it may be necessary to hire a market research firm to assist with data collection—and in some cases estimates may need to be used instead of hard data.

Using our four business concept examples, the grape grower would need to identify other growers who specialize in Spanish grapes such as tempranillo and grenache both within his/her appellation and outside the appellation. The tannat luxury winery would need to identify other wineries selling tannat wines in target markets, as well as other unique luxury wines that would compete with tannat. The wine bar/retail shop owner would need to visit other similar retail stores and wine shops in Boise, Idaho as well as similar competitors such as restaurants, bars, and grocery stores that may sell wine. Finally, the winery software inventor would need to identify all other suppliers that provide similar types of services—even if it is not software related.

Once the competitor data have been gathered, it is useful to display the data in a strategic group map in order to gain a clearer understanding of the competitive environment. In general, price/quality make up one axis while market scope makes up the other axis. The size of the circle indicates total annual revenues. Figure 2.3 is a potential strategic group map for the luxury tannat winery example.

In the luxury tannat winery example, they have targeted a part of the map that is currently unoccupied—high-priced/high-quality tannat wine for global markets. Their target annual revenue is $10 million. Their largest competitor is A, who is selling tannat wine globally, but at a lower price and quality level. Other competitors that they must consider are B through D, because these players are also in the global market, but at different price points. Regional players producing tannat wine at low price points are not direct competitors, but it pays to watch E, who is currently selling regionally but at a very high quality and price point. E could decide to expand to global markets. E is also a ripe acquisition target for competitor A, who may want to acquire a luxury brand.

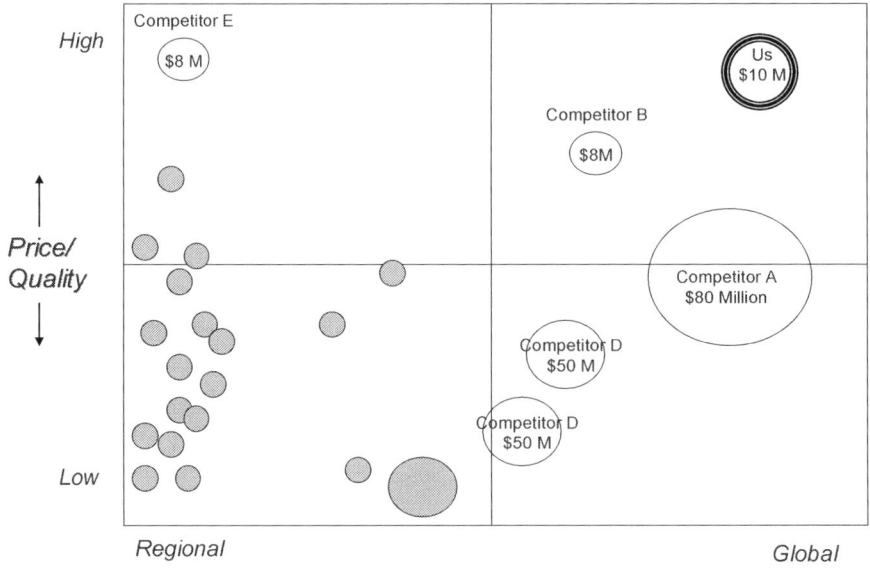

Figure 2.3. Example of strategic group map for luxury tannat winery.

By analyzing the strategic group map and the position of similar competitors, you can make an informed assessment about the level of competition. If many competitors appear on the map grouped together in the same area, that usually means it is not an attractive strategy due to intense competition. However, if there are very few competitors and/or large spaces on the map where you intend to operate (and your market analysis proved there is a need for your wine product or service), then the results of your competitor analysis are positive and you can rate yourself higher on the 1 to 5 point scale.

Distributor Analysis is a review of potential distribution channels you can use to deliver your product or service to the market. In many industries, this is not a critical analysis, but in the wine industry—especially for wineries—lack of a good distribution system could cripple the business before it even gets started. This is especially true in countries that have strict regulations on wine sales, such as the US, Canada, and Sweden. In these countries it is not always possible to approach an end-consumer or retailer and sell them wine directly. Instead, wineries must sell to distributors, importers, or government agencies, who will place their product. Unfortunately, there are many stories of small wineries that produce excellent wine, but could never get good distributor representation in the markets in which they wanted to operate. This caused sales estimates to be much lower than expected and severe financial strains to be placed upon the winery.

Therefore, a new wine business, such as the tannat luxury business would need to determine if they could get distributor representation in the eight key global cities in which they want to position their wine. They would most likely have to find importers in each country and make sure they could deal with the different regulations in each market. It is also important that they carefully evaluate the reputation of the distributors, importers, or other agents they intend to use. Unfortunately, there are some that do not follow through on promises, or do not have the contacts and relationships with on- and off-premise retailers to place the wine.

For the grape grower, distribution analysis is not as critical. In most cases, he/she will pick and ship the grapes to the customer wineries. However, he/she will need to ensure that the shippers provide temperature-controlled trucks or containers. The wine bar and retail shop in Boise will need to identify importers and distributors they can work with to purchase the wines they need. Because they are wine buyers rather than sellers, this is not as difficult and their main concern is finding people with whom they can develop long-term trusting relationships. Finally, the winery software provider also will not need to be as concerned with distribution as he/she will most likely use standard shipping companies to deliver products.

As with the other analyses, once the data have been gathered and assessed, a rating from 1 to 5 should be determined. If the score is 3 or above, it is time to move to the next analysis.

Macroenvironmental Analysis is a big picture analysis of five major factors that can impact a business. These factors are: 1) economic conditions, 2) technology, 3) societal values, 4) demographics, and 5) legislation. Though these issues can sometimes appear to be too large and fuzzy to analyze clearly, it is useful to ask questions about each factor in order to help focus the business concept as well as to determine if there are any major roadblocks to launching the business. Table 2.3 lists some of the key questions to ask for this analysis, including ratings for the current situation and potential to change in the future.

Again, the executive team that is completing the strategy development process should have an in-depth discussion

Table 2.3. Macroenvironmental Analysis Questions

Factor	Key Questions	Current Situation (1=Very Negative; 5=Very Positive)	Potential to Change in the Future (1=High Potential to Change; 5=Low Potential to Change)
Economic	What is the impact of economic conditions on the business concept?	1 2 3 4 5	1 2 3 4 5
Technology	Are there new technologies that could negatively impact the business?	1 2 3 4 5	1 2 3 4 5
Societal Values	Are the values of the target consumer changing in a way that could negatively impact the business (e.g., move against alcohol consumption)?	1 2 3 4 5	1 2 3 4 5
Demographics	Are the demographics of the target consumer changing in such a way to harm the business (e.g., growing older, younger, different nationalities/race, etc.)?	1 2 3 4 5	1 2 3 4 5
Legislation	Is there changing legislation and regulations that could help/hinder the business concept?	1 2 3 4 5	1 2 3 4 5

of these macroenvironmental issues to determine if the business concept has a good chance of succeeding. For example, in launching the luxury tannat wine, if the economy goes down in the eight major cities in which the company wants to target, how much will this impact sales? History has shown that, in general, wine sales may not decrease in an economic recession, but people do buy less expensive wine. Therefore, launching a high-priced luxury wine during a recession may not be a great idea; however, if the economic conditions are better in certain countries than others, then they may be able to reduce the risk.

Though there are not always clear answers with some of the macroenvironmental questions—because it is difficult to forecast the future regarding some of these issues—it is still useful to have the conversation so that everyone on the team engages in strategic thinking on big picture issues that can impact the business. Furthermore, a review of these factors may also reveal opportunities, such as changing legislation, demographics, and societal values that may open up new types of businesses that were not possible or popular in the past. For example, in the US wine sales have increased because of changes in direct shipping legislation, the growth of the Millennial demographic sector that drinks more wine that previous generations, and societal values that now focus on the positive aspects of food and wine.

The fifth external analysis is *Porter's Five Forces Analysis*, which is familiar to anyone who has ever taken a class in strategy. Developed by strategy expert Michael Porter (1985), this analysis asks that you examine five competing forces in your business environment and evaluate them as *weak, moderate, strong, or fierce*. If all five forces appear to be very strong or fierce, it is probably not a good business environment in which to enter. However, if just one or two of the forces are strong, and the others are weak or moderate, chances for success and achieving attractive profits are higher. Table 2.4 illustrates the five forces applied to the luxury tannat producer

Based on this analysis for the luxury tannat winery, three of the forces are rated weak to moderate and two are rated strong to fierce. This suggests a cautious recommendation to move forward with the business concept. The fact that there are not that many competitors in the luxury tannat market and that currently there is very little threat of new competitors bodes well for the business concept. Furthermore, the fact that supplies for the winery are not a huge issue (because they have their own vineyards) is very positive. The big challenge here is convincing buyers, who in this case are importers, distributors, and restaurant owners, that putting a high-end luxury tannat wine on their expensive wine list will be intriguing to their diners—and that it will produce sales. The substitute products force is always very high in the wine industry, because buyers may elect to drink beer, spirits, coffee, tea, soft drinks, water, juice, and other beverages. Only in the case of dedicated wine lovers who believe a meal without wine is a sin do you have a weak level of force on substitute products with Porter's analysis.

Finally, if after completing all five of the external analyses the strategy development team has determined that the business concept is viable and has a good chance of achieving attractive profits, then the next step is to engage in the Internal Analysis phase.

Internal Analysis Phase

Internal analysis is an inward look at the business concept. It involves a review of current operations, competencies, personnel, and other resources. If the company is just beginning and doesn't have current operations, then a review

Table 2.4. Example of Porter's Five Forces Applied to Luxury Tannat Winery

Force	Indicators of STRONG Force	Our Situation	Rating
1. Rivalry Among Competitors	• Many competitors • Very aggressive price cutting and discounting of wine • Slow market growth • Heavy acquisition activity	Five major competitors Not occurring Moderate growth No M&A activity currently	Weak to moderate
2. Entry of New Competitors	• Many new candidates • Low entry barriers (doesn't cost much to start a new business) • Rapid industry growth • Profit potential is high • Existing industry members have a strong incentive to expand into new geographies	No No – capital costs high Moderate growth Moderate potential Perhaps	Weak to moderate
3. Substitute Products to Wine	• Many good substitutes available at lower prices • Low switching costs for consumer • High-quality substitute available	Yes Yes Maybe – different varietal to tannat	Strong to fierce

Table continued on next page

WINE BUSINESS STRATEGY

Table 2.4 continued

Force	Indicators of STRONG Force	Our Situation	Rating
4. Supplier Power	• Only a few suppliers of grapes	Yes, but plant to purchase our own vineyards	Weak to moderate (because will have own tannat vineyard)
	• Only a few suppliers of other products	No—plenty of suppliers	
	• High cost in switching to alternative suppliers	No	
	• Supplier provides a differentiated input enhances the quality	No—only in grapes	
	• Some suppliers threaten to integrate forward	A few vineyard owners want to start a winery	
5. Buyer Power (In this case, buyers are importers, distributors, and restaurant owners)	• Only a few buyers	Yes—want to target top restaurants in eight cities	Strong to fierce
	• Buyer switching costs to competing brands are low	Yes	
	• Buyers can demand discounts	Yes	
	• Buyer demand is weak or declining	No—growing for tannat	
	• Buyers can wait to buy	Sometimes	
	• Buyers threaten to buy wineries	No	
	• Identity of buyer adds prestige to our list of customers	Yes—we need our wine in top restaurants	

of forecasted operating standards and/or outsourced operations, such as the use of custom crush operations for a new winery, is included. It is less time consuming than external analysis, yet requires a very honest appraisal of internal operations. Again, there are many different tools that can be used for internal analysis, but in the wine industry three key analyses should be conducted to determine if internal operations are sufficient to sustain the business and give it a competitive edge. These are: 1) Value Chain Analysis, 2) Critical Success Factor Analysis, and 3) SWOT Analysis.

Value Chain Analysis is identifying the primary and supporting activities that allow the company to produce a product or service for customers and analyzing the costs of these activities in comparison with other companies. In the wine industry a generic value chain includes supply procurement, operations, distribution, marketing/sales, service, and profit margin. This is illustrated in Figure 2.4.

The first step in value chain analysis is to diagram the primary and support activities in the value chain for your business. For example, Figure 2.5 illustrates a value chain

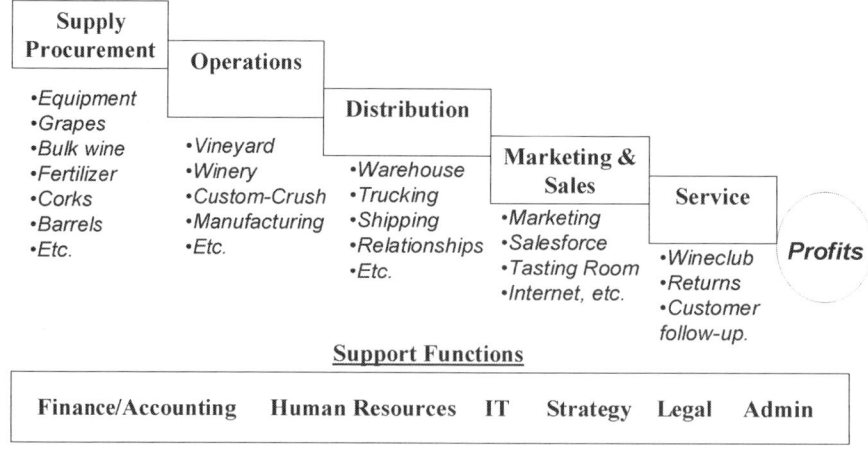

Figure 2.4. Generic wine industry value chain.

for the grape grower who plans to supply tempranillo and grenache grapes from the Sierra Foothills.

After the value chain is diagrammed, the next step is to estimate the cost to implement each primary and supporting activity. For example, what are your supply procurement costs? What are the annual vineyard operations costs? How much are you paying to deliver your grapes to winery customers? If this is a new business, then data on these costs need to be gathered in order to create an estimate.

Once cost estimates are obtained, then it is highly useful to analyze the value chains of your major competitors to see if they are different from yours. In the case of grape growing, there are not many differences in value chains, with the exception of farming practices such as organic verses traditional and establishing a branded verses nonbranded vineyard. Exceptional customer service can also make a difference with grape growing, as those growers who developing long-term and positive relationships with winery customers often have a higher rate of success.

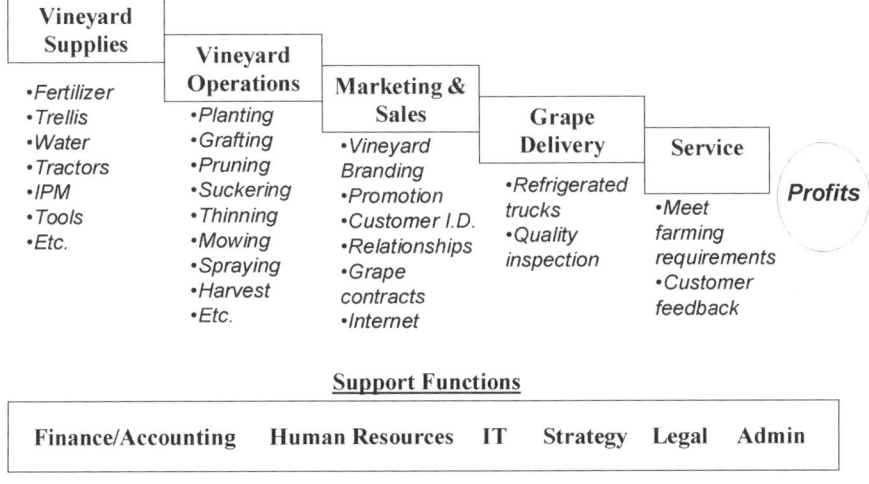

Figure 2.5. Example of grape grower value chain.

Large companies will engage in benchmarking practices in which they hire an objective third party to gather these data and then share the data (without use of company names) between all who participate. In this way, you can obtain an average cost of activities within the value chain and compare them to your cost estimates. If this is not possible, joining associations or networking can often provide similar estimates—especially in the wine industry where many small wineries, grape growers, and wine suppliers will work together to promote a region and share information so that all may prosper. It is also useful to discover if fellow competitors are outsourcing activities, because this may also be something you want to consider. This is often the case with start-up wineries, which may purchase grapes and rent a custom crush facility, rather than make the huge capital investment necessary to purchase a vineyard and fully equipped winery.

The final step of value chain analysis is then to analyze your value chain costs in light of the competition. From this, you can develop an action plan to address inconsistencies and cost disadvantages. The process also highlights areas in which you may have a cost advantage, which can be listed as a strength when you complete your SWOT analysis. Major methods to reduce cost disadvantages include adopting best practices of your competitors; eliminating low-value-added activities, relocating high labor cost activities to new geographies where costs are lower (not possible with a vineyard); adopting cost-savings technologies; outsourcing where it makes sense; and innovating new solutions to reduce costs.

The second internal analysis is called *Critical Success Factor Analysis* and can be described as identifying and implementing factors for your industry that separate winners from losers. The first step is to review your value chain and pinpoint critical processes that can spell the difference between disaster and success. These are usually different depending on the business category. Using our grape-growing example from above, we see that with grape growing there are several critical processes, including planting/grafting the right varietal for your appellation to match consumer trends; obtaining a sales contract with winery customers; meeting the farming requirements of the contract at a competitive cost; and harvesting and delivering the grapes at the optimal brix/acid/pH level for quality inspections. If these processes are handled correctly, there is usually a high chance of business success. Unfortunately, due to the lag time in planting or grafting the correct varietals to match market needs, this is not always an easy process to predict. The best that can be done is to annually complete the external analysis process to stay abreast of market changes.

Once the critical success factors (also called CSFs) have been identified, they are placed in a table and weighted in terms of importance. The weighted scale is usually scored in percentages, with 10% being low and up to 80% being very high in importance; however, they must add up to 100% (Table 2.5). Then you analyze how well you are doing in implementing the CSFs compared to your competitors. Again, you may have to hire a consultant to help you obtain the data, or informal networking may provide a clear idea on how well your competitors are implementing the CSFs. Generally a rating scale of 1 to 10, with 1 being very weak and 10 being very strong, is used for the calculation. Then you simply multiple the weighted CSF values times the 1 to 10 rating and add them up. The business achieving the highest rating shows that they have a strong competitive position, whereas lower ratings signal a weaker position. Once again, this information tells you where you need to make adjustments in order to become more strategically competitive in your marketplace.

Table 2.5. Critical Success Factors for Grape Growers in Sierra Foothills

Critical Success Factors (CSFs)	Weight	Your Company	Competitor 1	Competitor 2	Highest Potential
Long-term grape contract	25%	8 = 2	9 = 2.25	4 = 1.0	10 = 2.5
Hot grape varietal to match consumer demands	20%	6 = 1.2	8 = 1.6	5 = 1.0	10 = 2.0
High-quality product (brix/acid/ph)	20%	7 = 1.4	5 = 1.0	4 = 0.8	10 = 2.0
Farming costs	30%	4 = 1.2	8 = 2.4	7 = 2.1	10 = 3.0
Reputation (brand)	5%	4 = 0.2	7 = 0.35	2 = 0.1	10 = 0.5
Total	**100%**	**6**	**7.6**	**5**	**10**

The last internal analysis is *SWOT Analysis*. This is a process that is familiar to many people; however, most don't complete all of the steps of SWOT, which can make the process less beneficial. The first step is to list the Strengths (S), Weaknesses (W), Opportunities (O), and Threats (T) for a business. Much of this information is obtained through your value chain and critical success factor analyses. Table 2.6 lists some of the items to consider when identifying this information.

In the wine industry, there are certain areas that are more important than others. For example, an important strength is to have a unique product or service because the global market is so highly competitive. Being a small family winery that produces the same four varietals as thousands of other family wineries is not a strength, but a small family winery specializing in a specific varietal, or a unique farming or winemaking practice, is distinctive and can be a strength. Another specific issue in the wine industry is the threat of bad weather, pests, and water shortages. Many other industries don't have to consider Mother Na-

Table 2.6. SWOT Considerations for Wine Businesses

Strengths	Weaknesses	Opportunities	Threats
• Unique product/service	• Lack of financial resources	• Rising consumer demand for your product/service	• Increasing competition
• Key personnel	• No core competencies	• New markets	• Weather
• Attractive customer base	• Undifferentiated product/service	• Positive regulation changes	• Pests
• Financial resources	• High costs	• Product line expansion	• Water shortage
• Core competency	• Weak reputation/brand	• Online sales	• Changing Legislation
• Economy of Scale	• Lack of talent	• New technologies	• Market slow downs
• Good Brand name	• Behind on quality	• Acquisition or merger opportunity	• Shift in buyer desires
• Cost Advantage	• Behind on technology	• Partnerships	• Cost increases
• High quality	• Lack of distributors or partners	• Falling trade barriers	• Supplier consolidation
• Partnerships		• New consumer segments	• Distributor consolidation
• State of the art technology			

ture to the extent that the wine industry does. This creates a whole new category of threats and extra defense costs, such as crop insurance, which other industries don't have to consider.

After the SWOT lists are created, the second step is to draw conclusions about the items on the list. Table 2.7 lists some of the key questions to consider, such as which strengths are the most critical to our success and which opportunities can we really pursue given our financial resources? These types of questions should be considered in depth before

Table 2.7. Steps 2 and 3 of SWOT Analysis for Wine Businesses

2. Draw Conclusions
   Strengths
   Are our strengths matched to the Critical Success Factors for our industry?
   What strengths are missing?
   Least and most important?
   Weaknesses
   How serious are these weaknesses to success?
   What are our chances of overcoming weaknesses?
   Least and most important?
   Opportunities
   Which opportunities are most relevant for our wine business?
   Do we have the resources to capitalize on these opportunities?
   Threats
   Are these threats alarming, or can we defend against them?

3. Action Plan
   Strengths
   Which strengths can be used immediately? How?
   Which should be strengthened? How?
   Weaknesses
   Which weaknesses must be corrected now? How?
   Opportunities
   Which opportunities should be top priority?
   How can we address now?
   Threats
   What actions should we take to guard against these threats?

moving on to the third step of SWOT, which is to create an action plan to address the conclusions. Many strategy teams do a good job at creating lists, but do not create and implement action plans to address the identified issues.

Once the strategy development team has completed the SWOT analysis and crafted an action plan to address identified issues—with clear accountability and deadlines—they are ready to move onto the third phase: Strategy Development.

Strategy Development Phase

The strategy development phase is usually the most exciting and creative as it involves the actual writing of a strategy document. This includes mission, vision, values, long-term goals, and the specific strategy (game plan) the business will use. Now the strategy development team has completed the external and internal analysis phases, they have verified that their business concept is viable and has a good chance of achieving sustainable profits due to some specific competitive advantages. So now the first step is to modify and/or polish the original business concept into a clear mission statement, and then develop a vision and clarify values. The mission, vision, and values are not just for customers, but are also intended to communicate company direction and philosophy to employees, owners, suppliers, and the community.

A *mission statement* describes the purpose of the business or WHAT the business does. Ideally it should be simple and clear. After reading it you will know what the company is in business to produce and/or provide in terms of service. Some mission statements will also include the geographic location or market for the products or services, as well as information about what the company represents. Following are several examples of mission statements in the wine industry:

- **Beckstoffer Vineyard:** To be the highest quality grape grower of Northern California coastal premium wine grapes through the advancement of modern business and viticultural technologies—doing it our way! To realize exceptional returns from farming and grape sales while building an "estate" in vineyard properties. (http://www.beckstoffervineyards.com/about_history.cfm)
- **Fetzer Winery:** At Fetzer we strive everyday to ensure that the wines you enjoy are of exceptional quality and value, while managing our impact on the environment. Working in harmony with nature and with the utmost respect for the human spirit, we are committed to the continuous growth and development of our people, the quality of our wines, and the care of our planet. (http://www.fetzer.com/fetzer/wineries/philosophy.aspx)
- **Bistro Wine Bar:** To bring to the Springfield area a restaurant that will provide excellent food and wine at a reasonable price in a comfortable but refined atmosphere. (http://www.enotes.com/handbook-business-plans/bistro-wine-bar/business)
- **Amorin Cork:** To maintain market leadership and encourage innovation in all its businesses. It is also dedicated to strengthening the Portuguese economy by taking an active role in developing global opportunities for commerce, while being an active steward of the land. (http://www.amorimcorkamerica.com/assets/forms/Amorim-Media-Kit_April-2007.doc)
- **Foster's Group:** Is a premium global multibeverage company delivering a total portfolio of beer, wine, spirits, cider, and nonalcohol beverages. Our products inspire global enjoyment and are enjoyed by consumers all over the world. (http://www.fosters.com.au/about/profile.htm)

A *vision* describes where the company is going and what it wants to be in the future. Usually a vision statement describes an ideal time some 10 to 20 years in the future. According to John Kotter (1996), a clear vision statement

Table 2.8. A Company's MVV: Mission, Vision, and Values

Mission	WHAT business we are in?
Vision	WHERE we are going?
Values	HOW we will act?

paints a graphical picture of the future, is directional and focused, and is feasible and desired. It should be able to be explained in less than 10 minutes and ideally will include a short slogan or tag line, such as Henry Ford's vision of "a car in every garage." Some people in the wine industry have altered this statement to "a bottle of wine on every family dinner table three nights a week." Examples of vision statements in the wine industry include:

- **E.&J. Gallo:** To become the most innovative global marketer and distributor of wines. (http://jobs.gallo.com/whoweare/Corporate.asp)
- **Virginia Wine Industry:** By the Year 2015 the Virginia wine industry will double its market share within the Commonwealth and reach measurable sales on a national level. (http://www.virginiawines.org/strategicplan04.html)

Company values are the beliefs and practices that guide the conduct of the company as it pursues it mission and vision. In general, a business should agree on four to eight key values that are identified by company founders and employees. Some companies engage in a shared values creation exercise in order to gain buy-in and commitment from all employees. Others will have the executive team identify the values, and then communicate them to employees, customers, suppliers, and other key stakeholders. Values should not be pulled out of a hat or copied from other companies. Instead, they should be deeply felt and believed by the executive team, who should also role model implementation of the values.

Common values include integrity, teamwork, respect for the environment, customer focus, safety, and quality. In many cases these terms are modified to reflect terminology used by the individual wine business. Following are examples of some wine business values:

- **Constellation:** People, Quality, Entrepreneurship, Customer Focus, Integrity. (http://www.cbrands.com/CBI/constellationbrands/Careers/Values/)
- **Diageo:** We are passionate about consumers; We value each other; Freedom to succeed; Proud of what we do; Be the best. (http://www.diageo.com/en-row/About-Diageo/OurValues/)

In crafting the mission, vision, and value statements, extra time should be taken to ensure that all members of the executive team buy into the process. This is usually one of the most critical pieces of the strategy development process and can take several days of deep discussion in order to gain consensus and support. However, this is extremely important as these statements drive the purpose and direction of the company, as well as how people behave. Ideally, executives and other employees will later be evaluated in the performance management system on how well they exhibit and implement company values.

Finally, when the strategy development process is complete, the mission, vision, and values are usually published on the company website, as well as included in all major company brochures, literature, and employee communications. Many businesses will create wall posters to display mission, vision, and values in the main reception area and offices.

The next step of the strategy development process is to *write long-term strategic goals* for the business. These may or may not be shared with the public, but are usually communicated to employees. Long-term strategic goals—also called objectives in some companies—are the performance targets a company wants to achieve. Usually they are 3 to

5 years out—matched to the length of the strategy. These long-term goals are used to set the annual objectives during the annual planning meeting.

Long-term goals may be both financial and strategic in nature. Examples of *financial objectives* include:

- percent increase in annual revenues
- desired profit margins
- percent return on capital employed
- increases in earnings per share.

Examples of *strategic objectives* include:

- desired percentage of market share
- strengthening the company brand name
- achieving supply chain management leadership
- new product development
- technological leadership
- achieving lower overall costs than competitors.

In general, these are not as difficult to develop and obtain agreement on compared to mission, vision, and values. However, the executive team does need to discuss them in detail and come to consensus, as these long-term goals help to determine the company strategy.

The last portion of the strategy development phase includes *writing the strategy* document, which is the comprehensive plan explaining how the company will achieve its mission, vision, values, and goals. In general, this is one to two pages in length and ideally includes the type of strategy the company plans to pursue as well as what specifically it plans to do to address each of the five areas on the wine business strategy star.

Major Wine Business Strategies

Various strategy experts have identified different ways of describing generic business strategies, yet these are not

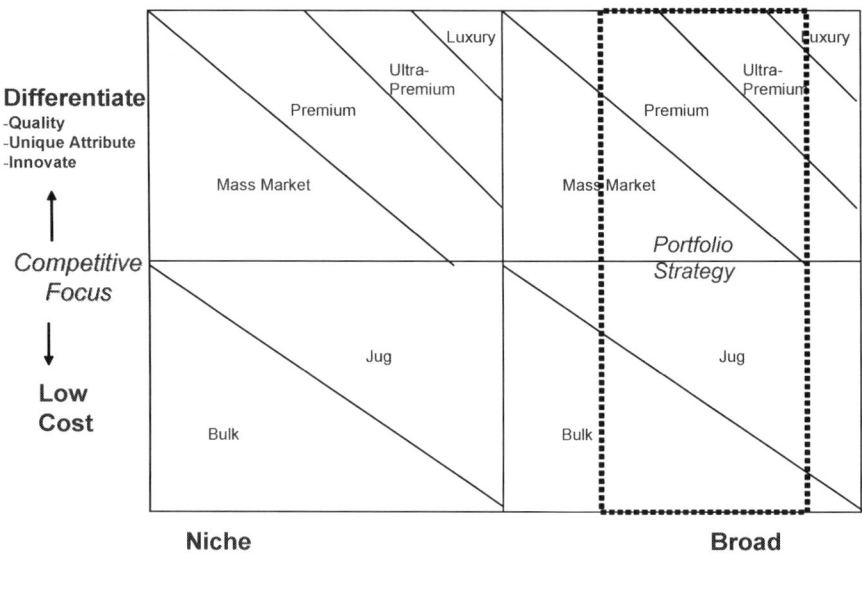

Figure 2.6. Five major strategies adapted to the wine industry.

that simple to apply to the wine industry because of all of the different price points, diverse product portfolios, and appellation issues. Therefore, for the purpose of this chapter, these concepts—based on Porter (1985) and Thompson and Strickland (2003)—have been adapted to the five major strategies illustrated in Figure 2.6. These are: Low-Cost Niche, Differentiated Niche, Low-Cost Broad Market, Differentiated Broad Market, and Portfolio Strategy.

Market Target Axis

When reviewing the matrix in Figure 2.6, it is useful to begin with definitions for each axis. Along the bottom of the matrix is Market Target, which ranges from niche to broad. A *Niche Market* in the wine industry can be defined as a small wine business that produces a product or service for a small geographical area, such as region, state, or province. If it is a winery, in general it produces

10,000 cases or less. A niche market can also mean a focus on a very specific type of market, such as Asian women or people with incomes of $5 million or more per year. A *Broad Market* is a larger wine business that has a target market that is at least national in scope (verses regional) and may be multinational. The larger the wine business and the more countries in which they operate, the further they are placed on the right side of the market target axis. In general, if the business is a winery producing 1 to 3 million cases, they would usually fall closer to the middle of the matrix, whereas large global wineries of 5 million cases or more would fall further to the right on the axis. Broad market can also mean that the customer base served by the wine business is very diverse.

Competitive Focus Axis

This side of the matrix illustrates the competitive focus the wine business is emphasizing. A *Low-Cost* focus is a wine business that strives to deliver the lowest cost for a product or service. In terms of wineries, these types of businesses usually specialize in bulk or low-priced jug or box wines. In the wine industry, a *Differentiated* focus usually emphasizes one of three major areas: quality, unique attribute, or innovation. Each of these are described below in more detail.

- A **quality focus** is an emphasis on providing the customer the best product or service in comparison with competitors. For a winery, this is a more challenging strategy to pursue, because all wines must meet certain quality specifications in terms of taste in order to even participate in the market. Furthermore, many customers equate quality with consistency of taste—which of course cannot always be provided due to vintage variation. The same holds true for grape growers, who may or may not have a good year with Mother Nature. Therefore, the differentiated quality focus is difficult to pursue for a winery unless you either have a large operation with excellent blending competencies or have such a prestigious brand with an allocated customer following that they understand and appreciate vintage variations.

Some examples of wineries that have done an excellent job implementing the Differentiated-Quality focus are Veuve-Cliquot Non-Vintage, Rosemount Yellow Label, and Kendall Jackson Chardonnay Vintner's Reserve. All of these wines have a loyal following because consumers expect they will get the same high level of quality each time they purchase the wine.

- A **unique attribute** focus is an emphasis on a particular attribute that the wine business does very well—usually better than other competitors. For example, a focus on a signature varietal, a special style of winemaking, or even copyright-protected packaging that is distinctive fits this differentiated strategy. The key is to provide something unique to customers that no one else can do as well as you. There may be copycats, but you are considered to be one of the five best in the world in producing your specific product or service. In some cases, a special appellation—if it is famous and small enough—can serve as a unique attribute. The Champagne region, Rutherford Bench in Napa Valley, or Pauillac in Bordeaux can capitalize on this because the size of the region is limited and wineries that operate there have a special competitive advantage just because of their location. Likewise, famous vineyards, such as La Tache, fit this definition. Other unique attributes are: historic buildings; oldest business; most awards; highest consistent ratings; environmental stewardship; unusual vineyard programs; etc.

Some examples of wineries that have done an excellent job implementing the Differentiated-Unique Attribute focus are the five first growths in Bordeaux, Silver Oak in

California, which only produces cabernet sauvignon from Napa or Sonoma aged in American oak, and Mendocino Wine Company, which is the first winery in the US to be carbon neutral.

- An **innovation focus** is an emphasis on being the leader in producing innovative products and services. This strategy is not pursued as much in the wine industry as it is in other industries. It calls for a large research and development budget, as well as continual monitoring of customer trends and hiring or developing an innovative workforce. For a winery, examples might be first to market with a new kind of packaging or a unique marketing promotion. Wineries using this strategy often develop new and unusual brands in-house—rather than acquiring a brand—and even develop new varietal blends or styles. The downside of an innovation strategy is that others quickly try to duplicate what you've done, so you must capitalize on being first to market and gaining a wider market share.

Some examples of wineries that have done an excellent job implementing the Differentiated-Innovation focus are Bonny-Doon in Monterey, California, which has been the first to venture forth with new varietals and light-hearted labels; Lindemans in Australia, which has launched an innovation strategy of bottling wine from different countries around the world under the Lindeman's label; and some of the Champagne houses that have launched Champagne in small bottle with straws for the night-club market.

Putting it All Together

So in putting together the boxes of the matrix, the four strategies of Cost Niche, Differentiated Niche, Low-Cost Broad Market, and Differentiated Broad Market become obvious. However, there is also a fifth that is rather unique to the wine industry and this can be called the *Portfolio*

Strategy. As illustrated in Figure 2.6, it is a Broad Market strategy that encompasses both the Low-Cost and Differentiated quadrants. This is the strategy pursued by most of the very large global wine players such as Fosters, Constellation, LVMH, Concho Y Toro, Gallo, Diageo, and others. They serve global markets at multiple price points, and they can do this competitively because of the economies of scale they achieve from being so large. This allows them to negotiate favorable pricing with suppliers because they order such large quantities of barrels, bottles, corks, and other supplies. They have also developed excellent relationships with distributors and importers in countries around the world. The fact that they operate at all price points allows them to serve multiple consumer segments and to be flexible in their promotion focus based on economic swings and other changes in the global economy. Indeed, most have other product divisions besides wine, such as beer and spirits, or even perfume and luggage, as is the case with LVMH. In this way, if one sector is down, they have others they can rely on to ride out the financial tides.

Obviously not everyone can emulate the Portfolio Strategy because of the vast financial resources and scope needed to implement it effectively. That is why half of them are publicly traded companies in order to obtain the funding for expansion. This type of public scrutiny also requires that they have very cost-efficient operations and hire talented leaders.

So in reviewing all five strategies (Table 2.9), it soon becomes apparent to the wine industry expert that the most commonly pursued strategy is that of Differentiated Niche. This is because there are thousands of small family wineries around the world and most do not pursue a Low-Cost focus. Instead they focus on the varietals that grow well in their region and sell them locally but at a higher bottle price than jug or box wine. A few that manage to excel at

Table 2.9. The Five Strategies Defined

Strategy	Definition
Low-cost niche	Sell product or service at lowest cost to a small and/or targeted market.
Low-cost broad	Sell product or service at lowest cost to a large and/or diverse market.
Differentiated niche	Sell a differentiated product or service based on either highest quality, a unique attribute, or innovation to a small and/or targeted market. Most common strategy in the wine industry.
Differentiated broad	Sell a differentiated product or service based on either highest quality, a unique attribute, or innovation to a large and/or diverse market.
Portfolio strategy	Sell multiple products and/or service at all price points to a very large globally diverse market. Emphasize economies of scale.

the Differentiated Niche strategy—whether that be with an emphasis on quality, unique attribute, or innovation—may grow to be larger and sell their wine in global markets, thus moving into the Differentiated Broad strategy.

The least popular strategy for a winery is Low-Cost Niche, but in most every wine region of the world there are a few wineries and some cooperatives that do pursue this strategy and do it well. The Low-Cost Broad strategy on its own is even rarer, but there are a few companies that implement it perfectly. In the US, The Wine Group, which owns the box wine brand Franzia and other jug wines, has done well in this arena. Likewise, The Bronco Wine Company, located in the Central Valley of California, prides themselves on never selling a wine for more than $10. In France there are some very large cooperatives that also fit this strategy.

Beware of the Muddled Winery Strategy

Though it is true that most small wineries pursue a Differentiated Niche strategy, many do not do it very well. This is because they are not clear on how they are different from competitors. In most wine regions of the world, you will find hundreds of small family wineries all producing the same varietals from the same appellation in very similar styles. They have not clearly differentiated themselves on quality, a unique attribute, or being innovative. Instead many copy their neighbor.

At the same time, in each region, there will be a few wineries that stand out. This is because they are very clear about what makes them different and they use this in all promotions, as well as in internal policies and practices. It is these wineries that should be studied as role models because they have clear strategy implementation.

Selecting the Best Strategy for Your Wine Business

The best way to select the most appropriate strategy for your wine business is to review all of the data from your external and internal analysis. Then identify something that you are very good at doing (and passionate about)—and that no one else is doing or doing very well—and select that as your strategy. This could mean that you do select a Low-Cost Niche strategy, because no one else is doing it—and you know you can do it well! Or you may decide to focus on a new varietal that no one else is producing. Whatever you choose, make sure it can be clearly differentiated and that you have or can obtain the skills and resources to implement it well.

The Eight Complementary Strategic Options

In addition to selecting one of the five wine business strategies, you may also want to consider adopting one or more of the eight complementary strategic options that support the five strategies. These are as follows:

1. **Alliances & Partnerships:** Can you form an alliance or partnership with another company that will help you implement your strategy better?

2. **Acquisitions & Mergers:** Would it make more sense to purchase or merge with an existing business to implement your strategy faster? This is a very common option for the Portfolio Strategy.
3. **Outsourcing:** Should you consider outsourcing certain aspects of your business in order to be more efficient and focused? This is very common for small start-up wineries that often outsource cellar activities.
4. **Forward or Backwards Integration:** Does it make sense for you to add on operations that you currently outsource. For example, if you are a winery and purchase grapes, it may eventually make sense for you to purchase a vineyard, which is considered backwards integration. Likewise, instead of using distributors to sell your wine, you may want to hire your own sales force (where legal), which is considered forward integration.
5. **Product Line Expansion:** Does it make sense for you to expand your product line? Perhaps you are doing really well selling a luxury priced wine, but decide to produce a second label at a lower price? Likewise, if your signature grape is syrah, you may want to begin to offer a sparkling syrah line.
6. **First Mover:** Is there an opportunity for you to be a first mover in a new market or category? If so, do you have the resources to implement effectively? The advantages of the first mover are increased brand recognition and larger market share initially—until the fast followers arrive.
7. **Fast Follower:** Does it make sense for you to adopt a fast follower strategic option where you quickly copy what others are doing successfully and move into their market as the second or third competitor in the category? The advantage of the fast follower is you don't have to invest in the upfront market research and promotion costs.

8. **Build to Sell:** Does it make sense for you to build your wine business and then sell it for a profit? Then you can move on and start another business. This is a common option for wine brands that have become very famous. They are an acquisition target for Portfolio Strategy companies. The owner can sell the brand and then start a new company.

Writing the Strategy Document

Once the mission, vision, values, long-term goals, generic strategy, and complementary strategic options have been selected, then it is time to write the strategy document, which includes all of these elements, plus the specific game plan to implement it. This is where it is time to take a walk around the strategy star and describe what exactly you will do to address each of the five points of the star (see Figure 2.2).

Customers: Based on your strategy, who are your target customers: distributors, retailers, direct consumers? How do you plan do communicate with them: face-to-face, tasting room, sales force, Internet, etc.? What type of marketing and sales strategy will you implement? How will it be different that your competitors? What are your customer goals? What types of customer service will you offer (e.g., 24-hour response to all customer requests; lack of red tape, etc.)?

Operations: Based on your strategy, what type of internal operation will you have? What are the key work processes you will need to implement? Where will you be based? How will you work with suppliers? What will you outsource? What are your quality standards? What are your operational goals? How can your operation be the most effective and efficient in supporting your strategy? What type of organizational structure will you use?

Employees: Based on your strategy, what types of employees do you need to hire to run your operations and service customers? What skills and educational background will they need? Where can you hire them? What training will you offer? How will you motivate and retain them? What types of innovative reward systems can you provide that will support strategy implementation? What human resource policies need to be implemented? What type of leaders do you need? What type of company culture needs to be developed in order to support your strategy? Who will report to whom? How will you promote and develop future leaders?

Distribution: Based on your strategy, how will you distribute your products and services? What is your philosophy in working with distributors? How can you build positive working relationships with distributors? What type of regional, national or global distribution system do you need? What regulations do you need to be aware of in providing your wine business product or service?

Information: Based on your strategy, what are the key types of information that need to flow easily between your operation, customers, suppliers, distributors and other parties? What types of information systems are needed to make this happen? What types of technology do you need to implement? How many meetings should you hold with employees, customers, etc.? What formats will you use for communication (e.g., email, vlogs, newsletters, etc.)? How will you reduce the levels of bureaucracy so you can quickly and effectively meet customer needs?

Format for the Strategy Document for Start-up Winery

There is no one correct way to write a strategy document because they are different for each business. However, in general they are usually 1 to 2 pages in length and include all of the components that have been described in the preceding section on strategy development (Figure 2.7).

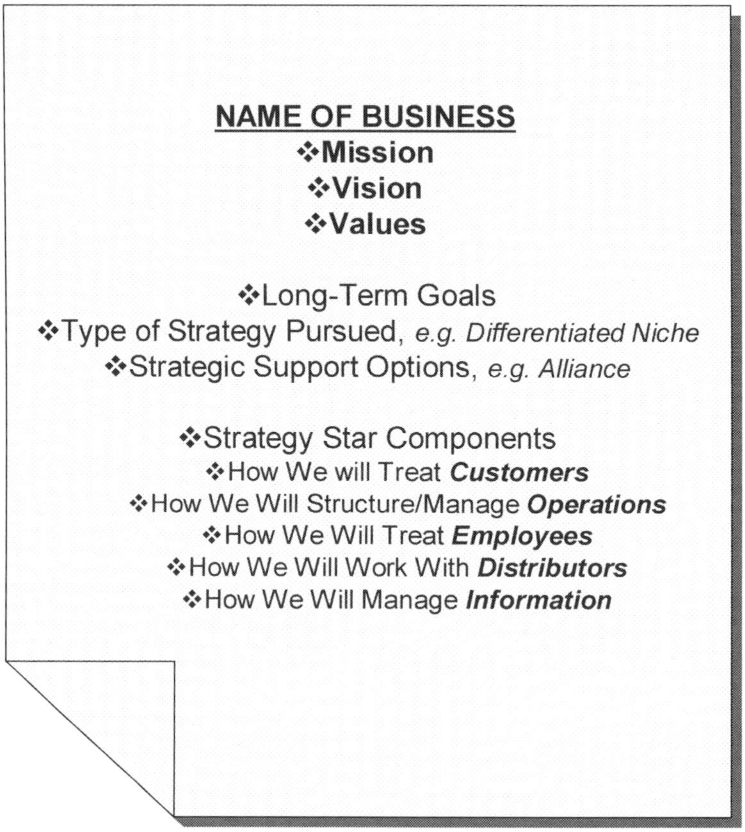

NAME OF BUSINESS
❖ Mission
❖ Vision
❖ Values

❖ Long-Term Goals
❖ Type of Strategy Pursued, *e.g. Differentiated Niche*
❖ Strategic Support Options, *e.g. Alliance*

❖ Strategy Star Components
  ❖ How We will Treat **Customers**
  ❖ How We Will Structure/Manage **Operations**
  ❖ How We Will Treat **Employees**
  ❖ How We Will Work With **Distributors**
  ❖ How We Will Manage **Information**

Figure 2.7. Example of contents of a strategy document.

STRATEGY IMPLEMENTATION AND EVALUATION PHASE

Strategy Implementation and Evaluation begins with an action plan for each component on the strategy star. The action plan includes a timeline, goals, and clear accountability for people responsible for the various processes. In order to begin this phase, it is highly recommended that you involve the people in your organization who are responsible for each of these functions. Therefore, for the customer point of the star, you will want to involve the head of your marketing and sales function. For the operations point of the star, you will want to involve your operations manager. If your business is large enough, you may also have a director of human resources, a distributor relationship function, and a technology officer. However, as most small wine businesses do not have these functions, then divide up these responsibilities among the employees you do have. You may also outsource these functions, in which case you can work with your supplier on building an implementation plan.

In setting action plans for each of these areas, you generally want to create both a long- and short-term implementation plan. The long-term plan is usually 3 to 5 years out and includes clear objectives for each function—derived from the long-term goals on your strategic plan. The short-term plan is your annual plan for the coming year, which will describe what you hope to achieve in the next 12 months towards the long-term goals. For example, using the luxury tannat winery example, the 5-year goal for the customer point of the star may be to place your wine in 80% of the 100 key restaurants you've identified in the eight major city markets. For the first year, you are hoping to achieve 20% of this goal, because you realize you have to spend most of the year doing relationship building with those key accounts, as well as finding distributors and importers to help you distribute the wine in those cities.

Obviously this phase of the strategy development process can take some time. In fact, it is actually the longest of the four phases because after the action plan is developed, the next step is to begin implementing it—which of course, takes the rest of the year. An important part of implementation, however, is ongoing evaluation. Because a good action plan always has deadlines, key metrics for goals, and specific accountability built in, at a minimum you will want to conduct quarterly, if not monthly, reviews of progress. Again, all key employees who were involved in developing the plan and have accountability for implementing it should be involved in the evaluation process.

The format for the ongoing evaluation will vary by wine business, but many companies prefer to hold quarterly meetings where they review progress against goals as an executive team. Then if there are areas where goals fall short, as a team the group can discuss and agree on revisions to the plan, changes to the goal, or other needed actions. In cases in which the team surpasses the goal, they may want to accelerate or increase their original plan. With a new strategy, these ongoing evaluations should be viewed as continual improvement and learning opportunities, rather than a negative setting in which people are publicly embarrassed if they have not met their goals. This is because it is not always certain how a new strategy will unfold—it may give rise to positive results much faster than planned or do the opposite. A wise business owner, with the support of his/her executive team, recognizes this and celebrates and learns from both successes and setbacks.

THE STRATEGY REASSESSMENT PHASE

Strategy reassessment is a miniversion of the strategy development process and is usually conducted each year as part of the annual executive team meeting. Many businesses make this part of their ongoing evaluation process. The purpose of strategy reassessment is to verify that the current strategy is still relevant and to make any needed updates or revisions.

Therefore, at least 1 or 2 months before the strategy reassessment meeting—which is usually held as part of the annual executive planning meeting—someone in the business should be given the responsibility of conducting a mini-external and internal analysis. If there is no one internally who can do this, some wine businesses will hire an external consultant to gather the data and assist in facilitating their annual planning meeting.

The data collection process doesn't have to be nearly as extensive as with a new strategy development process. However it should involve gathering the most recent data on wine consumers, market conditions, competitor positions, social and political trends, changing regulations, and other important external considerations that may impact the wine business. For the internal portion, gathering data on current operational goals and efficiencies, employee competencies, and any key changes you know about your competitors' internal operations—especially critical success factor indices, is important. This information should then be compiled in a report and presented at the annual meeting in order that the executive team can review and make decisions regarding any needed revisions to the strategy and both long- and short-term goals. Obviously, at this meeting the action plan for the next 12 months is also developed with clear goals, metrics, deadlines, and accountability.

Conclusion

In conclusion, strategy development and implementation in the wine industry is not that different from other industries, because the process steps are basically the same. What is different are the thousands of players in the global wine industry from mom and pop wineries to new online wine shops to winery conglomerates like Constellation. This makes for a very fragmented global industry, and therefore it is more difficult to gather competitor and customer data. Also, because there are so many businesses, they operate at all levels of business sophistication, with many very unclear about their business strategy.

At the same time these differences give rise to more opportunities to develop innovative strategies in the wine industry. This is because innovation is often driven by passion and a strong interest in a subject. This is another unique difference of the wine industry. Most people who work in wine do so because of their passion for the product and the beautiful places on earth in which it is nurtured. This type of passion, coupled with a strong understanding of business strategy development, could result in some amazing new wine businesses in the future. At the same time, lack of understanding of the importance of a clear business strategy could also cause some wine businesses to exit the scene. Therefore, the time to prepare is now.

LEARNING OBJECTIVES:
- Identify vineyard selection issues
- Describe business considerations when planting a vineyard
- Explain the annual vineyard management cycle
- Identify important issues in viticulture business
- Gain insight from Beckstoffer Vineyards

CHAPTER 3

THE BUSINESS OF VITICULTURE

Richard Thomas
Professor Emeritus Santa Rosa JC

Mark Greenspan
Viticulturist, Owner of Advanced Viticulture, LLC

David Beckstoffer
President & CEO, Beckstoffer Vineyards

Liz Thach
Wine Business Professor, Sonoma State University

The sight of a vineyard stretching green and verdant along a hillside during harvest is uplifting to behold, especially one that is heavy with fruit and has the potential of creating good wine and a healthy financial reward for the vineyard owner. However, the green of the vineyard doesn't always translate to positive cash flow, as multiple variables come into play regarding the business of viticulture.

This chapter describes some of the basic viticulture issues from a business perspective. It begins with an overview of vineyard selection considerations, including information on appellations, the major types of grape varietals, regulations, and zoning. From there a discussion of the planting issues is presented, including prepping the vineyard, installing irrigation and trellising, and planting the vines. Next, an overview of the annual vineyard management process is provided, with a description of the seasonal growth cycle of the vine as well as farming processes to support it. Finally, some of the current issues facing viticulture in the New World, as well as future issues, are presented. The chapter concludes with an example of how a world-class vineyard is managed at Beckstoffer Vineyards.

Vineyard Selection

In selecting a vineyard location, the most important considerations are *climate and soil*. Though wine grapes can be grown in many parts of the world, there are certain locations that are considered to be ideal. These are referred to as the five Mediterranean regions of the world, namely: 1) the European Mediterranean region (Italy, France, and Spain); 2) the North Coast of California (near Napa, Sonoma, Lake, and Mendocino counties); 3) the South Coast of Australia (near Adelaide); 4) the Central Coast of Chile (between Santiago and Talca); and 5) the central coast of South Africa. These regions all have warm dry summers and cool wet winters, which are excellent climates for growing grapes.

Soil is also important, in that grapes thrive best in slightly rocky soils with a certain balance of acid and nitrogen. For example, soils that have too much clay are more difficult for vines to grow. On the other hand, soils that are too rich in nutrients do not stress the vine enough, so the quality of the fruit is not as intense. Therefore, consideration of soil is very important in vineyard selection. However, in certain cases, it is possible to augment the soil with appropriate nutrients and additional soil amendments to make it more palatable for wine grapes.

The Issue of Appellation

An obvious business consideration regarding vineyard selection is the cost of the land. Vineyard real estate costs are usually dictated by *appellation*. An appellation is a designated wine grape growing region of the world that is defined by soil, mountain ranges, bodies of water, and weather. In the US appellation is often referred to as an AVA, or American Viticultural Area, and must be approved by the government. For example, in Sonoma County, there are currently 13 approved AVAs, ranging from Rockpile, Dry Creek, and Alexander Valley in the north, to Russian River near the coast, and Sonoma Valley and Carneros in the southern part of the county. Grapes from vineyards in these specific appellations can usually demand more money, because the appellation has been verified as an ideal place to grow a specific type of grape varietal. For example, Pinot Noir grapes from the Russian River can usually demand more money per ton than Pinot Noir grapes from other AVAs in different parts of California.

Famous appellations in other parts of the New World include some of the following: 1) Australia is well known for the Barossa Valley, McLaren Vale, and the Adelaide Hills; 2) New Zealand is famous for the Marlborough region; 3) Chile is well known for its Maipo, Rapel, and Maule valleys; 4) Argentina is best known for the Mendoza region, where more than 90% of its grapes are grown; 5) South Africa is famous for the Cape region; and 6) Canada for the Okanagan Valley in British Columbia and its Ontario region for excellent ice wines.

Obviously the more famous an appellation, such as Rutherford in the Napa Valley, the more expensive the price of land. For example, recent prices for Napa vineyard land range from $75,000 to $300,000 per acre. In the Barossa Valley of Australia undeveloped vineyard land ranges from $20,000 to $70,000 per hectare (1 hectare equals 2.47 acres), but that doesn't include the cost of water, which would amount to another $5,000 to $7,000 per hectare. In Bordeaux, France, vineyard land has been sold at ranges of $200,000 to $600,000 per hectare. Though vineyard land fluctuates based on the location and the amount of grapes on the market, land in famous appellations will always be sold at a premium.

Which Grape Varietal To Grow?

Appellation also dictates, to some extent, the type of grape varietals to grow. This is because the climate and soil in a specific appellation are usually appropriate for growing certain varieties of grapes. For example, the Napa Valley is famous for its Cabernet Sauvignon grapes; and Dry Creek Valley in Sonoma County is famous for its Zinfandel grapes. This is because each of these appellations has the perfect mixture of soil, sun, and moisture to nurture these specific types of grapes to a state of high perfection. Indeed, appellation is so important that in some countries, such as France, the government actually dictates the varieties of grapes that can be grown in each appellation.

Though it is possible to grow other types of grapes in an appellation that is known for only one or two varieties of grapes—by modifying the soil and other scientific methods—"Mother Nature" still dictates, for the most part, where specific grapes should be grown. Table 3.1 illustrates some of the specific grape varietals that are famous in New World countries.

Therefore, when selecting a vineyard site, it is important to be aware of what type of grape variety one wants to grow, because the variety can dictate where to buy land. The

Table 3.1. Famous Grape Varietals in New World Countries

Country	Grape Varietal
Argentina	Malbec
Australia	Shiraz
Canada	Vidal Blanc
Chile	Carmenère
New Zealand	Sauvignon Blanc
South Africa	Pinotage
US—Napa	Cabernet Sauvignon
US—Sonoma	Zinfandel

choices in varieties are also complex, as there are more than 5,000 different types of grapes in the world. The most famous ones, however, descend from *Vitis vinfera*, a Mediterranean vine used for most of the famous vineyards and wines in the world. Of these, the most well-known white is Chardonnay, the Queen of Grapes; and the most well-known red is Cabernet Sauvignon, the King of Grapes. Other famous white grapes include: Sauvignon Blanc, Johannisberg Riesling, Viognier, Chenin Blanc, Pinot Gris (or Grigio), Gewürztraminer, Sémillon, and Muscat. Other famous red grapes include: Pinot Noir, Merlot, Malbec, Syrah (or Shiraz), Cabernet Franc, Zinfandel, Sangiovese, Tempranillo, Barbera, and Petite Syrah.

Grape Regulations

In the New World, wines are usually listed by their varietal name on the front of the bottle. However, the percentage of grape variety that goes into the bottle is usually determined by government regulation within the country. In many New World countries, at least 85% of the grapes must be of the variety listed on the label; however, this varies by country. For example, in the US at least 75% of the grapes must be of the variety listed on the bottle. If the bottle says "Chardonnay," then at least 75% of the grapes must be Chardonnay grapes, but the other 25% may be Chenin Blanc, Sémillon, or another variety. However, if the bottle also states that the grapes are estate grown and bottled, then 100% of the grapes must be grown on the estate. Another grape regulation that is becoming more common is when a wine bottle lists a "vineyard designate" of the name of the vineyard in which the grapes were grown. If this is listed on the bottle, then at least 95% of the grapes must come from that vineyard.

This system is different from the original system established in the Old World wine countries. In most of these countries, the grapes that go into the bottle are labeled based on the appellation from which they originate. In this case, it is expected that the consumer is familiar with the type of grapes and the style of the wine from that region. For example, in France if someone buys a bottle of Bordeaux, they know that if the wine is red, it will contain a primary blend of the five red grapes that are grown in Bordeaux, namely: Cabernet Sauvignon, Merlot, Malbec, Petite Verdot, and Cabernet Franc. If the wine is white, it will contain a primary blend of Sauvignon Blanc and Sémillon grapes.

Zoning Regulations

A final issue with vineyard selection has to do with the land zoning regulations in the area in which the vineyard will be established. Most New World countries require that permits must be obtained to establish a new vineyard, and that the land must be zoned for agricultural use. In addition, there are environmental issues to consider, such as the amount of water needed for the vineyard, how the water will be recycled and cleaned if chemicals are used in the vineyard and the water drains into other areas, as well as erosion issues. For example, in many parts of California, there are hillside vineyard ordinances that require new vineyards with a slope of more than 30% to complete an engineering evaluation and implement a plan for erosion control.

Other zoning issues have to do with obtaining permission to cut down trees, and/or replant trees if taken down due to vineyard establishment. For example, in South Australia, many of the native Eucalyptus trees were chopped down years ago when agriculture first came to the McLaren Vale. Now the government has implemented a program to replant many of the native trees near the vineyards and old orchards. Related to this is the impact on wildlife. Special fencing may be required to keep deer and kangaroos outside of vineyards so that they cannot eat the leaves. In addition, gophers, which may attack roots, and certain species of birds, which may eat the grapes, must be considered.

A final vineyard zoning issue has to do with neighbors, who may have houses near the vineyard. In California, land must be zoned for agriculture use, and neighbors living next to a vineyard are often concerned about the noise of tractors or other equipment, as well as drifting sulfur dust or other chemicals that may be used for pest control within the vineyard.

In conclusion, based on all of the issues that must be considered when selecting a vineyard, it is important to obtain a vineyard real estate agent and other expert advisors to assist in the selection and permitting processes.

Planting the Vineyard

Once a vineyard sight has been selected, and clearance on appropriate permitting has been verified, the land can be purchased. Many vineyard owners need to obtain a loan to purchase and develop the land. Banks and other financial institutions will require a soil sample report, a cost–benefit analysis, and a marketing plan—ideally a signed contract from a buyer to purchase the grapes upon maturity.

The Soil Sample Report

It is important to analyze the soil in the vineyard by taking samples of soil from various parts of the vineyard site, as well as conducting a visual appraisal of soil structure and texture. The sample is then sent to a lab for analysis. The resulting report will describe the type of soil (e.g., clay, sand, loam), the pH level of the soil, calcium, magnesium levels, water holding capacity, and other important information. In addition, it will suggest the type of grape variety and rootstock that will perform best in the particular soil and appellation, as well as any additional nutrients that should be added to the soil, such as lime. Finally, it will provide information on the amount of water the vines will need. This information is used to help calculate the cost of establishing the vineyard.

Cost–Benefit Analysis

Developing a cost–benefit analysis for a loan on a vineyard usually requires an expert in finance and accounting, and is covered in later chapters in this book. However, the basic elements of this on the *cost side* include: cost of the land, cost to prepare the soil, cost of the grape vines, trellising, irrigation system, fencing, other equipment, installation, and vineyard maintenance. The *benefit side* of the equation is the "potential yield and price of the grapes over the years." In general for yield, most grape varieties start producing a 10% yield at 3 years of age; 30% yield at 4 years of age; 70% yield at 5 years of age; and then 100% from age 6 to approximately age 30. From age 30 on, most grape vines produce less fruit, but at usually the same or higher level of quality. For this reason, most tables illustrating net returns begin at year 4 and extend for 30 years.

Regarding the price side of the equation, most New World wine countries measure this as price per ton or tonne (1000 kilograms or 2200 pounds in US: 1 kilogram = 2.2 pounds)

THE BUSINESS OF VITICULTURE

or *gross dollar per acre or hectare* (2.47 acres = 1 hectare). The price fluctuates based on the supply within an appellation as well as global supply of a specific grape; however, it is possible to calculate an estimate of the price per ton range based on historical prices within a region. This is usually what is provided as part of a cost–benefit analysis as part of a loan process to purchase vineyard land, and is calculated in best, average, and worst case scenarios.

The Marketing Contract

Because grapes are an agricultural product, and are dependent on the good will of nature, as well as the cyclical nature of grape supply (see Figure 3.1), it is important for growers to secure a long-term contract with a winery before

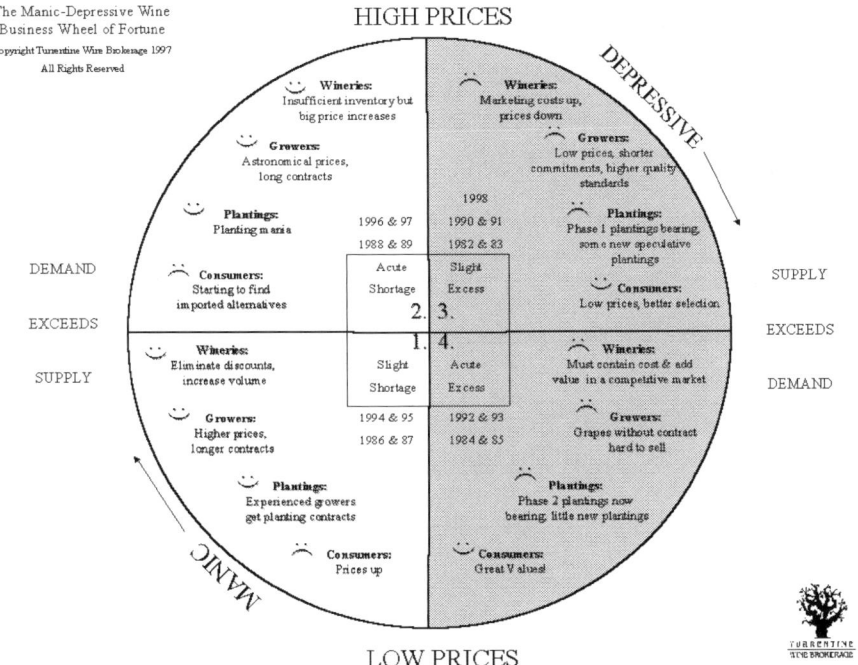

Figure 3.1. Cyclical nature of wine industry. Reprinted with permission of author, Bill Turrentine, Turrentine Wine Brokerage, LLC, 2003.

planting a vineyard. This negotiated and written contract establishes the price the winery will pay for the grapes over a period of time, usually 3 to 5 years. It also describes how the vineyard must be managed, including amount and type of irrigation, fertilizer, thinning, and harvest specification. Any deviations from the contract may result in a reduction of the agreed upon price, or even cancellation of the contract. Most financial institutions will not provide a vineyard development loan without a written contract.

If a grower doesn't have a long-term contract, he or she may choose to sell their grapes on the spot market, which may fluctuate quite broadly, depending on the global grape supply. There are some years when growers actually make more money selling their grapes on the spot market, but if they are bound to a written contract this is not possible. The trend, however, is for growers and wineries to establish long-term buying relationships that are of a win–win nature for both parties.

Preparing the Land

Once the vineyard has been purchased and any needed financing has been secured, the land can be prepared. This includes some type of "ripping" or "slip-plowing" of the soil so that the grape vine roots can penetrate the various layers within the soil. In addition, all of the old roots and large rocks are removed from the field. A special type of tractor is usually required to do this type of work. Prepping of the land in this manner usually occurs in late summer, and then a thin layer of straw or hay is placed over the soil so that it can rest and erosion is prevented until planting of the vines in early spring.

Laying Out the Vineyard

Determining the number of rows in a vineyard, as well as the spacing between rows, is part of the design, or laying out, of the vineyard. This is done differently based on the needs and philosophy of the grower. Traditional vineyard row spacing in the New World has been wide (12×8 feet in the US) so that tractors can travel between the rows. This is similar in Australia and New Zealand, as they perform much mechanical harvesting and pruning of their vineyards. However, current trends are towards narrower row spacing, such as 6×8, 6×9, 6×10 ft, or 1×1 meters (vine×row spacing).

In the Old World, most vineyards were traditionally designed with tighter spacing (1×1 meter), because many were established earlier and based on hand harvesting. Philosophies are also different, however; many New World countries believe they can achieve higher yields and higher quality by planting the vines further apart, but with higher trellising. On the other hand, many Europeans argue that even though each of their vines yields less because they are so stressed due to tight spacing, their quality is higher (vines need to be stressed to produce higher quality). They also argue that they produce just as much—because they have more vines per acre/hectare. Regardless of the choice, it comes down to the particular philosophy and needs of the grower, and now many different types of spacing are found around the world.

Planting the Vines

Once the vineyard layout has been determined, and the grape variety to be planted has been selected (e.g., Zinfandel), there are two other important decisions to make regarding the type of rootstock and the age of the vine to be planted. Regarding *rootstock*, this is "the root of a grape variety to which a fruiting vine is grafted." Some grape vines are planted on their original roots, but a large majority are grafted to a heartier rootstock that is resistant to soil-borne diseases or pests, and/or because they will grow well in certain types of soil. Some of the more common rootstocks are St. George, 110-R, 1103 Paulsen, 101-14, and 5C.

Many of the rootstocks are from vines that are native to North America, which have been grafted with the *Vitis vinifera* grapes from Europe. This combination creates a hearty vine, and was used to replant many of the vineyards of France in the late 1800s after they were ravaged by phylloxera, a root insect related to aphids. Interestingly enough, one of the few countries in the world that is still planted to the original *Vitis vinifera* rootstock from centuries ago is Chile. This is because Chile is more isolated from the rest of the world, and the diseases and pests that have infiltrated the vineyards of other countries have not entered Chile. There is some concern on how long this may continue, however.

The second planting decision after choosing a rootstock is whether to use "green vines" or "dormant benchgraft vines." Green vines are newly grafted vines that are less than 1 year old. They were grown in a greenhouse from a piece of grafted rootstock and have green leaves on them. Dormant benchgrafts were also grown in a greenhouse as green vines, but then were planted in the ground for 1 year and then removed in the winter when they were dormant. They look like a stick and have no leaves—similar to any bare-rooted plants available at nurseries.

There are pros and cons to using either choice. The advantage of green vines is they are less expensive (US$2 to $3 per vine) and can be planted later in the season (as late as early summer). The disadvantage is they require more water, are less hardy, and require 1 more year before they produce grapes, in most cases. The advantage of dormant benchgrafts is they can be planted earlier in the season—late winter to late Spring (March–May in the Northern hemisphere; July–September in the Southern hemisphere); generally they take 1 year less to produce fruit, are hardier, and require less water. The disadvantage is they usually cost at least US$1 to $2 more per vine than green vines.

The holes in which to plant the vines usually must be dug to at least 2 feet in depth. In large vineyards, special equipment is used to do this. However, in smaller, steep, or rocky vineyards, this is still often done by hand using a pick-axe and shovel. Once the holes are dug, any additional soil or nutrients may be added, and then the vine is placed in the hole and the soil is filled in around it. Often growth tubes, which look like plastic tubes, are placed over the young vines to protect them from the elements and pests (deer, rabbits, etc.). In this way, they begin growing safely within the tubes, which are removed when the vine is older and has climbed out of the tube. Obviously the cost of digging the holes, planting the vines, and installing the growth tubes is part of the business and cost issues regarding vineyard installation.

Installing Trellising and Irrigation

Another important viticultural business issue is installation of a trellising and irrigation system. Both of these are costly items but, once installed, will last for years. There are many different types of trellis systems in various price ranges that can be used. The most common in New World countries, and often the least expensive, is the traditional vertical shoot positioned (VPS) trellis. With this system, the vine grows up a central iron pole, and then is trained "to the cordon"—that is, the strongest shoots from the vine are tied horizontally to wire (either on one or two levels). From the cordons tied to the horizontal wire, additional shoots grow upward, leaving the fruit clusters (grapes) to hang down below the cordon.

Selection of the most appropriate trellising system often depends on the appellation and the amount of sun exposure to the canopy each day. Canopy management—using trellising systems to manage how much sunlight the grapes receive, thus impacting fruit quality—is one of the important areas of viticulture. This is why visitors will see many different types of trellising systems in the New World countries.

It is worth noting that the historical method for trellising was to leave the vine alone. In this case, it naturally grows in the form of a short "tree or bush." In Australia, these old vines are referred to as "bush vines." In the US they are called "head-pruned vines," in reference to the method in which they must be pruned. In ancient times, the most popular trellising systems were to let the vine grow up a wall or a tree.

Irrigation systems also differ by vineyard and country. Most modern vineyards use automatic timers to water the vineyard through "drip" irrigation lines, which are hung along the bottom wire of the trellis system. These irrigation systems often serve a dual purpose of watering and delivering a liquid fertilizer as needed. Historic vineyards relied upon the rain and fog for moisture, or used flood irrigation methods that usually involved diverting a local river to flood the vineyards at certain times of the year. This method is still used in parts of Chile and Argentina. In many appellations of France, it is against government regulation to irrigate vineyards, so growers there rely totally on "Mother Nature." This has led to the practice of a "vintage" year, meaning "the year in which the grapes got ripe," due to the correct amount of natural moisture and sun.

VINEYARD MANAGEMENT

Once a vineyard has been planted, it must be managed on a yearly cycle. This involves hiring workers to manage the process, as well as purchasing or leasing the appropriate equipment and supplies. Because the vineyard is governed by nature, its schedule is also adjusted to the seasons. Following is a high-level outline of the annual vineyard management process.

Spring: Bloom (Northern Hemisphere: March–June; Southern Hemisphere: September–November)

In the spring, the dormant vine goes through bud break, in which the first green leaves shoot from the wood. There is very rapid growth of the leaves and multiple shoots streaming from the main cordons at this time. The vine can actually grow up to 1 inch in length per day. Each of these shoots will form two clusters of small flowers (similar to a lilac flower). These will "bloom" when the individual flower "throws the petals" and some tiny green balls are left on the vine. These will later grow into the grape cluster. During this time of year, the major concern is both frost and rain, which can damage both the leaves and fruit cluster. Therefore, various types of technology, such as heaters, large wind machines, and sprinklers are available for use in the vineyard to combat this threat. In addition, there is potential danger from insects and disease, which may call for pesticide management. Finally it is important to "sucker" the vine during the spring, or cut off any new shoots that grow near the roots or trunk. This allows the energy within the vine to focus on the fruit clusters.

Summer: Veraison (Northern Hemisphere: June–September; Southern Hemisphere: December–February)

During the summer, the vine will pour all of its energy into the fruit cluster in an attempt to ripen it. As the sun continues to shine on the vine, the fruit cluster will go through a process called *veraison*, in which each small grape will become larger, change to its mature color, and begin the ripening process. (All grapes are green when first formed.) Colors can range from pale yellow, light green, red, purple, and an inky blue-black—depending on the grape variety. During this season, it is important to thin the leaves as part of canopy management to ensure the fruit cluster receives enough sun and that too many vines don't drain energy from the grapes. In addition, it is often necessary to "thin" the clusters by cutting off the excess fruit clusters, so that more energy can be directed to the larger and healthier ones. This also helps to ensure the quality of the grapes for intensity of flavor. Finally, continued pest and disease management is necessary, such as watching for "powdery mildew," which results from too much moisture.

Autumn: Harvest (Northern Hemisphere: September–November; Southern Hemisphere: March–May)

In autumn, the grape clusters become ripe enough to be harvested. For this to occur, the *brix* (sugar level within the grapes) must reach a certain level. This is usually between 22 and 26 degree brix (percent sugar), but also depends on the grape variety and region. For example, in the Marlborough region of New Zealand, it usually doesn't get hot enough for the grapes to reach this level of brix. Therefore, they often harvest their famous Sauvignon Blanc at levels ranging between 17 and 21 brix. Another consideration is the acid and pH level of the grapes. All of this calls for much measuring of the grapes, which is usually done by hiring workers to pick small samples of the grapes and analyze them in a laboratory. Once it is determined that the correct level of sugar, pH, acid, and flavor is reached, then the harvest begins.

Harvest is an incredibly busy time of year in a vineyard and winery. Workers often work very long hours to ensure that all of the grapes are picked at the perfect time. Most of this work occurs early in the morning, when the grapes are firm and the sun hasn't warmed them to a higher temperature. The grapes are either hand-picked by workers or harvested by machine, and then rushed to the winery in large trucks. The concern is to get the grapes to the winery to be crushed while they are still fresh.

In Australia, where they don't have a large agricultural labor force, they use machines to harvest over 90% of their vineyards. This is actually more economical, and allows them to produce good wine for a lower price than some other countries. In some of their vineyards, they actually crush the grapes into must in the vineyard by machine, and then freeze it—before transporting it thousands of miles by truck to the winery. Crushing in the field and keeping the grapes cold with dry ice (CO_2 cap) allows them to maintain the quality level of the grapes, yet also produce an economical wine.

A final concern during harvest is the weather. Rain is about the worst thing that can happen during harvest, because it causes the grapes to become water logged and lose flavor. This is bad for quality. On the other hand, weather that is too hot or too cold can also cause problems during harvest, as the grapes either become too ripe too fast or do not ripen in time. It is for this reason that growers around the world watch the weather reports anxiously during harvest. It is also the reason that Mediterranean regions, with their hot dry summers, are so ideal for grape growing.

Winter: Dormancy (Northern Hemisphere: December–February; Southern Hemisphere: June–August)

After harvest, the vines slowly begin to "go to sleep." First their leaves begin to change colors, into golds, yellows, oranges, and reds. Then the leaves fall to the ground, and the vine is left as a black skeleton of itself. It rests and lets the winter rains soak into its roots. The grower finally gets to rest also for a month of two after harvest, but then must begin the pruning season. In many parts of the New World, this is done by hand, so workers again must go to the fields and individually prune each vine so it can grow again in the spring. Pruning a vine correctly takes special skill, and is also dictated by the type of trellis. Again, in certain countries, and especially Australia, most vines are pruned by machine, where "hedge or box pruning" of vines is very popular. Finally, fertilizer may be applied to the vines during the winter so that they can begin their growth cycle in the spring, strong and vibrant.

The Issue of the Maturing Vineyard and Regrafting

Some growers and winemakers prize grapes from old vineyards. Indeed, in many parts of the world there are vineyards that are well over 100 years in age still producing grapes. However, because yield decreases as the vine ages (though many say quality increases), growers concerned with higher financial returns will often tear out an old vineyard (30+ years) and replant. Though it will take them approximately another 3 to 5 years before the vine production reaches an acceptable level, many believe this is preferable to harvesting smaller yields.

Another issue has to do with regrafting of a vineyard that is producing a grape variety that is not as popular or is oversupplied. For example, when there was a surplus of Chardonnay on the market several years ago, many vineyards grafted their vines over to Sauvignon Blanc or Pinot Gris. Though this can be costly in terms of labor—each vine has to be regrafted by hand with a graft from the new variety—it is a faster way to convert a vineyard than tearing out vines and replanting. Both of these are business decisions that need to be analyzed carefully in terms of cost–benefit, as well as forecasting future market needs.

How Many Bottles of Wine Does Each Vine Produce?

A mature grape vine can produce approximately 10 to 12 pounds of grapes (22–26 kilograms). This translates into five bottles of wine at 750 milliliters each, or 1 gallon (3.7 liters). Another way to view this is approximately 589.5 grapes are needed to produce one bottle of wine. This varies, however, because different grape varietals have different sizes of grapes.

If analyzed from the per ton perspective, 1 ton of grapes equals approximately 160 gallons of wine. Since there are five bottles of wine in each gallon, then 1 ton yields approximately 800 bottles of wine. This is a useful equation when determining production for a total vineyard. For example, if a small 2-acre vineyard generally yields 4 tons per acre, then the total bottle production of the vineyard would be 6400 bottles (2 acres × 4 tons × 800 bottles). As a case has 12 bottles, the approximate case production for the vineyard would be 533 (6400/12).

IMPORTANT ISSUES IN VITICULTURAL BUSINESS

In examining the business of viticulture today and in the future, several important issues arise. Each of these can impact the profitability of vineyards in various New World countries, depending on the scenarios that continue to occur now and in the future.

New Technology in the Vineyard

The advances made in technological equipment for the vineyard over the last decade or more have been prolific, and have resulted in increased product quality and cost savings for those companies that use them. The majority of the technologies enable vineyard managers to collect data that can be stored, easily retrieved, and analyzed in a database, and facilitate more efficient, information-based farming decisions. The acquisition, evaluation, and application of diverse agricultural data types over time and space is often referred to as "precision farming," and is already in use in other agricultural industries. Currently, major groups of vineyard technology can be grouped into three categories: 1) automated weather monitoring and telemetry/control equipment, 2) GPS, or global positioning systems 3) database systems, and 4) sensor technology. Each of these is described in the following paragraphs.

Weather Monitoring Equipment is used to monitor humidity, temperature, wind, and solar radiation in the vineyard. The equipment is usually set up on large poles at various locations within the vineyard, and collects information 24 hours a day. The stations transmit weather data frequently using radio signals to a computer base station. Other systems have moved to a Web-based software platform, where users may access their current and historical information from any computer or portable Internet device from anywhere in the world.

The base station (or Internet-based) software determines critical information regarding the progress of disease and insect pest populations (which develop according to weather patterns) and regarding how much water is being used by the vineyard. Employing the vineyard pest and disease information, the vineyard manager can then examine the data to determine the appropriate amount and timing of pesticides and fungicides that are needed. This is an improvement from traditional practices where large vineyards were sprayed on a set schedule, because there was no data available indicating the need for these inputs. This can result in large cost savings, because those materials need only be applied when necessary and not on a constant interval, as was done conventionally. Savings are realized in material costs, labor costs, and equipment maintenance costs. Vineyard sustainability is achieved through the judicious use of such vineyard protectants.

Vineyard water use information from the weather stations allows growers to irrigate according to the needs of the vineyard, and not simply by a set schedule or a gut feeling. The information provided is a quantity of water that is evaporated from the vineyard, for which the grower can replace all or a portion of through his irrigation system. Scheduling irrigation using this information results in cost savings from reduced water costs, pumping costs, labor costs, and system maintenance costs. When combined with drip irrigation technology (where water is applied in controlled amounts to each vine) the water status of the vineyard can be manipulated to provide the best quality of wine from each location. Excessive irrigation degrades fruit and wine quality and overstimulates vine growth, while, on the other hand, insufficient irrigation leads to vine stress, reduced productivity, and poor fruit maturation.

The weather stations have other benefits as well, including alerting vineyard managers to frost danger in the Spring or Autumn, so that immediate action can be taken to protect the crop (e.g., use of heaters, sprinklers, wind machines, etc.). Many a vineyard manager has been awoken in the middle of the night by a phone call generated automatically by a weather station alerting him that temperatures are approaching levels that can destroy their crop if not protected. Current technology allows for text messages to be sent to the user, alerting him or her to the specific location of the weather event.

Additionally, the weather data is stored in a computerized database, and can be analyzed from a historical perspective, which allows vineyard management to track local climate and microclimate characteristics and make more informed decisions about how to farm the vineyard in the future.

Weather stations are not being used only for weather monitoring. The state of the art includes stations that not only monitor weather, but can monitor any other type of electronic sensor. Irrigation flow sensors, soil moisture sensors, water depth sensors, etc., may be attached and monitored by the weather stations and the information provided along with the other weather data. Furthermore, the weather stations have the ability to actuate irrigation valves, frost control fans, and anything else that may be turned on or off by a switch.

The weather station technology has grown so much that tracking of weather conditions is just a start. The units act like little "robots" in the vineyard that can be controlled automatically or by remote control.

Global Positioning Systems (GPS) are small satellite receivers that can be handheld, placed on the roof of a harvester or tractor, or used from an airplane. The receiver collects data from satellites rotating around the earth. The data reports the exact coordinates of where the receiver is located on the earth to within 3 to 4 feet (higher accuracy devices are also available). This technology has grown so much that it is now a part of the consumer market, finding its way into automobiles, cell phones, and even golf bags! This detailed geographical data is very useful for vineyards in several ways. For example, a handheld device is useful when a viticulturist or technician is monitoring a vineyard for disease or pest damage. Pest and/or disease severity may be assessed throughout a vineyard and entered into the device at any number of locations. The severity indices are mapped to each pinpoint location using the GPS signal. When the viticulturist returns to the office, the information can be downloaded to a computer and a severity map can be generated using the set of observations just collected. The vineyard manager can use this map to generate an application map of, for example, a fungicide to control bunch rot. Rather than spraying the entire block with the material, only the portions of the vineyard having significant levels of bunch rot need be treated, thus saving time, labor, and material costs. Taking it one step further, an application map can be generated and uploaded to a unit mounted to the sprayer or tractor that is also equipped with a GPS unit. The application map can be used to automatically control the amount of material being sprayed based on the tractor's location in the vineyard.

On a harvester, a GPS can be useful in tracking the location in the vineyard where the highest and lowest grape yields are harvested. For example, mechanical harvesting machines can be outfitted with sensors that automatically weigh the fruit as it is being harvested in each part of the vineyard. This data is combined with the GPS location to identify which locations in the vineyard produce the highest and lowest yield, as well as to identify patterns of yield variability within the vineyard.

GPS technology is often used on airplanes when taking aerial photographs of vineyards. The aerial images provide an assessment of vine size throughout the vineyard and are geographically referenced to the same locations as the yield information and can be overlaid onto one another. Using special geographic software, the vine size and yield information can be analyzed to identify areas of the vineyard that are overcropped and/or undercropped. Vineyard managers can then assess what might be causing this variability, and take actions to balance the vineyard yield, such as pruning styles, bunch thinning, fertilizer, or irrigation— depending on the situation.

GPS technology is being used in conjunction with *vineyard databases* and GIS (geographical information systems) to facilitate visualization and management of spatially variable data. But the databases themselves have developed into essential components of a data management system.

Tremendous amounts of data are tracked within vineyard operations, not only about viticultural conditions, but labor hours, equipment and other resource allocation, pesticide applications, infrastructure maintenance, etc. As regulations become more stringent, keeping logs on paper or even on spreadsheets becomes inefficient. Vineyard databases have been developed to assist the viticulturist and vineyard manager to not only log the data, but to retrieve the data at any time in the future.

Computer-based databases are certainly the foundation for these systems, but the real workhorse of the technology has been the Personal Digital Assistant (PDA). The PDA, a small portable computer, may be carried into the shop, yard, or vineyard where data is entered using a stylus, barcode scanner, RFID (radion frequency identification), or even a camera. PDA technology has advanced such that the units are very low cost. To add field ruggedness, the units may be placed into hard shell cases. Alternatively, there are several field-grade PDAs that incorporate a PDA into a hard, rubberized case that is intended to be "drop proof." GPS receivers may be used with the PDAs, either by plugging them into standard receptacles on the PDA or through a wireless Bluetooth data connection.

Once the data has been collected in the PDA, it can be synchronized with a central database, which may be located on a personal computer or on a server. PDAs may use a hard-wired Internet link or may, alternatively, synchronize using a wireless Internet link. The wireless technology, especially when using a cellular-based wireless connection, allows for rapid synchronization of data between field personnel and the farm office. This provides the potential for almost real-time data acquisition from the field to the vineyard managers.

Sensor Technology is used to measure the characteristics of a grape. Newer sensors use reflected light in visible and near infrared light bands to measure grape composition. Grape samples are taken in locations throughout the vineyard during the ripening phase of the vineyard season. These samples are analyzed using special equipment that immediately determines not only the brix and acid level of the grapes, but many other characteristics that are important to producing the best wine from each vineyard. This data allows both the vineyard manager and the winemaker to make adjustments to their "farm plan" regarding irrigation and fertilizer amounts, as well as canopy management techniques, to ensure that the mature grape is of the highest quality. Most importantly, though, the information provides real-time information as to when the grapes should be harvested. Harvesting at optimal ripeness for flavor and texture of the wine is the single most important decision of the season. With the rapidity of measurement made possible by advanced sensor technology in combination with GPS technology, the assessment of grape maturity may be done rapidly throughout a block to determine regions of a block having different levels of ripeness. Harvest of portions of blocks can then be performed to capture fruit at optimum ripeness while awaiting harvest of less ripe sections of the vineyard.

It should be clear that each of the types of technology do not exist only in their own realm. Rather, they all link up to form a complete information management system. The current and future trends in technology are that synergies are forming among the individual technologies, integrating vine and grape sensors with spatial technologies (GPS and GIS), weather and climate data, and recording the data in a large multiuser database.

In conclusion, all of this new technology in the vineyard not only helps to increase the quality of the grape harvest, but also results in cost savings due to savings in materials and labor. Large vineyards around the world are already using most of these technologies, and expect to apply newer technologies as they are created. Eventually, it may be possible to gather specific data from an individual vine in a 500-acre vineyard across the globe and act upon it accordingly. All of this begs the question, however, of what will happen to the small family vineyard still farmed in a traditional manner without the luxury of expensive technology. It is possible that many of these smaller vineyards will be acquired by the larger vineyard corporations. It is also feasible that some of them will continue to farm in a traditional manner, in which the human eye and hand will nurture the vine instead of a machine. The potential exists, however, as cost of technology continues to fall, that some of the technology will be adopted even by smaller growers.

Labor Versus Mechanization Issues

The increase in technology use within the vineyard raises the question of the fate of the vineyard worker, who has traditionally managed, pruned, suckered, and picked grapes by hand. Now, in the larger vineyards, much of this work is done by a machine that can both prune and harvest the vines. In addition, new trellising is being used not only to facilitate the machine harvester, but also to reduce the amount of time spent on leaf pulling, so workers are no longer needed to do this either. It is only in the small, prestigious vineyards, or those planted on hillsides, that traditional vineyard workers are still needed. Even though it has been proven several times over that mechanically harvested wines are of the same quality as hand-harvested wine, many winemakers still retain a bias for hand-harvested grapes in high-end wines.

Interestingly enough, many of the vineyard workers in the US and Chile are reporting that they would rather do other types of work anyway. Working in the vineyards was something their parents and grandparents had to do, and it is no longer fashionable for many of them to work the vines. This trend, which is resulting in a shortage of vineyard workers in the US, is also driving a need to adopt more mechanization in the vineyards. Those workers who do remain must become "knowledge workers" and learn to use the new computer and satellite technologies that are infiltrating the vineyards.

Organic and Biodynamic Vineyards

Another trend that is receiving more attention in viticulture is that of organic and/or biodynamic farming. Originating in Europe, many consumers there are demanding products that are grown without pesticides. Organic farming is usually defined by a government body, and if food products are grown and produced according to their regulations, they can be labeled "organic." In the wine industry, *organic viticulture* is basically growing grapes without the use of synthetic pesticides or fertilizers. The only exception to this is sulfur, which is used to combat powdery mildew, and is considered organic because it is a natural substance. Organic viticulture has been adopted by both small and large vineyards in several areas of the world. It is also being used as a consumer marketing tactic to appeal to those customers who will only purchase organic products.

Biodynamic farming is more extreme than organic farming, and is more philosophical in nature. Based on the work of Rudolf Steiner, it is defined as "a science of life-forces, a recognition of the basic principles at work in nature, and an approach to agriculture which takes these principles into account to bring about balance and healing" (Biodynamic Farming and Gardening Association, 2008). Biodynamic farming promotes the use of organic fertilizers and farming based on the cosmic rhythms of the earth. Currently biodynamic viticulture has only been adopted by a few vineyards in Europe, the US, New Zealand, and other countries, but has been growing in popularity.

ISO Certified Vineyards

Related to the trend of organic viticulture is another trend that is starting to be driven by consumers in Europe and New Zealand. This is ISO certification for vineyards. ISO, which stands for the International Standards Organization, certifies business processes that have been documented and proven to result in consistent product manufacturing practices, as well as continuous improvement. Within a vineyard, the certification is called *ISO14001* and can be described as an environmental management system that requires vineyards to document inputs and outputs, and analyze the effects these have on the environment. It also supports sustainable agricultural practices. To date, several vineyards in Europe, as well as seven in New Zealand (Sileni Estates, Ata Rangi, Palliser Estate, Martinborough Vineyard, C.J. Pask, Vidal Estate, and Mission Estate Winery) have been certified. The growth of this trend will most likely depend on how important it is to consumers.

Growth of Vineyards Around the Globe

A final trend to consider for viticulture is the growth of vineyards around the world. Many vineyards in Eastern Europe are coming into full production. In addition, India and China are planting large vineyards. For example, in the last 10 years, China has invested much money into developing large vineyards in the northern area of their country, near Mongolia. They have received much consulting support from the French and Australians, and appear to be able to produce and harvest grapes at a very economical price, due to their low labor costs. Currently it is not clear that the wine quality is sufficient to be exported globally, but the rate of wine consumption is climbing steadily within the country.

The question of what will happen to grape and wine prices, with so many vineyards being developed throughout the world, is one that needs to be asked. With wine consumption still at a small growth rate in most parts of the world (or decreasing in France and Italy), what changes and new trends will this increase in wine grapes have in the future?

Vineyard Management at Beckstoffer Vineyards

Beckstoffer Vineyards was started in Napa Valley, California, in the 1970s by Andy Beckstoffer. Uniting a passion for wine and a keen knowledge of business, Andy began purchasing prime vineyard land in the Rutherford Bench appellation of Napa. As his reputation for growing high-quality wine grapes grew, many of the top wineries approached him to purchase grapes. Today Beckstoffer Vineyards has become the largest independent vineyard owner in the North Coast of California, owning and operating approximately 3000 acres of vineyards in the Napa Valley, as well as Mendocino and Lake Counties.

Viticultural practices at Beckstoffer Vineyards are driven by a clear mission and strategy. The stated mission is "To be the highest quality grape grower of Northern California coastal premium winegrapes through the advancement of modern business and viticultural technologies—doing it our way! To realize exceptional returns from farming and grape sales while building an 'estate' in vineyard properties." The company has grown steadily since its inception, frequently taking advantage of opportunities to expand its operations in the North Coast. Each of its three regions is managed under a clear strategy. In the Napa Valley, the emphasis is on producing the highest quality grapes using the best processes, people, and technology. The emphasis here is not to be the low-cost producer, but to be the best and produce the highest quality grapes. The strategy in Mendocino and Lake Counties is also to produce high-quality grapes, but with an emphasis on the cost side of the equation as well, because these regions currently do not command the high prices seen in Napa appellations.

Much of the success of Beckstoffer Vineyards lies in its sustainable farming practices and use of advanced farming techniques. The company has a strong commitment to sustainable farming and has successfully integrated practices including cover crops, biological pest controls, optimized water management, habitat preservation, and erosion control practices into their farming plans. They also own several organically certified vineyards in Mendocino County.

Beckstoffer uses some of the newest vineyard technologies to monitor and analyze their vines, including weather monitoring stations, GPS-based mapping, geographic information systems (GIS), and plant moisture sensing tools. The data are collected, analyzed, and used as a tool to make viticultural and farming decisions on matters such as irrigation scheduling, pest control, nutrient applications, and other vineyard management issues. They also employ modern farming technologies such as automated irrigation systems, pulsating sprinklers for frost control, and hand-held computers to increase farming efficiency in the vineyard. Mechanical harvesters are used in parts of Mendocino and Lake Counties to help reduce costs, but all of their Napa Valley vineyards are hand-harvested at the request of their winery customers.

Much emphasis is placed on progressive management practices and employee relations at Beckstoffer Vineyards. Decisions involving vineyard investments and redevelopments are assessed using computer financial models and discounted cash flow analyses to calculate returns. Custom-built agricultural accounting software enables managers to receive monthly reports that compare actual costs, labor productivity, and equipment use to budget. Additionally, much importance is placed on positive employees relations, with strong efforts to ensure that employees are well trained on viticulture and safety practices, as well as how to operate vineyard equipment. Employee surveys and performance reviews are scheduled on a regular basis to provide opportunities for constructive feedback from employees on working conditions, and frequent employee events promote the family atmosphere that is important at Beckstoffer Vineyards. Furthermore, all employee communication is provided in both English and Spanish to support the hundreds of Spanish-speaking field workers. Three separate office locations are established for the more than 200 full-time workers, and competitive pay and benefits are provided for all employees. Finally, Beckstoffer employees take a lead in participating in community activities and in supporting local associations, such as the Farm Bureau, Napa Valley Grape Grower's Association, and California Association of Winegrape Growers.

Beckstoffer Vineyards operates with a viewpoint of "partnership with the customer." Beckstoffer will create a "farm plan" for each vineyard block, and then discuss the plan with winery clients so it can be customized to fit their specific winemaking needs. The quality and history of certain Beckstoffer vineyards have become so well known that several wineries designate the specific Beckstoffer vineyard of origin on the wine label.

Indeed, "keeping the customer in mind first" is one of the success philosophies of Beckstoffer, and is used in much of their decision making. They look at their winery clients as partners rather than merely buyers and see themselves as an integral part of the final product, not just a materials supplier. Prices of many of Beckstoffer Vineyards' grapes are, in fact, indexed to the final price of the bottle of wine that they go into, thereby sharing the risks and rewards with their winery partners.

At Beckstoffer Vineyards, vineyard acquisitions and redevelopments are, to a great extent, based on "knowing the market." Knowing when to buy, when to plant, and what to plant can be the difference between success and failure in a business where there can be several years between the time you purchase a vineyard and the time you produce your first grape.

In conclusion, Beckstoffer Vineyards is an excellent example of a vineyard operation that is focused on the triple bottom line. It operates with practices that are economically viable, environmentally friendly, and with a clear focus on customers, employees, and community.

LEARNING OBJECTIVES:
- Identify business decisions for the six major steps in the winemaking process
- Explain the pros/cons of using multiple labels for winemaking
- Gain insights from Penfolds winemaker, Peter Gago
- Identify new technologies in winemaking
- Describe current and future issues impacting enology

CHAPTER 4

THE BUSINESS OF ENOLOGY

Linda Bisson
Professor of Viticulture & Enology, University of California, Davis

Roy Thornton
Professor of Enology, California State University—Fresno

Peter Gago
Winemaker, Penfolds

Winemaking is a complex process. Not only is it formed from a marriage of science and art, but it is also shaped by the whims of "Mother Nature" and the financial dictates of business practices. This can create both opportunities and challenges for the winemaker, and potential clashes with the winery CFO, as the business of winemaking unfolds to create the glories of a fine wine that can generate a positive financial return for the winery.

This chapter provides an overview of the major business issues regarding enology (the study of winemaking). It includes a review of the basic winemaking process and associated cost issues. In addition, it explores the decision-making process regarding first, second, and multiple labels. Finally, this chapter provides an overview of some of the new technologies and methods being explored in winemaking, as well as future issues impacting enology. The chapter ends with a description of the role of the winemaker, with special emphasis on the tension between the "art" and "business" of winemaking.

Business Decisions in the Winemaking Process

From a business perspective, the basic winemaking process can be broken down into six major steps. These are: 1) harvest and crush; 2) fermentation; 3) aging; 4) blending, stabilization, and finishing; 5) bottling and labeling; and 6) storage. The decisions and methods used in each of these steps can vastly impact the total cost of production, as well as the final qualities and styles of the wine. It is for this reason that a winery needs to be very clear on the type of wine they want to produce and the market segment to which it is directed. Are they in business to produce high-end luxury or artisan wines at very high price points; do they want to produce midprice wines—the "fighting varietals"; or are they in the jug, or value wine business? Table 4.1 describes some of these price categories. Another option is to produce wine at two or more of these price points and create second or multiple labels to distinguish the wines.

A newer trend is to identify the preferred flavor profiles of specific consumer segments and craft a wine designed to please the palate of the consumer. Australia has invested much money in this type of research, and has been successful at producing fruit-forward, less-oaked wines that

Table 4.1. Standard Wine Price Categories

Jug	up to $3
Popular premium	$3–7
Fighting varietals	$7–10
Classics	$10–14
Premium	$15–25
Ultra-premium	$25 and up

are very popular with consumers around the world. Now a few other wine countries are beginning to follow this trend, which is different from making wines that please the palate of the winemaker or wine critics or adhere to a strict regional definition of "terroir." Clarity on strategic direction, however, is critical in making business decisions about the winemaking process. Wine style strategy drives everything from where the grapes are sourced, to the types of fermentation tanks, barrels, and packaging used. However, once a decision has been made, a wine business can move forward in the winemaking process with clarity of costs, rates of return, and other business issues.

Step 1: Harvest and Crush

The process of negotiating and contracting for grapes was described in Chapter 3, but because the price of the grapes can be the most expensive component in a bottle of wine, it is important to reiterate this point. In a luxury wine, the winemaker will want the highest quality of grape possible, and will dictate farming and harvesting requirements to the vineyard. This may include how much water, fertilizer, thinning, etc., is used, as well as at what flavor, brix, and acid levels the grapes should be harvested. All of this impacts the total cost of the grapes, and dictates the cost of a bottle of wine. However, it is important that the costs of each step of the wine production process be known and calculated on a per bottle basis to guide winemaking practices and keep per bottle costs in the desired range.

There is an old adage in the American wine industry that states if you divide the price per ton of the grapes by 100, then you will know the minimum you need to charge for the bottle of wine. For example, if you paid $2800 per ton, then you need to charge at least $28 per bottle. Costs per ton generally reflect costs of farming, and many of the practices employed for the production of grapes destined for luxury wines increase the cost of viticultural input. Obviously this doesn't hold true in every case, or every country, but it illustrates the impact of the price of grapes on a bottle of wine. An alternative to per ton pricing is per acre pricing. In this case the costs of farming are covered directly by the purchaser. This allows the winery to fully understand the cost of vineyard operations and to decide if they are really warranted. It is also linked to payment on quality, rather than tonnage.

There are also business issues associated with the crush process. This is mainly tied up in the cost of the crush equipment and labor. A small winery may choose to outsource the winemaking process to a "custom crush facility," or third party, that will crush, ferment, age, and bottle the wine for them. This is a common strategy for beginning wineries that may not have the capital to invest in crush equipment, or may not have the proper permits to establish a bonded winery.

However, there is a point when it becomes more economical to invest in crush equipment. This usually includes a grape drop bin hopper, sorting table, destemmer, crusher, press (can be one piece of equipment), hoses, cleaning equipment, pumps, and other associated items. The type of available equipment varies. For example, batch or continuous presses can be purchased, and each contributes different attributes to the juices produced. Oftentimes for a new winery only one type of item can be purchased. The winemaking management team will have to decide on

THE BUSINESS OF ENOLOGY

what equipment will best meet their wine business strategy. Chapter 5 of this volume provides more detail on some of the pros and cons for the decision points on this topic.

The crush process is different, depending on the type of wine. White grapes are destemmed and crushed or can be sorted whole to be immediately pressed to yield juice. Red grapes are crushed, but fermentation proceeds in the presence of the skins and seeds in a tank. Juice containing skins and seeds is called "must." In some cases, red wine fermentations contain intact berries or clusters. Table 4.2 illustrates some of the basic differences.

Step 2: Fermentation

The fermentation process is critical to producing a good-quality wine. Fermentation is the process wherein *the sugar in the grapes is converted to alcohol and carbon dioxide*. For this to occur, yeasts must be present. There are yeasts naturally occurring on the grape surface or found on the surfaces of the winery that can grow in juice or must and conduct the alcoholic fermentation, which is then called a

Table 4.2. The Crush Process

White Wine	Red Wine	Blush Wine
1. Grapes are sorted	1. Grapes are sorted	1. Red grapes are sorted
2. Grapes are destemmed & crushed (usually)	2. Grapes are destemmed (usually)	2) Grapes are destemmed and crushed
3. Juice is drained from the grapes and skins are removed	3. Whole grapes and/or whole clusters, or crushed grapes with skins, are put in fermentation tank	3. Red grapes pressed and skins are removed
4. Juice is put in the fermentation tank	*Juice is pressed from grapes later	4. "Pink" juice is put in fermentation tank

"native flora fermentation," but many winemakers will also inoculate a wine with a *cultured yeast* strain. The main reason is that cultured yeasts give the winemaker more control over the fermentation process. This is because some native yeasts are unpredictable and may get "stuck," and the wine will not complete fermentation on time. They may also emit off odors. This leads to loss of tank space, requires manipulation of the arrested fermentation to get it to complete, and more careful monitoring of the fermentation, all of which can increase the final cost of the wine. Not only can it be costly to get the wine "unstuck," but off-characters may form, resulting in an inferior wine that must be discarded, treated, or blended out. Therefore, a sound management process may be to "take out an insurance policy," by using native yeasts on 10% of the juice or must, but inoculating 90% of it in a different tank with a good cultured yeast strain. This way, if fermentation starts to slow or develops an off-character, the winemaker can add yeast from the "insurance tanks." This is a good safety value that can keep costs in check and allows for the development of a good-quality wine.

Other obvious cost issues with fermentation are the types of tanks or barrels used. Most wineries ferment in stainless steel tanks with temperature controls. This way they also have the option to conduct a "cold soak" pre-ferment if they choose, and ensure during the heat of fermentation that the wine does not rise above 85 degrees (the point at which delicate flavors begin to burn off and fermentation may stop). However, some wineries prefer to ferment in large vats/foudres. This was the traditional method from the Old World, and is still used in some countries today. For example, some of the smaller wineries in Chile still use large oak fermenting uprights, and the Mondavi Winery in Napa switched to using specially designed large oak fermenting uprights for its high-end red wines. Though the uprights were quite expensive, Mondavi believes that they produce a higher quality wine based on their cost–benefit analysis.

During the fermentation process, most red wines will produce a "cap" of grape skins that float to the top of the tank. This cap contains a lot of the tannins, flavor, and color or *anthocyanins* that create the unique quality of the red wine. These components must be extracted from the skins in the cap during the winemaking process. Therefore, winemakers usually "break the cap" several times a day to mix with the rest of the wine. This allows a better mix in the tank moisturizing the cap and aeration to help the yeast. It also allows the heat of the fermentation to be used for extraction of cap components. This process, called "punching down," can be done manually with large paddles or with special equipment that is either built into the fermentation tank or lowered over the tank. In one variation of this technique, the entire cap is submerged in the tank. If a more temperate approach is needed, one can more gently irrigate the surface of the cap to extract components without risk of mechanical rupture of cap components. This can be done with a pump by pumping fermenting juice from the bottom of the tank, the racking valve, over the top of the cap. This process is called "pumping over." It is also possible to take all of the juice to another tank, allowing the cap to settle to the bottom, then return the juice to the original tank. This is called "rack and return." Which process is used depends upon the varietal, the level of extraction achieved, and the style of wine being produced.

The fermentation process can last from a couple of days to several weeks, depending on the type of wine being made. The process is finished when most of the sugar has become alcohol, which causes the yeast to die. The resulting alcohol level usually ranges from 8% to 16% (MacNeil, 2001). Obviously, the length of the fermentation impacts the cost of the wine, as does the need for refrigeration. In addition to ethanol and carbon dioxide, fermentation also produces heat. If too much heat is produced by the yeast, the wine can lose volatile characters and take on a "cooked" character. Fruity white wines and blush wines take less time to ferment, but may require more cooling. They generally cost less to produce than red wines because they are typically not aged as long. Some wines will go through a second fermentation, which is called *malolactic fermentation*, and is described in more detail in the aging section.

When the wine is finished fermenting, it is racked from the tanks (with red wine it is also drained off the skins and pumped to a different tank) and pumped into either another tank or oak barrel for aging. When red wine fermentation is complete, this is called "free run" wine. The remaining skins and wine are usually gently pressed again to get additional wine, usually with higher temperatures, which may be mixed with the free run wine or used for a different label. The remaining material—bits of grape debris, skins, stalks, seeds—is now called pomace and is usually used as mulch and placed back in the vineyards, though some enterprising wineries will sell it to other companies.

Acetic acid bacteria can also grow on the surface of or in the wine, producing acetic acid and ethyl acetate, giving wine a vinegar flavor or volatile acidity. The wine must be protected against oxygen to prevent this from occurring. *TCA taint* is produced by mold growing on wood and wood products and interacting with chlorine-containing compounds. TCA is a volatile compound that can infect wine by spreading through the air. It causes a dank, musty smell in the wine that can be detected by consumers in very low concentrations. TCA also affects natural corks randomly and can cause up to 1 in 12 bottles to have off notes. If this invades a winery, it is very costly and time consuming to get rid of. That's why the best wineries are very clean, and require much labor to maintain.

Following is a quote from Simon Blacket at Wolf Blass Winery in the Barossa Valley of Australia (L. Thach, personal interview, July 2003). He describes some of the sophisticated winemaking techniques regarding fermentation, as well as cost-savings measures they use in the winemaking process:

> We've upgraded the new Wolf Blass winery with the newest and best wine-making technology in the world. For example, each of controls on these stainless steel fermentation tanks is linked to a computer network. A winemaker who is traveling in Asia, can log onto his or her laptop to check the progress of the fermentation in this very tank, and make adjustments to the temperature from thousands of miles away.
>
> When it is finished, this facility will have the capacity to handle the equivalent of 80,000 tons of grapes....Every thing that goes into the wine making process here is recycled in some form, and if possible, used to gain revenue. For example, we actually sell our treated waste water to the local golf club. Everything here is operated with the utmost efficiency for both quality of product and cost-savings, which helps drive revenue.

Step 3: Aging

Aging of wine is done to allow the flavors to more fully develop; sometimes to permit the wine to take on some "oak" flavor (when newer oak barrels are used); to allow polymerization reactions and softening and stabilization of the flavor and to encourage all of the elements to balance out. It can last from a few weeks to several years, and can greatly impact the cost of a wine. The costs are associated with the amount of time a wine is "tied up in the aging process," as well as the cellar space and overhead costs for storing it. In addition, wine can be aged in less expensive stainless steel tanks, or oak barrels that range in price from US$300 to $1000. The price is dependent on where the oak is from (which can range from Romania to the Nevers Forests of France), oak age, as well as the cooperage and process used to make it.

Cooperages, which craft the oak barrels, use a variety of methods. In addition to determining the type of wood from which the barrel will be made, they also "toast" the inside of the barrel with fire and smoke. The old-fashioned and more expensive method is to place the barrel over a real fire. A less expensive method is to use a torch gun or ceramic heaters. The level of "toasting" determines what type of flavor will be imparted to the wine—ranging from a light toast to a heavy, spicy mocha toasted flavor. Some barrels are produced using hot water or steam rather than a flame to shape the staves.

To complicate matters further, an oak aging barrel will impart oak flavors for only about 3 years. After that, there is less flavor imparted and the barrel is usually used for storage, for aging where oak extractives are not desired, or is sent to the flower garden. Luxury wines are usually made in very expensive barrels, which are often used for 1 to 2 years only. Some winemakers will blend wines made from new and older barrels; and some will blend wines made from barrels from different countries. The type and age of the oak barrel can make a big difference on the final taste, quality, and cost of the wine.

A less expensive method for imparting "oak flavors" is to use *oak "tea bags,"* chips, or *oak staves*. They can be put inside a stainless steel tank or added to an older barrel that has lost its flavor. These methods are often used for less expensive wines.

> A potentially critical problem can surface during the aging process and the process can be affected if proper procedures are not followed in sterilizing and cleaning all of the equipment each time wine is moved from one container to another. If this is not done correctly, then certain yeasts

and bacteria can grow in the barrels and taint the wine. The most common of these is "Brett character" caused by growth of the yeast *Brettanomyces*. Brett character is the subject of much study currently; some claim that a little of it enhances the wine and gives it more "interest." Virtually everyone agrees that any of the "barnyard, horsey" Brett character results in spoiled wine.

The issue of how much oak flavor to add to a wine is one that has received much attention in the international wine press. Some countries, such as Italy, have traditionally used very little oak in their popular Chiantis and Sangioveses, whereas others, such as Spain, often age their high-end Tempranillos for 5 years or more in oak. Americans have been unjustly accused of "adding two-by-fours" to some of their California Chardonnays, whereas the Australians have countered this trend by using absolutely no oak in a number of their Chardonnays and labeling the wines as "unwooded." The issue of how much oak consumers like in their wine is one that is being researched in more detail through the use of flavor profiling. At the end of the day, it is the grape and terroir that drive the flavor; the oak only plays a supporting role.

Two other components of aging are *malolactic fermentation* and *racking*. Malolactic fermentation usually occurs naturally with red wines and some whites when they are barrel aged. It occurs when malolactic bacteria in the wine convert the "malic" (tart) acids to "lactic" (milk) acids. It often creates a creamier taste to the wine plus a buttery aroma while softening the acidity, which is usually considered desirable.

Racking is allowing the particles within the wine to settle to the bottom of the barrel or tank, and then drawing off the clear wine into another tank or container. Wine may be "racked" several times during the aging process. It not only helps clear particles from the wine, but also adds some air, which may enhance the mature flavor of the wine. The amount and timing of oxygen exposure during aging can be important to the softening and development of mature flavors. This exposure to air can occur naturally during movement of the wine between tanks or barrels or can be done deliberately and with no movement via the process of "microoxygenation." In this case, minute amounts of air are deliberately bubbled through the wine to increase the exposure of wine components to oxygen.

All of the components of the aging process impact the price and taste of a wine, and are also perceived as part of the "artistry" of winemaking. Creating the perfect flavor and balance of a wine depends, in part, on the aging process.

Step 4: Blending, Stabilization, and Finishing

Many would argue that step 4 in the winemaking process, especially the blending component, is where the true artistry of the winemaking process begins. In the blending phase, the winemaker blends together wine from various tanks or barrels to create a perfect harmony of taste and structure. France has always been lauded as the birthplace of blending, as anyone who tastes a great Bordeaux can attest. These wines are usually blended from five grapes: Cabernet Sauvignon, Merlot, Petite Verdot, Malbec, and Cabernet Franc. However, the blend is never the same from year to year, as the weather and other factors impact the composition of the grapes. Therefore, the winemaker must taste and blend different amounts of wine from each varietal to achieve the harmony of the whole "Bordeaux." Only a master winemaker can perfect this blending process.

However, in many New World countries, single varietals may be bottled, with 100% of the wine coming from the same variety. This is not always common, but it does occur. The type and percentage of wine that can be blended is also dictated by country regulations. For example, in most countries, if a wine is advertised as a single varietal on the label, at least 85% of the wine in the bottle must be of that grape varietal. However, it can come from different barrels or tanks, and be made from different types of oaks and yeasts. Therefore, there are endless permutations in the making of a fine bottle of wine, and every year it may be slightly different. Thus, many people refer to the blending process as the artistic heart of the winemaking.

Once the blending process has been completed to the satisfaction of the winemaker, many wines will also need further stabilization or flavor adjustment before going to the bottle. The processes of fining and filtering may be employed. This is not always necessary and depends upon the clarity of the wine post-aging and blending. The purpose of *filtering* is to clear up any small particles floating in the wine. Some settling will occur naturally so not all wines will need to be filtered, but if cloudiness remains suspended in the wine filters can be used that will remove the particles either via a mechanism of exclusion of the particles from the wine due to the small pore sizes of the filter or due to adsorption of the particles to the filter matrix.

Fining refers to the addition of agents to the wine. Some fining agents may be used to help particles agglutinate so that they can be removed more easily by filtration. Other fining agents are used to remove undesired components from the wine. Clays will bind to proteins and remove them. Some of the protein fining agents, like egg whites, will remove phenolic compounds and can make the wine less bitter or astringent. Fining agents can also be used to adjust the color of the wine. Fining is controversial, because, if overdone, it can strip the wine of flavor. Filtering has less of an impact on wine as it does not remove flavor or aroma components, but if ultrafiltration is used the macromolecular structure of the wine may be impacted. Both fining and filtering are focused on improving both the visual clarity of the wine and microbiological stability.

The wine must also be stabilized against unwanted reactions occurring postbottling. Protein removal discussed above is done to make sure a protein haze does not form in the bottle. It may also be important to prevent tartrate crystallization in the bottle. To do this, the wine is supercooled, which catalyzes the crystallization of the tartrate so that it can be easily removed from the wine. This way the crystals, which are mistaken for ground glass or small diamonds by many consumers, will not be present in the wine at the point of sale or after it is chilled in the refrigerator. It may also be necessary to make sure the wine is microbially stable. If there are residual nutrients it may not be possible to control microbial growth in the bottle with only an addition of sulfur dioxide. In this case, it may be necessary for the wine to undergo a sterile filtration as it is being bottled. This is especially true if there is any residual sugar in the wine. The "sight" of the wine in a glass is one of the five phases of wine appreciation, as described in Table 4.3.

Step 5: Bottling and Labeling

Many small wineries will outsource the bottling and labeling process, because it is an expensive process that is only conducted a couple months out of the year. Investing in a complete bottling line doesn't always make sense for a new winery, when they can hire someone else to do it. One of the more popular methods is hiring a "mobile bottling

Table 4.3. The Five "Ss" of Wine Appreciation

First S: Sight	to view color & clarity
Second S: Swirl	to volatilize the esters
Third S: Sniff	to smell the aroma
Fourth S: Sip	to taste (gargle in mouth)
Fifth S: Swallow/Spit	enjoy the finish

line," which arrives on a large truck and is set up outside the winery.

However, most larger wineries will have their own bottling line on premise, and may even offer bottling services to other wineries for a fee. Business issues regarding bottle and label type and design are often regulated to the marketing division of a winery, but it must be done with consideration of wine production costs. A standard bottle size and design are usually less expensive to process in terms of overall material and processing costs than a specially designed bottle that requires hand-labeling and perhaps etching. Again these are all decisions that must be considered as part of the overall wine strategy. Table 4.4 lists the major components for a wine bottle, and a range of costs (Perdue, 1999). Obviously the use of half-bottles, magnums, and other odd-sized bottles would have different prices ranges.

The bottling process is usually a very noisy one, as the bottles clank down an assembly line with workers who guide them through the filling, labeling, corking/closure, and quality control process. Workers must use hearing protection and be properly trained on the operation of the equipment as a safety precaution.

Step 6: Storage

The last major winemaking process that has business cost implications is that of storage. Most wines, once bottled,

Table 4.4. Sample Bottling Costs for Standard 750-Milliliter Bottle

Bottle	$0.20–$0.95 per bottle
Cork	$0.08–$0.35 per bottle
Foil	$0.00–$0.25 per bottle
Label	$0.03–$0.35 per bottle
Bottling, corking process	$0.18–$0.35 per bottle

are stored for a time so that the wine may "marry" or harmonize within the bottle. This is called *reductive aging* or *postbottling maturation* (MacNeil, 2001). The cost issues obviously have to do with the storage space and temperatures, as well as the delay in getting the wine to market. Many wineries will rent storage space at special warehousing facilities designed to store wine at appropriate temperatures, as well as ship it to distributors and customers, whereas others maintain in-house storage. New World wine countries often differ on their philosophy regarding bottle aging of wine. Chile, for example, tends to bottle age the majority of its high-end red wine for 1 to 3 years before release. Some of the older wineries in Chile, such as Santa Rita, have large underground caves and cellars with stacks of unlabeled bottles filled with wine, which, once aged, are dusted and labeled for shipping. Australia and the US bottle age some wines in shorter time segments, usually 6 months to 1 year. There are some exceptions, however, such as the famous Penfolds Grange, which is first aged in oak for up to 2 years, and then bottle aged for another 3 years before release.

THE USE OF MULTIPLE LABELS IN WINEMAKING

Many beverage industries will produce more than one type of beverage and use different labels or brands to differentiate types. Coca-Cola, for example, sells regular Coke, as well as Diet Coke, Caffeine Free Coke, and other brands. It even alters the taste of its regular Coke for different countries and cultures around the world, based on the preferred taste profile of its major customer segments.

The wine industry also uses multiple labels, but usually to distinguish its high-end, more expensive wines from the less expensive ones. This is particularly useful for a winemaker, who may discover that the quality level of a grape crop from a specific block is not good enough for the high-end label, so it is rerouted to their second-tier label. A good example of this is Opus One in Napa, California. They produce a very high-end "Bordeaux" style red wine, which retails for approximately US$160. However, they also sell a second label called "Overture," which retails for US$50 and is only available for sale at the winery. Many other wineries around the world follow this tradition; however, they do not always advertise to the customer that they produce the secondary label. It may be sold through different channels and not be associated with the winery.

The use of secondary or multiple labels is a business decision made by the management team and winemaker. Many small wineries will begin with only one label, but later on see an opportunity to procure grapes or bulk wine at a good price, and therefore will expand into a "second" product line. The advantages of doing this are the flexibility of cashing in on a good opportunity, as well as maintaining the high quality of the first label and not diluting the brand. The disadvantage is potentially flooding the market with a wine that is difficult to sell.

A growing trend in the use of second labels is that of "private labeling," in which a grocery store, restaurant, airline, or cruise ship wants to sell a wine with their own label. They contract with a winemaker to produce a wine that matches their image and price range, and then sell it at their establishment under a different label. In most cases, the name of the winery and winemaker who made it is never divulged. Large grocery stores in the UK, such as Tesco, and Safeway in the US are currently using this process. This is an excellent method for wineries to sell more wine and potentially enhance revenues.

The Role of the Winemaker at Penfolds

The role of the winemaker at Penfolds is threefold: 1) to ensure the grapes are picked at the optimal time to maximize the fruit quality, 2) to preserve those fruit characters by ensuring the length of time in barrels and the temperatures at which the wine is stored is optimized, and 3) to blend the right grapes and varietals to achieve the style in which the winemaker pursues. It is easy to make great wine if the grapes coming from the vines are of superior quality.

Winemaking has evolved over the last few years in that while there are still many differences within regions and states amongst winemaking techniques, there are fewer differences around the world than perhaps 10 years ago. For example, many of the regions around the world are using Italian filtration systems, German presses, and Australian rotary fermenters. The Old World, as defined in the wine industry, has adopted this new technology as well as the New World of wine. The technology does not replace the "art" of winemaking, but rather just allows fine winemakers around the world do what needs to be done on a much larger scale. This willingness to use new technologies has allowed great wines to be produced on a much larger scale than before. Advancements such as air-conditioned warehousing, filtration methods, computer-controlled fermenting, stainless steel crushers, and microoxygenation have all contributed to the ability to produce fabulous wines on a commercial scale.

Metaphorically, it is like a manufacturing a stereo. Once a person could buy a stereo in which all of components were produced by one manufacturer. Now the best quality stereo manufacturers purchase components from around the world from the suppliers who build the best components. When these are combined together in the final stereo system, the resulting sound quality is far superior to that produced by the old stereo manufacturers who made all of their own components. This is similar to winemaking when looking at all the technological components now used in producing fine wine.

Some of the future trends in winemaking may include the usage of enzymes, the prevalence of adding acids, increasing flavor by increasing alcohol to up to 15%, and, of course, the one that nobody wants to mention—genetically modified grapes. All of these future possibilities are under exploration today and some are utilized to different degrees. However, it would be premature to say they are the norm or standard in winemaking.

At Penfolds, the role of the winemaker involves a number of major objectives. First, it is the responsibility of the Penfolds winemaker to maintain the style differentiation of the various Penfolds wine ranges, while optimizing the quality of any given vintage. Penfolds traditional strength is its ability to blend across variety and area to maintain a "house style" for many of its wines (e.g., Grange, Bin389, Koonunga Hill). That said, it also makes many wines from single regions (e.g., RWT Barossa Valley Shiraz, Coonawarra Bin 128 Shiraz) and it also creates single vineyard wines (e.g., Magill Estate Shiraz, Clare Valley organics). To optimize the quality within each style, more time is spent in the vineyards batching grapes and keeping parcels separate. This leads to enhanced grower relationships—paying by the hectare instead of by the ton—and resulting in smaller yields, but higher grape quality. In other words, it is the responsibility of the winemaker to make the best possible wine within the style that the brand has delivered over time. Thirdly, and perhaps most importantly, as a Penfolds winemaker, one of the key responsibilities is to look ahead 5 to 10 years by trialing and experimenting with new styles and varieties. This keeps the brand innovative and fresh while respecting its history. Finally, as all winemakers will attest, there is a need to minimize mistakes and wine faults. While it takes fabulous grapes to make great wine, there is still an obligation to overdeliver at all price points and optimize grape resources accordingly.

New Technologies in Winemaking

The study of winemaking is a fascinating one, as there are ongoing attempts to identify new methods and technologies to enhance the taste of the wine. This has resulted in increased quality and consistency of winemaking around the world. Some of the new technologies that are being emphasized today are described.

Chemical Profiling of Wines

Wines are a complex mixture of compounds and it is often difficult to determine which compounds are associated with, or responsible for, specific flavor or aroma attributes of the wine. Many wineries are using sophisticated spectroscopic chemical analyses to profile all of the chemicals in their wines to start to tease apart the ones that are most important for their style and consumer profile. In this case the specific compound does not have to be rigorously identified, and in general many attributes are the sum of our simultaneous perception of multiple chemicals, just the pattern of chemicals. Winemakers can then follow the changes in concentrations of these components and use that information to guide winemaking practices.

Reverse Osmosis and Spinning Cone

Though these technologies to remove alcohol from wine have been used for several years now, they are becoming more prominent as consumers increasingly ask for lower alcohol wines. In addition, with warmer vintages, grapes sugars have been rising and so there is sometimes a need to reduce alcohol after the wine has been produced. Reverse osmosis employs the use of a large generator-type machine that passes a percentage of the wine (usually 10%–20% of the total wine lot) through a filter. The filter separates the water and alcohol from the color, flavors, and other key elements. The water and alcohol combination is then distilled so that only the water remains. The water is then recombined with the other components. This smaller percentage is then blended back into the rest of the lot of wine, thus lowering the total alcoholic content of the wine. Spinning Cone Technology involves a series of double metal cones within a large cylinder. Again a certain percentage of wine (usually 20%–25%) from a lot of high-alcohol wine is poured in the spinning cone. It passes through the cone twice, with the flavors, color, and other elements being extracted on the first pass, and the alcohol extracted on the second pass. The dealcoholized portion is then blended back into the remainder of the lot, which lowers the total alcohol content.

Microoxygenation

With this process tiny bubbles of oxygen are released into the bottom of the fermentation tanks. This creates reactions that modify tannins, remove green characters, and stabilize color. The more mature taste of the wine is considered desirable. What is exciting about the process is that it softens the wine by reducing astringency/phenolics and shortens the time the wine must age in barrels. For example, a wine that may have been barrel aged for 18 months in the past may now only need to age for 12 months. Therefore, microoxygenation can save a winery money and increase quality. The aim of this technique is better *tannin management*. Tannins were previously thought to just be related to wine astringency and mouth feel. It is now known that tannins play a more important role in the sensory and taste perception of wines so there is strong interest in learning how to manage the types of tannin present in a finished wine. Microoxygenation is only one of the tools that allow manipulation of tannin content. There are a few downsides to the process, however, and winemakers must be careful not to over do it. Oxygen can have many effects on a wine including stimulation of the growth and metabolism of unwanted microorganisms. This can cause some backlash effects in the taste of wine and result in increased costs in the long run.

Co-pigmentation

Another exciting new discovery is called co-pigmentation. The color of wine is often not just due to the concentration of the pigment molecules or anthocyanins but to the types of compounds these compounds physically interact with in the wine. This complex color is called a co-pigment. Co-pigmentation of a wine can be manipulated by manipulating the concentration of the noncolored species that form co-pigments. This involves blending different types of wine to achieve more stable and interesting colors, as well as unique tastes. For example, some winemakers are experimenting with a blend of the Viognier grape and adding it to red wines to help stabilize the color of the wine. By doing this, they have discovered that Viognier creates a more interesting color and taste. The process of co-pigmentation is expanding, and winemakers are becoming more creative and thinking outside of the box as they blend new and exciting wines. Since color is so integral to the perceived quality of wine, other methods for the stabilization of color are also being explored, such as understanding the factors leading to the development of stabilized colored polymers important to the longevity of a wine. As with tannin management, the formation of these types of polymers may be stimulated by oxygenation of the wine, depending upon the types of color molecules and other reactants present.

Flavor Profiling

A fifth process that is receiving much attention of late is called flavor profiling. In flavor profiling enologists first determine the characters most attractive to consumers for a specific varietal. Enologists are spending more time identifying specific impact flavors or characters for a given varietal and determining how those characters can be enhanced or modified. This helps to determine the optimal time to harvest to ensure the best flavors will stay in the wine and not disappear during the winemaking process. Flavor profiling links the impact characters of a varietal to the palate preferences of specific consumer segments. For example, in California, there is ongoing research on the preferred wine flavor profile of the Hispanic market. Currently this market segment does not consume much wine, but is the dominant consumer group in the state. Therefore, wine researchers are attempting to discover what wine flavors they prefer and link that data to the grape data. From this information, they can craft a wine that will be popular with this customer segment.

Another exciting feature of the flavor profiling research is that it will not only allow winemakers to produce wine that specific consumer segment like, but will also result in data about the grape varietal, vineyard, and harvest that will increase consistency and quality. No longer will it be a gamble regarding when to pick the grapes, because the winemaker and vineyard manager will have the ideal flavor data for the grape and know when it forms, stabilizes, and disappears. This type of information may also allow winemakers to eliminate certain steps in the winemaking process, such as extended or cold maturation. This in turn can reduce costs and maintain or increase quality. The true benefit of such flavor profiling is a better understanding of the characters and potential for making the best wine possible from the fruit available to the winemaker, and for a reasonable cost.

CURRENT AND FUTURE ISSUES IN ENOLOGY

There are several current and future issues that impact the business of enology. One of the most controversial is the development of a *global wine brand*. There have been rumors in the global wine industry for several years that one of the major wine players will eventually be able to achieve this—though it has not occurred to date. If a true global wine brand were to develop, it would be similar to the Nike model, in which a tennis shoe is designed, but the material and labor to make it are sourced from around the globe. The consumer doesn't necessarily know from where the rubber, cloth, shoelaces, design, etc., originated from; instead, they focus on the final branded product, for which they are willing to pay a premium.

If this model were to play out in the wine world, someone would create a wine that had a consistent taste year after year. They could source the grapes from any location in the world that met their quality and price requirements. They could produce it in any country that had low production costs, but produced at a high-quality level. Their only concern would be the consistent taste, quality, and price.

This is very counter to the current method of making wine, which relies on the concept of *terrior*: the soil, weather, tradition, terrain, method, and people who make the wine. Wine has always been a "local" craft, even though it is sold globally. The conventional thinking is that the consumer should be able to "taste the place in the glass." Therefore, the development of a global brand would break all tradition. However, there are some intriguing benefits with the concept. The winery that developed such a brand could achieve excellent return on investments, because they would have the flexibility of sourcing their grapes from any place in the world that fit their brand flavor profile. Furthermore, if they could deliver a strong brand with consistent taste, with good advertising they could develop a loyal following similar to Coke. Whether or not this will come to pass is still to be determined.

Another issue or trend in the business of enology is that of *customer focus and sustainability*. For many years winemakers have been accused of making wine they like versus what the average customer prefers. This is changing, and wineries need to be on top of what consumers like and want. They need to explore consumer preferences, experiment with new varietals, and help develop the novice wine consumer, rather than making fun of them because they prefer sweet, nontannic wines. Australia is already doing an excellent job of this.

Related to this is the need to focus on *"green"* or *sustainable winegrowing* practices that are environmentally and socially friendly, as well as economically viable (see Chapter 17 for more information on this trend). More and more, consumers want to know that their wineries are using "green and clean" practices to sustain the environment, such as recycling waste water, using natural fertilizers and pesticides, reducing energy use, and supporting wildlife and vegetation habitat. Consumers are becoming more sophisticated and want products that are produced using true green technologies, not ones that are just a different shade of brown. This poses challenges to winemakers worldwide to understand what new practices and technologies hold clear benefits for the environment versus simple replacement of one environmental pollutant with another. Wineries have the potential of not only being "off the grid" or nonconsumers of commercially generated energy sources but to be net contributors to the energy needs of their communities.

Related to this issue is the one of making *organic wine*. The number of organic and biodynamic wines that have come on the market in the last several years has increased dramatically. Many of these are in the European markets, as consumers there want to know that their wines have been grown and made using organic standards. This is an issue that some New World countries have already addressed (e.g., New Zealand) but that others may need to pay more attention to in the future. The term "organic" is not synonymous with "sustainable" however, and it will be important for all wine production to be truly sustainable.

A fourth issue or trend is the *quality/value equation and consumer profiling*. Many consumers believe that wine is overpriced. This is why many drink other alcoholic beverages. The great success of Charles Shaw wine in the US, and similar brands in other New World countries, attests to the fact that consumers will drink wine, but they want to believe they are getting a good price for the quality. Much of the research suggests that the novice consumer starts at low-end wine and trades up, yet many wineries still set their prices beyond the average consumer target. Wineries need to do a better job at profiling their customers at the different market segments and then producing products that meet their needs.

A final issue, which has been mentioned in other chapters, is that of *ISO certification*. This is a process whereby wineries document all of their production processes and confirm that they follow them exactly to ensure quality results. To achieve ISO certification, external auditors visit the winery to evaluate whether or not everything has been documented and is accurate. The benefit of doing this is that, in the future, some retailers and consumers may demand this type of certification. Therefore, wineries that get a jump start on documenting practices may come out ahead in the future.

Interestingly enough, in California, state regulations are so strict that most wineries are already documenting processes to conform with these requirements. In addition, those countries that are encouraging wineries and vineyards to adopt sustainable wine-growing practices are also facilitating the process towards ISO certification, because of the rigorous review and documentation of existing management practices.

Conclusion

In conclusion, the business of enology is a fascinating but complex matter. There are multiple decisions that must be made regarding customer needs, company strategy, costs, and revenue generation. In addition, there is always the variable of "Mother Nature," who may bless the vineyards one year with abundance and quality grapes, and then either rain, hail, or not warm them with sun the next. The winemaker's job is never stagnant. Circumstances are always changing, but there is a satisfaction in the profession, because, like a fine wine, a winemaker does grow better with age. With each season, with each vintage, the winemaker learns new ways to adapt, and this continually growing knowledge and experience is critical to the bottom line of a wine business.

LEARNING OBJECTIVES:
- Define supply chain and supply chain management
- Describe the process of determining your core competencies
- Explain the commodity strategy and supply chain decision matrix
- List all of the winery equipment and supplies to consider for supply chain management
- Describe various quality control systems that impact supply chain management
- Identify supply chain management issues and trends for the future

CHAPTER 5

WINE SUPPLY CHAIN MANAGEMENT AND QUALITY CONTROL

Thomas Atkin
Associate Professor of Business, Sonoma State University

Jon Affonso
Winemaker, Rail Bridge Cellars

The wine business has experienced huge changes over the last two decades in terms of advances in technology, globalization of markets, and stabilization of political economies. Managers have realized that material and service inputs from suppliers have a major impact on their organization's ability to meet customer needs. This has led to an increased focus on the supply base and the organization's sourcing strategy.

As a result of these changes, companies now find that it is no longer enough to manage only their own organization. They must also be involved in the management of the network of all upstream firms that provide inputs (directly or indirectly) for the organization. In addition, they must be involved with the network of downstream firms responsible for delivery and after-market service of the product to the end customer as well. From this realization emerged the concept of the "supply chain." The wine industry has followed suit in this need to focus on managing its supply chain to reduce costs, improve quality, and meet customer needs.

This chapter provides an overview of how the wine industry interacts with its supply chain. It begins with a definition of supply chain and its management function. It then describes the process of determining your core competencies to determine which aspects of your business process to outsource to suppliers, and then reviews a commodity strategy to assist in doing this. Next the chapter reviews the major supplies for the wine industry, and the various quality control systems that impact supply chain management. Finally, the chapter concludes with a review of supply chain management issues and trends for the future.

WHAT IS A SUPPLY CHAIN?

The *supply chain* encompasses all activities associated with the flow and transformation of goods from the raw materials stage (extraction), through to the end user, as well as the associated information flows. Material and information flow both up and down the supply chain (Handfield & Nichols, 1999). Supply chains are essentially a series of linked suppliers and customers that handle the product until it reaches the consumer and beyond, if recycling or disposal is involved.

Supply chain management is the integration of these activities through improved supply chain relationships to achieve a sustainable competitive advantage. The path to sustained growth involves managing costs across these multiple enterprises. Figure 5.1 illustrates a supply chain, with the "plan" portion depicting the "management" aspects.

Each organization has to define its own position in the supply chain. The wine industry exhibits a wide variety of supply chain designs, from the complete control demonstrated by the vertically integrated supply chain of larger wineries, such as E.&J. Gallo, to the fragmented marketplace of the smaller wineries. Several issues make the wine supply chain distinctive. First, the production of wine is very fragmented, with a large number of small wineries operating very independently. For example, there are currently over 1,000 wineries operating in Sonoma and Napa counties alone. This has resulted in a lack of coordination between producers and suppliers.

Second, the distribution system is unique due to the laws controlling the sales of wine in each state. This body of law requires a three-tier system that demands that wineries sell only through approved distributors in each state and limits sales direct to customers. This has led to a situation where a large number of wineries are competing for the attention

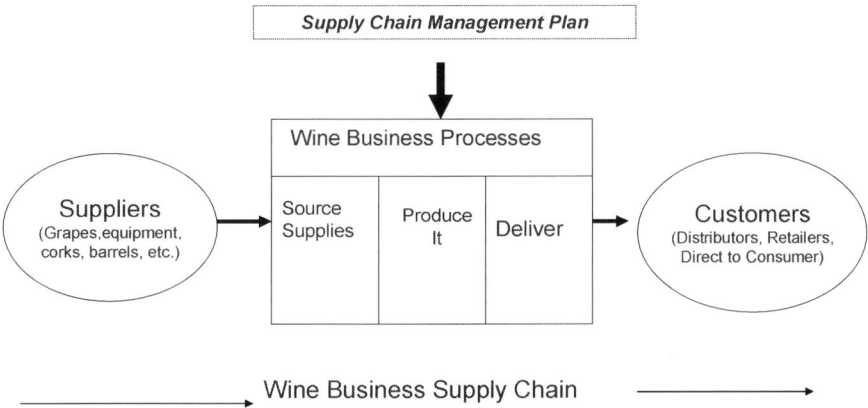

Figure 5.1. Supply chain management.

of a handful of distributors who have become very powerful. These two factors, combined with the vagaries of an agricultural product, make the wine supply chain different from all others.

Determining Your Core Competencies

The development of a supply chain is essentially a series of make versus buy decisions. Any and all aspects of wine business activity can potentially be outsourced. The grapes can be grown on the private estate of the winery or they can be obtained from a grower's vineyard. The bottling of the wine can be handled in-house or performed by a mobile bottling operation. Custom crush facilities can perform any and all winemaking operations from crushing the fruit to fermenting, ageing, bottling, and even marketing the final wine. Firms will develop different supply chain strategies depending on their corporate mission and competitive advantages.

First, each organization has to decide whether to provide the product or service in-house or to outsource it. In order to decide, a firm must determine their core competencies. What is it that they do better than their competition? In addition, what materials give the firm a strategic advantage over their competitors? These two factors constitute strengths within the company and are activities that should remain inside the company. The current thinking is that the firm should concentrate on developing or enhancing these strengths. These are the things that are critical to the end product and that the firm has a distinct advantage in providing. These advantages may be due to economies of scale, higher quality, or a favorable cost structure.

Similarly, the firm should also look for weaknesses in areas where they do not produce an item as effectively or efficiently as their suppliers or competitors. In this situation, a company can develop the item into a core competency by bringing in skilled experts and investing in equipment to produce the item in-house. This is often expensive and time consuming. As long as the material is not a core competency, direct quality control is not a serious issue, and if it is less expensive to buy it than make it, then outsourcing is the preferred solution. Many firms have recognized that anything that is not part of a firm's core competence is a candidate for outsourcing. If a supplier can do a better job on a product or service, then it makes sense to buy from the supplier rather than make it in-house.

Commodity Strategy

A commodity strategy is the next step in deciding how to deal with all of the inputs that are required to provide a bottle of wine to a consumer. Inputs can generally be classified across two dimensions: the profit contribution of the input and the supply risk (Monczka, Trent, & Handfield, 2002). These two dimensions combine to form a 2×2 matrix, as shown in Figure 5.2.

Using this matrix as a guideline, we can see how best to manage all of the required inputs. Each cell of the matrix calls for a different set of techniques for obtaining it. For instance, the most valuable input for a winery to obtain is the grapes. This would fall into the strategic category so the winery would communicate extensively with those suppliers and continuously review performance. Some of the techniques that can be utilized for each category are listed below and an analysis of the key inputs for wineries appears in the following sections.

Major Supplies in the Wine Industry

The following sections describe the five major supply components for a winery. These are: 1) wine/grape source, 2)

	STRATEGIC – High supply risk and high profit contribution	LEVERAGE – Low supply risk and high profit contribution
HIGH Profit Contribution	• Limit suppliers • Develop partnerships • Involve suppliers early • Closely manage price/cost • Continuously review performance	• Multiple sources • Maintain competitive pressure • Seek waste elimination • Monitor performance
LOW Profit Contribution	ACQUISITION – High supply risk and low profit contribution • Ensure supply • Focus on service • Frequently review • Minimize acquisition cost	MULTIPLE SOURCES – Low supply risk and low profit contribution • Use multiple sources • Procurement cards • Automate • Give minimal attention
	HIGH – Supply Risk	LOW – Supply Risk'

Figure 5.2. Supply chain decision matrix.

winemaking equipment—both fixed and variable costs; 3) storage; 4) distribution; and 5) inventory control management. A discussion of the many subcomponents of each category and its decision points is included.

Wine/Grape Source

Buying Bulk Wine

The first make versus buy decision in the development of a winery's supply chain is to determine if they are going to produce the wine or buy it on the bulk market. Most wineries categorize the production of their style of wine as a core competency. Other companies have vast distribution channels, and still others specialize in unique packaging or are able to react quickly to changes in consumer preferences. Any of these core competencies creates a situation where it might be better to buy some or most of their wine in bulk, from another producer. The main factor in buying bulk wine is the availability of the wine. How specific of a wine style and type is the company looking for? Is the company simply looking for red or white table wine or are they looking for an Old Vine Zinfandel from bench land areas of Dry Creek Valley in Sonoma County? The specificity of the wine relates directly to the availability, which is the dominating factor in the market value of that wine.

Price is not determined by how much it costs to produce the wine. Price is really only an expression of what the market will bear. This of course creates a certain amount of volatility in the market, causing the supply to become an ocean of inexpensive good-quality wine or a desert of poor wine. The wine industry has formed wine brokerage companies that specialize in the buying and selling of bulk wines. They are excellent sources of information as to what the market has available as well as forecasting what will be available. They help in reducing the level of price volatility; however, it is still a free market system.

Producing Wine

Because most wineries consider wine production in their style to be a core competency, it is suggested that most wineries should produce most if not the vast majority of their wine. This allows the production to remain in-house where the company can retain constant control over the quality of the product.

In an ideal world, the entrepreneur would have enough money to build the perfect little jewel of a winery to accommodate exactly what the winemaker had in mind. That million-dollar investment to start up a winery is out of reach for most of us. There are a couple of ways of avoiding that investment, however.

One of the evolutionary changes that occurred in the wine industry over the past 20 years is the business relationship called "custom crush." It is the leasing or renting of excess space in a winery facility to individuals who want to make their own wine without having to make the financial investment needed for a winery. This gives smaller wineries the capacity to grow and allows them access to state-of-the-art equipment. These facilities offer a wide range of services, including:

- crushing and fermenting,
- winemaking and tank storage,
- barrel storage,
- analysis,
- filling and packaging,
- regulatory compliance,
- business services.

"Shared premise" is a similar arrangement in which an established winery shares the use of its facilities with other winemakers. By alternating the use of winery space and equipment with an established producer, a grape grower or brand builder can become a vintner without the major investment. This arrangement offers more stability in winemaking style and more control over costs and pricing than custom crush.

The capital expenditure for building a winery is significant, and it is often difficult to produce wine without some financial backing. In order to obtain backing, you may have to show that your product is viable in the marketplace. Custom crushing facilities make it possible to offer a wine in the market before the winery is built. By outsourcing the production of your wine, you can avoid or at least postpone the enormous capital cost of your own facility and the equipment needed to crush, ferment, press, age, and bottle your wine. This allows the winery to produce a product, earn some revenue, and prove to the bank that its product is worthy of financial support.

With financial backing, a production facility can now be built. To develop a new winery, at minimum a facility, pressing equipment, food grade tank storage, and climate control systems and hoses are required. More often than not, grape crushing/destemming equipment, filtration equipment, pumps, as well as barrels, are additional items, all of which are fixed costs used to produce wine.

Buying Grapes From a Grower
Now it is somewhat obvious to suggest that producing wine should be a winery's core competency. Is the same true of the suggestion that a winery should produce its own grapes? This is not as clear. There are several reasons why outsourcing would be a good, if not a preferred, option for obtaining this strategic resource.

Many private growers produce high-quality grapes and sell them to wineries. The vast majority of wineries in the US purchase some grapes from private growers because the growers have a unique expertise and a close understanding of the grape vines on their property. They often have more experience and their close proximity to the vines makes them better equipped to respond to infection or infestation. In many cases, the growers are more focused and better suited to produce quality fruit than the wineries.

Due to the temperate climate in most winemaking regions of the world, vineyard land is in high demand. This increases the average price per acre to enormously expensive amounts. In addition to this, the winery must consider the large capital cost of the farming equipment, the trellis system, the irrigation system, labor, and the vines themselves.

The down side to purchasing fruit is that there is an inherent adversarial relationship between growers and wineries. Growers want more money per ton and higher yield. Wineries want less money per ton and lower yield for quality. Due to the importance of the commodity, a close strategic relationship is required. Constant negotiations are extremely important between growers and wineries to ensure a mutually beneficial relationship. If the relationship is close and mutually beneficial, there are few problems. However, if communication breaks down or the specifications are not clearly stated, there is a possibility of producing grapes that do not meet the requirements of the winery. This lack of communication often results in wineries that are forced to purchase inferior grapes or growers losing an entire year's income. It is for this reason that the winery should be in constant contact with the grower throughout the growing season as well as after harvest to discuss the quality of the fruit. This not only increases the likelihood that the fruit is maintained correctly during the current growing season, but it allows for improvement by both parties for the future.

Growing Grapes From Estate Vineyards
While the initial capital expenditure of planting a vineyard is high, in the long run it is often more cost-effective to plant your own vineyards. It is important not to grow too quickly so that the company is able to absorb the debt and still remain healthy. Unless someone began as a grower before they started their winery, many wineries do not initially own their own vineyards. In time, wineries gradually begin to purchase land and plant their own vineyards. Wineries often initially outsource the labor and expertise by hiring a vineyard management company instead of purchasing their own equipment and employing the labor. As the winery continues to plant more and more vineyards, it will be more cost-effective for the winery to purchase its own equipment and hire a vineyard manager and labor to maintain them. While in the long run it is more cost-effective to grow grapes, the main advantage is that it eliminates the adversarial relationship between the grower and the winery. This allows the winery to have absolute

control over how the fruit is grown, which further helps to achieve the fruit specification and quality the winery is looking for.

Winemaking Equipment: Fixed Costs

There is a large amount of equipment to purchase if the winery has decided to produce the wine in-house. First, a building will be needed somewhat close to the grapes so the grapes won't suffer damage in transit due to temperature. The following items will be needed to outfit that building to produce wine.

Tanks

Tanks are necessary for fermentation and storage of wine. Although most are constructed of food-grade stainless steel, some wineries still use cement or wood tanks. Newer technologies are developing food-grade plastic tanks. Stainless steel has become the worldwide standard because of its strength, long life, and the ability of its surface to be easily sanitized. The cost of tanks can vary depending on whether they are fitted with a jacket (to control temperatures) or other specific features that have been added (such as cleanout doors). Jacketed tanks are double walled to allow a hot or cold liquid to pass around the tank without being in contact with the wine. The liquids are able to reduce or increase the temperature quickly to stabilize the wine. Currently, a good rule of thumb is that tanks will cost about US$4.00 per gallon. Therefore, a 6,000-gallon tank represents an investment of about US$24,000.

Newer tank technology is helping to improve the quality of the wine and speed up the fermentation process. The most recent designs serve to reduce fermentation time by one half through modern methods of pumping over and breaking the cap. Tank designs such as Selector System allow cap wetting in which the cap is steeped, remixed, and disintegrated. This improves the extraction capacity of color, bouquet, and taste and avoids the formation of solids. It also means less time spent in the tank so the tanks can be used more efficiently. Additional features allow the tank to empty completely so laborers don't have to crawl inside to perform dangerous cleanups.

Barrels

Barrels differ from tanks in that they are not only a storage vessel but they are an aging tool that dramatically affects the flavors, tactile impressions, and the aromas of the wine. Barrels are described by the origin of the wood used (i.e., French, American, Eastern European), the grain size of the wood, the volume it stores, and the amount of toast (how long the inside of the barrel is exposed to fire). The grain size also dictates the amount of oxygen that passes through wood into the wine, which dictates how fast a wine will age. All of these factors directly influence the amount and type of oak extraction that influences the wine. Every cooper is different and each barrel type imparts different characteristics. These factors will have a large influence on the style of wine and the winery should take care to purchase the correct barrels and a diversity of barrels to create the style they want to achieve.

Like everything else, barrels are very expensive. The type of oak and the expertise in building it are the two main factors that determine the cost. Barrels are still largely assembled and toasted by hand, which makes them labor intensive. Most barrels are either French or American oak, but many cooperages are sourcing some of their wood from Eastern Europe and even China. While French oak is generally twice as expensive as American oak, French oak may or may not be the type of oak that fits your style. On the other hand, while cost is a factor, the style of the wine should be the dominating factor in deciding which barrels to purchase rather than simply buying the least expensive barrel.

Another difference between barrels and tanks is that barrels have a much shorter life span. On average a barrel will last 3 to 7 years depending upon the winery's style. This means that new barrels must be constantly cycled into the program as older barrels are culled out. This creates a more complex management dynamic that must be monitored. The shorter the life span of the barrel, the larger the influx of new barrels and the higher the expense. This shorter life span makes the item almost a variable cost, but because the item is used for several different wines and different vintages it is often considered a fixed cost.

Of course, like most commodities, there are some price breaks for larger volumes. However, because most wineries purchase barrels from several different coopers, their volume of barrels is not as large as if they bought all of their barrels from one cooper. Therefore, there is less advantage for the supplier to provide significant price breaks and thus the savings are not very significant.

Oak additives (i.e., chips, beans, or dust) have been developed as substitutes for barrels. They come in French or American varieties and at any desired toasting level and grain size. These additives can be added directly to the wine or come in muslin sacks that act as tea bags to infuse the oak into the wine. Also, barrel innerstaves can be placed inside older barrels to increase their life span. This allows the use of older barrels while imparting compounds from the new oak. In addition, tank innerstaves can be used in the same fashion to impart oak character in stainless steel vessels rather than barrels. All of these alternatives can have a substantial cost savings.

Presses

Technology has taken hold in the process of pressing the grapes. The hydraulic vertical basket press and the horizontal bladder press are by far the two most common types of presses. The hydraulic vertical basket press operates by placing the grape skins into a slotted cylinder with a hydraulic lid that presses down from above. The wine is then allowed to flow between the slots and is collected in a pan at the base. The horizontal bladder press is the more modern version and it exerts the pressure by inflating a bladder mounted to the inside of the cylinder. Wine is then allowed to flow from slots in the cylinder on the opposite side. Because of the relatively low pressure exerted on the pomace and gentle mechanical treatment of the skins, the solid content in the juice is low. Automated control systems on both models drive the process to allow maximum yield and decreased stress on the juice.

Another type of press, the continuous press, is essentially a screw auger that presses the pomace toward a small outlet. This is efficient because it saves loading and unloading of the container, but a big disadvantage is that it causes tearing and breaking of the skin's tissue.

Cellar Equipment
There are several pieces of equipment that are used to varying capacities and they vary drastically depending upon the wine's style and price point. Items like crushing/destemming machines, filtration devices, pumps, hoses, and fittings are all commonly used in wineries. However, how and if they are used is more a discussion of winemaking than supply chain management. Suffice it to say that they are all expensive fixed costs with relatively long life spans. These pieces of equipment must first be decided upon based on wine quality and then their cost can be factored into the development of the facility.

Bottling Line
Setting up a bottling line is expensive but there are choices available. The bottling activities only take place part of the time so many smaller wineries use custom crush facilities to bottle their wine. A very creative solution is the mobile bottling line, which is essentially a trailer that can be

moved from winery to winery for short filling runs. These facilities can also be used if the winery has exceeded its in-house bottling capacity.

Winemaking Equipment: Variable Costs

Winemaking Supplies

There are hundreds of different products that contribute to the production of thousands of different wines. It is beyond the scope of this chapter to discuss all of the different products and how they are used, but some of the more common supplies are listed in Table 5.1.

Many of these products have several different substitutes and can be purchased from a few or even one supplier. This consolidation gives the winery a little more leverage when it comes to pricing. There is a significant advantage for a supplier to sell a single winery a large volume of materials, which translates into the ability to offer significant price breaks. The better a winery is at reducing the number of suppliers providing items, the more efficient their supply chain will be, the better customer service they will receive, and the lower the price per unit will become.

Table 5.1. Common Winemaking Supplies

Yeast
Yeast nutrient
Diammonium phosphate
Potassium metabisulfate
Tartaric acid
Fining agents
Settling enzymes
Caustic cleaning agents
Citric acid
Lactic acid bacteria
LAB nutrients
Lab analysis
Diatomaceous earth
Filters

Bottles and Alternative Containers

Tradition from Western Europe has dictated the form of most wine bottles. The three most common shapes are: the Bordeaux shape (long cylinder with a short shouldered neck), the Burgundy shape (shorter cylinder with a long unsholdered neck), and the Alsace shape (longer and narrower cylinder with a long unsholdered neck). Traditionally, certain wines are used in certain bottles. Therefore, the actual shape of the bottle helps described the wine it contains. Red wines typically are packaged in a dark green or dark brown glass to protect the wine from the damaging effects of light. The vast majority of bottles have a volume of 750 milliliters. However, smaller 375-milliliter and larger 1.5-, 3-, and 6-liter bottles are common and other sizes do exist. The neck opening design of most bottle types has standard dimensions to fit the standard cork size.

Glass bottles are by far the norm but substitutes are available. Glass is susceptible to breakage and the quality of the wine deteriorates if not consumed fairly quickly. Plastic bottles made out of polyvinyl chloride (PVC) or polyethylene terephthalate (PET) are being used more commonly with wine, especially in the small 187-milliliter size on airlines. Also Foster's has released some of their major brands, such as Lindemans and Wolf Blass, in 750-milliliter PET, which appear to be well-accepted by customers. One downside to PET is a shorter shelf life for the wine, so it must be consumed rather quickly upon distribution.

The bag-in-box container has a cubicle shape that is economical to ship, maximizes shipping space, and can survive a fair degree of rough handling. The box contains a vacuum-sealed bag that prevents oxidation so it can keep the wine fresh for up to 4 weeks after opening. Similar to bag-in-box is the cardboard brick option, which is low cost, has a good oxygen barrier and longer shelf life, but is not as well accepted by customers. Currently used frequently by the milk and juice industry, it creates some negative connotations in the minds of wine customers—some of who fear children may accidentally drink the wine, thinking it is juice.

Another option is to store the wine in easy-to-open cans. Similar to beer and soft drinks, this container might make the product more accessible but will drastically affect the perception of quality of the product. Another issue is the infusion of aluminum flavor into the wine. The cans would have to be Teflon coated to prevent the product from being in contact with the metal. However, this may be especially useful on airplanes and in the catering industry.

Whichever container is used, the winery will be purchasing a very large quantity and will have some leverage with the manufacturer. While leverage can be used, it is not recommended going too far as it might begin to affect customer service. Many glass companies work with cardboard suppliers and they will package the bottles into the winery's finished boxes as part of their service before they arrive at the winery. This greatly increases efficiency of the bottling line as the cases are emptied and then refilled with the box they arrived in. It also eliminates the use of unnecessary packaging material. The glass company might also work closely with the winery to improve efficiency of materials by operating on a JIT basis. This assists in cash flow by not tying up money in inventoried supplies. Because bottles can be produced in large amounts fairly quickly and the material is easily recyclable, most glass suppliers have to keep very strict quality control programs to stay competitive and this is another key element to good customer service.

Labels and Foils
Labels and foils are devices that can really make a wine stand out on the shelf. They are a key component of the marketing effort for any winery. Foils are simple tin cap-

sules that cover the top of the bottle. They provide little practical function but drastically improve the appearance of the product. Labels inform the consumer as to the content of the bottle. Some wording on the label is required by federal regulation and needs to be scrutinized to ensure that all information is in compliance. However, there is still plenty of space for brand identification and graphics. Close communication with the printer is needed during the design phase to achieve the desired presentation. Glue labels and pressure-sensitive labels are the most commonly used. The equipment and expertise needed to produce high-quality image is well beyond the core competency of the winery. It is not a loss of strategic advantage to outsource and it is still easy to maintain control of the design of the labels. It is important to choose a printer that will provide the range of materials and the assurance that the label will, in fact, work in conjunction with the labeling machine and the container utilized.

There are several manufacturers that provide these services but often the companies that offer the most service tend to be the most expensive. Therefore, the winery should determine how much service they are willing to pay for. A winery does purchase a significant amount of labels and can use them as promotional materials as well. If you can limit the number of suppliers, this again will allow the winery to maximize the amount of leverage it can have without affecting the customer service.

Corks

For centuries cork was the absolute closure of choice for wine. It seals the bottle as the liquid forces the cork to expand and some believe cork also allows small amounts of oxygen to enter the wine to assist in aging. In the 1980s the incidence of cork taint began to increase in a certain percentage of bottled wines. It was a taint that could, depending upon its intensity, cause a reduction in fruit character of the wine or even cause an outright moldy smell similar to wet cardboard.

Beginning in the 1990s several companies began to manufacture and sell synthetic corks that mimicked a cork as far as the ability to seal a bottle of wine. Questions have arisen concerning their ease of opening and the ability to maintain a tight seal over several years' time. But, once thoroughly tested and evaluated, a number of wineries deemed synthetic corks quite acceptable replacements for true bark cork and they are now widely used. Other wineries have opted to remain with the traditional natural cork closure due to tradition and consumer perception.

The screw-cap, or Stelvin closure, is also a popular alternative to cork—especially for white wines. Technological advances have enabled screw caps to perform well in preserving the freshness of the bottled wine and it is much more convenient for the consumer. It is now used widely around the world, especially in New World wine countries such as Australia, New Zealand, and South Africa. A recent entrant in the closure arena is a glass closure that seals well, is reusable, and customers consider to be quite attractive. One downside is the higher cost, but it is being adopted by wineries for their more expensive white wines—especially in Germany. Other closures, such as the MetaCork and Zork, offer creative new ways to seal wine and reduce the hassle of opening the bottle. As innovation continues to rise in the closure arena, it is expected that new alternatives will be made available in the future.

In most cases, the decisive factor in a decision on closures revolves around consumer acceptance. Will the consumer react negatively to the use of an alternative closure? How important is the issue of cork taint to consumers?

Suppliers in the cork industry have made strong efforts to increase the quality of their products. They have begun to vertically integrate the supply chain so that more control of the harvesting and production of the corks can be achieved. Batches can now be tracked from the forest to the warehouse. More sanitary processing methods and facilities have been established. Peroxide is now used instead of bleach and the bark is stored on concrete rather than dirt during curing. All of these services are controls to look for and should be demanded as a criterion of purchase. Again, many suppliers provide these services and it allows the winery some leverage.

Storage

Bottle Aging

Even though the majority of wine purchased in the US is consumed within 48 hours after purchase, wineries must still be concerned about bottle aging. Bottle aging is any chemical reaction that goes in the wine while it is in the bottle. Wine is often filtered at the time of bottling to inhibit the wine from spoiling. As a result, the wine is "shocked," which suppresses wine's flavors and aromas for a month or two. Once the wine recovers it begins to express the characters of the fruit and what the winemaker crafted into it as well. In time the wine will continue to age as the tannin ages and the fruit characteristics will become more prevalent. The wine will ultimately become more complex and interesting to the senses.

Environmental Conditions

With the amount of evolution the wine undergoes in the bottle, a winery must be concerned with the bottle environment from the time it is bottled to the time when it is consumed. Extreme heat, extreme cold, light, and humidity are all important factors to control in wine storage. Temperatures around 90°F and light can rapidly accelerate the aging process. Heat will dry out the closure and allow the wine to leak, oxidize the wine, or cause the wine to cloud.

Extreme cold at around 30°F can slow down the aging process or cause the wine to precipitate crystals. It is important that wine storage be between 50°F and 70°F with an elevated level of humidity to prevent the cork from drying. It is for these reasons that trucking, retailers, and consumers often go to lengths to maintain this environment.

These conditions should be maintained from case storage to consumer. This means the winery should ensure the case goods storage for the winery, the transportation of the wine to the distributor, the case good storage at the distributor, the transportation from the distributor to the retailer, and the retailer case storage are all climate controlled. It is a tremendous amount of refrigeration.

Distribution

Three-Tier System

In the US, the three-tier system is a by-product of prohibition that forces all alcohol sales to travel from a producer, through a distributor, and on to a retailer before it can be sold to the consumer. Distributors are organizations that either act as a broker or take possession of the product and act as an agent of the winery. As an agent they are supposed to actively promote and sell the wine. However, as time has passed, most of the distributors have consolidated, leaving only a handful, and they often have to sell hundreds of brands. It is difficult to ensure the kind of customer service that is commonly provided with other service providers when dealing with so many brands. In the end many distributors are not able to actively promote every brand so the winery is forced find other ways to gain attention.

Options in this area are quite limited due to the three-tier system imposed by government regulation. But even within this system there are opportunities to promote the wine. These include winemaker dinners, tasting rooms, active public relations, wine clubs, private tours and/or parties at the winery, and special auction lots. These activities will help to create a bond with the distributor, the retailer, and even the consumer, all of which makes it easier to move the wine through the system. These should be core competencies within the winery. There are services that will allow you to outsource these promotional aspects. However, the whole purpose is to build a direct relationship with the distributor, retailer, and consumer. By outsourcing this service, it places an intermediate between them and decreases the level of closeness. It is far more effective to service these entities directly and personally. Even with the benefit aside, the cost is not expensive and the result is very profitable.

Internet
The Internet provides an additional channel for reaching consumers. Direct communication with the end user is established and the wine is delivered to their doorstep. This allows the winery to develop a long-term relationship with its customers and keep them informed about new vintages. Small package carriers typically take care of the delivery, making it very convenient for the customer. If the wine is purchased in the tasting room at the winery, tourists can now ship wine to the residence anywhere in the US. The wine club management can also be outsourced, but the purpose is to have a direct relationship and an outsourced service will be an additional layer.

Tasting Rooms
Direct access to customers is also provided by wine tasting rooms. Customers can sample the wine before purchasing it. This can often be the beginning of a longer relationship that can be enhanced by wine club membership and Internet promotions.

Wine Shipping
One function that can be outsourced is the warehousing and shipping of the wine. Many wineries use specialized public warehouses and coops to store their product and prepare it

for shipment. These suppliers have also developed expertise in handling the paperwork and government reporting requirements that accompany the shipping process. They also provide the proper environmental conditions for transporting and ensuring wine quality is maintained. Using refrigerated trailers, heaters, and insulated quilts can provide temperature control during transportation.

Inventory Control Management

Monitoring the inventory of case goods allows for visibility of the entire stock of product that is ready to be sold. This ensures that the little or no product is lost or stolen. This is a service that a storage warehouse will provide, but the level of accuracy is often not as good as if the winery does it. This is another area that should be a core competency of the winery. There is software available that makes this much easier, but a specialist is preferred to ensure efficiency and accuracy.

A specialist will monitor rates of depletion in all markets where the wine is sold. This information can be passed on to marketing and even the production department to act as a type of feedback loop. If sales are up or down in specified markets, it might allow marketing to reallocate their promotion to areas where sales are down from areas where sales are high. It might also indicate to production which wines are preferred and what production levels they should predict for the next season. Again, this is all proprietary information that is essential to the winery's strategy, which further underlines why this service should be one of the winery's core competencies.

QUALITY CONTROL SYSTEMS IMPACTING SUPPLIES

Cost of Quality

Given that a large portion of the action takes place outside the winery, it is important to be able to manage activities taking place at the supplier. A good example is the supply chain for corks. For years wineries just accepted the fact a certain percentage of wine would suffer from cork taint. Wineries would try to identify tainted lots of cork and reject them, but little effort was made to force suppliers to change their handling methods and reduce the level of taint. Cork producers have recently reacted to the problem by instituting new methods of sorting and cleaning the cork. Producers are now making it right the first time rather than letting poor-quality product make its way to the customer.

Many of these issues are tied to corporate goals in achieving high quality. Dr. Joseph Juran developed the concept of cost of quality. He was able to show managers that poor quality was actually more expensive than making things right the first time. Products made with high defect levels are costly to the company in terms of product that has to be thrown out, loss of future sales, shipping goods back, inspection of incoming goods, and materials consumed in destructive testing.

Organizations have found that, in order to achieve better quality, they have to take an active role in the measurement and management of activities at the supplier. The first requirement is to clearly communicate the expectations and specifications for the item. In addition, many firms are reducing the size of their supply base to facilitate closer relationships.

Supplier Certification: ISO 9000

It is critical that a winery selects and maintains only those suppliers capable of providing the highest quality products. Certification is a process in which suppliers' manufacturing and quality processes are thoroughly evaluated to make sure they are in "control." Certification is implemented in stages beginning with supplier-provided information. A team will then visit the supplier to assess their management capability, cost structure, information systems, and total quality management philosophy. A cross-functional team usually performs this audit. A positive outcome means that the supplier is able to do more business with the customer. After a period of time, incoming inspection of the supplier's goods may no longer be necessary because systems at the supplier are well developed. The primary purpose is to work only with suppliers capable of continuous quality improvement.

This is a huge effort, however. In the absence of the resources to perform such an audit, a smaller winery may want to rely on an audit framework like ISO 9000. This is an international set of standards that tests the qualifications of the applicant to ensure conformance quality. Many purchasing managers recognize ISO 9000 standards as a necessary foundation for the implementation of total quality management.

Hazard Analysis—HACCP

Another process that examines supply quality issues is a Hazard Analysis and Critical Control Point (HACCP) system. It includes mapping manufacturing processes within the winery and identifying potential hazards. This includes materials and supplies, and is therefore quite useful from a supply chain management perspective. A documented monitoring system and preventative action plan is established so that each critical control point is monitored to ensure it stays in control and that appropriate actions are taken if it is out of compliance.

Business Excellence Model

The model is a proactive one that focuses on adopting a philosophy of continuous improvement and self-assessment, rather than forced compliance. It measures nine aspects of business management, including three that impact the supply chain: strategy and planning, resource management, and quality systems and processes. It is a comprehensive system and very useful for large wine companies to consider adopting.

Trends in Supply Chain Management

Stricter Supply Tracking—Bioterrorism Compliance

There is much discussion about increased scrutiny of ingredient labeling and tracking of different wine lots for product safety and traceability reasons. The Public Health Security and Bioterrorism Preparedness and Response Act in the US is just one example of a new law that has required stricter compliance in terms of tracking all winery supplies going into each lot of wine. This includes documenting all supplier information, original manufacturing data, how a specific supply is used in the winery, and continued tracking and documentation of the wine into the distribution and retail chain. In this way, if a consumer gets sick from drinking the wine, or finds a small particle of glass or even an insect in a bottle of wine, the bottle can be traced back to the specific winery lot from which it originated. In addition, each ingredient in the product—from every yeast to fining agent—can be traced back to its original supplier.

Although ingredient labeling on the wine bottle has not yet been required by law in most countries, the Office Internationale de la Vigne et du Vin is diligently reviewing this issue and encouraging its member nations to be proactive in documenting and tracking this data. Indeed, in the US, Bonny Doon has become one of the first wineries to voluntarily supply ingredient information on their wine labels. These types of issues are only going to increase in the future.

Continued Emphasis on Total Quality Management

One of the keys to achieving customer satisfaction through total quality management is an emphasis on monitoring quality at the source. This means that wineries will have to continue to develop systems for measuring supplier performance. These systems can be used to help make supplier selection decisions and determine where to commit supplier development resources. For instance, software has been developed that allows the winery to track activities at their grower's vineyards.

Information Sharing

Information is the lifeblood of any supply chain. A free flow of information is necessary to minimize inventories and maximize customer satisfaction. An example is the sharing of information between wineries and distributors. As we saw earlier, wineries can help to manage inventories by tracking depletion rates while distributors can help wineries by providing consumer profiles and analysis of sales information. Systems such as electronic data interchange (EDI) and vendor managed inventory (VMI) will help to accomplish this.

More Outsourcing

Firms will continue to outsource activities that are not part of their core competency. In the extreme, this could even result in the virtual winery, a winery that utilizes suppliers for everything. One local winery sells private label wine to a large retailer. The juice is imported from Chile, fermented and bottled at a custom crush facility, and shipped directly to the distributor. No bricks and mortar.

Longer Term Contracts

The percentage of long-term contracts continues to increase. Firms often combine supply base reduction efforts with the use of longer term agreements to reduce transaction costs and gain the benefits of closer relationships, such as improved quality and design assistance. Reference contracts between wineries and grape growers are typically multiyear contracts that stipulate certain activities to assure quality and consistency. Price changes can be tied to the county crush report. An evergreen contract renews itself automatically unless one of the parties cancels or renegotiates it. Per-acre contracts give the winery a large amount of control over vineyard farming decisions.

The Rise of RFID Technology

"Mega retailers like Wal-Mart, the largest company in the world, and Costco, the largest retailer of wine in the U.S., are rapidly adopting sophisticated supply-chain technologies and are requiring their thousands of suppliers to do the same" (McCallum, 2007)

RFID technology is now being rolled out which allows retailers like Wal-Mart to better track their supply of products and increase efficiency. RFID potentially allows individual wineries to track the origin, varietal, and age of wine in each barrel or tank as it moves through the production process. It can also help wineries comply with FDA regulations regarding the retention of all records concerning the production of wine for 2 years.

RFID technology uses an RFID tag, which consists of a microchip attached to a substrate, to store a unique identification number. When attached to an item, the unique identification number differentiates that individual item from any other. The RFID tag uses an antenna to communicate with an RFID reader using radio frequencies encoded with the item's identification number. Once retrieved by the reader, or transceiver, this identification number can be used to access a database that contains information specific to that item. The information in the database can include a description of the item, production date, lot number, shipping history, or any other pertinent information.

There are several *benefits to wineries stemming from the implementation of RFID*:

1. First, because RFID allows for real-time communication of product information, a supply chain that utilizes RFID tags at an item level can instantly provide an item's inventory count and where the inventory is located in the supply chain.
2. If items going through production are already RFID enabled, production throughput data is much easier to obtain. The system does not require a direct "line-of-sight" to read the information as is required by bar codes.
3. Another benefit of RFID implementation occurs when a company requires remote data collection. For instance, when cargo is being shipped overseas, GPS equipment can work with RFID tags to communicate the exact location of an item regardless of where along its ocean route it may be.
4. Lastly, RFID technology allows companies to react more quickly to market changes. As a result of real-time data collection, companies utilizing an RFID-enabled supply chain can immediately identify potential shipment delays or insufficient inventory levels and react accordingly to avoid complications.

RFID has only been used to a limited extent in the wine business. One winery in particular has used RFID technology for over 10 years. It has allowed them to follow works-in-process and finished goods throughout their facility. The payoff has been increased inventory accuracy and full asset visibility. Since 2004, the system has incorporated all components, including packware, corks, and labels (Gary, 2006).

This was especially helpful when international sales picked up and the number of SKUs tripled. The languages, alcohol contents, disclosure requirements, and product sizes were different for international markets, which created a new SKU for each variation. This would have quickly overloaded the barcode equipment and inventory system that the company had in place. However, with their new RFID system, managers could not only identify and track each SKU separately through the supply chain, but they could group SKUs together by lot numbers, date codes, or grape variety in the computer system to increase efficiencies within their winery. Should a certain lot number or date code need to be pulled from inventory, the SKU grouping would allow workers to identify the product very quickly with a minimal amount of human intervention.

The implementation of the system was expensive, however. It required converting several large warehouses into wireless storage area, installing radio frequency readers on all lift trucks, conducting a massive training session for all of the employees, and entering over a million cases of wine into the system.

Although RFID technology has been used in some form for decades, in the last 5 years cost reductions have been achieved as well as advances in tag miniaturization, mass production, and signal reach. Improvement in the readers has also helped to make RFID an appealing replacement for line-of-sight bar coding. A standard active tag with no additional hardware (i.e., weatherproofing or shock protection) costs between $1.50 and $2.00 (Gary, 2006).

Wal-Mart (who now stocks over 50 different wine brands in certain stores) announced that it was implementing a new set of guidelines for its suppliers. Wal-Mart believes that RFID can provide them with a way to always know, with little added effort, where a shipment is at all times and whether it will arrive in time to fill the shelves. Better yet, they believe that RFID will allow the inventory tracking system itself to contact the supplier with a low inventory alert days before it becomes an issue and for the supplier to send back a report with the exact location and EPC numbers of the replacement product for the inventory system to track.

These new guidelines dealt directly with the implementation of RFID technology, first for their top 100 suppliers, then for the remainder of their supplier base. Many small-to-middle sized suppliers view this request as a choice to either "comply or die." One type of RFID system that suppliers have implemented in order to meet Wal-Mart's minimum requirements is called a "slap-and-ship" system.

A "slap-and-ship" system consists primarily of a printer that produces RFID-enabled labels (Figure 5.3), a reader used to validate that the labels are properly encoded with an EPC, and a software package to link the components together. The supplier manually applies these labels on each case of product and on each pallet as it is being prepared for shipment. Because these labels can be designed to include a barcode on the front, many suppliers have leveraged the backwards compatibility with older distribution

Figure 5.3. RFID data flow system.

and shipping systems. The cost of a "slap and ship" integration package (excluding project consulting, installation, and design or testing) can run as low as $10,000.

There are several ethical issues surrounding RFID information. One concern is that the information may be intercepted by thieves if the tags are not properly deactivated by the retailer. Also, if a retailer reads the RFID tags from merchandise in a consumer's shopping cart while they are in the store, they can immediately gather valuable information on that consumer's shopping habits. Companies pay millions of dollars a year on marketing studies that tell them what products people buy and when, but retailers can obtain this information for free before the consumer even leaves the store. The retailer may use this information to display personalized advertising messages to the consumer suggesting other items they may wish to purchase. That could potentially be considered an invasion of privacy.

Conclusion

This chapter presents supply chain management in the wine industry as a dynamic endeavor. Firms in the wine business have many options available to them in designing a supply chain that will give them a competitive advantage. It takes constant monitoring and communication to keep the supply chain running as effectively as possible.

LEARNING OBJECTIVES:
- Define wine marketing and branding
- Identify the Five Ps of wine marketing
- Explain why branding is so critical and the conditions for successful branding
- Describe three wine market consumer segmentation methods
- Describe how to position a wine brand
- List the characteristics of a good brand name
- Explain how to build wine brand consumer loyalty
- Discuss the challenges of taking a national wine brand to international markets

CHAPTER 6

Marketing and Branding Wine

Linda Nowak
Wine Marketing Professor, Sonoma State University

Paul Wagner
Wine Marketing Professor, Napa Valley Junior College

Jean Arnold
President, Hanzell Vineyards

Marketing wine and creating a strong brand to differentiate your product is one of the most challenging activities in the global wine industry. Though growing grapes and making a fine wine require years of training, experience, and talent, trying to sell that same wine on the global market with the plethora of competing labels seems, at times, to be an almost impossible task. However, there are many successful wine brands on the market and most of those brands have been developed by employing professional wine marketing and branding strategies.

This chapter describes the process of establishing a wine brand. It begins with a definition of branding and marketing, and then describes the five Ps of wine marketing and how they are different from other industries. Next is an explanation of why branding is so critical and the conditions for successful branding. From there the chapter provides an overview of wine customer segmentation, and describes how to position a brand. Information on how to ensure consistency of brand image, the characteristics of a good brand name, and customer loyalty techniques are also provided. A section on the special challenges of taking a national wine brand to international markets is also included. Finally, the chapter ends with a description of how two wine brands were developed.

DEFINING MARKETING AND BRANDING

There are many ways to define "branding," but perhaps the simplest is this: *a brand is the end result of all marketing. It is the perception of the product and its name in the consumer's mind.* When all the advertising campaigns have run, when the promotions are over, when the consumers have tried the product (or not!), and have formed their own opinions, that final perception is the brand. When all is said and done, the true test of the effectiveness of your marketing is the price you are paid for the brand alone.

For some companies, the brand is a powerful tool. Many companies take an existing brand—a perception in the consumer's mind—and introduce new products or services bearing the recognized symbol of the brand. This allows them to take advantage of the positive perceptions of the consumer and tie them to new products and new sources of revenue.

On the other hand, some brands become a liability in the marketplace, for one reason or another. We all remember companies that have changed their name and brand to distance themselves from existing negative perceptions in the marketplace, and others that have changed their brand name because the new direction of the company was no longer captured by an old and out of date brand name. If the brand is the primary word you own in the consumer's mind, you want to make sure that it really does capture what you do, and convey your strengths.

The term "marketing" *is used to describe all of those activities that create a perception around a brand.* The more obvious elements of marketing are the advertising and public relations activities that try to convey meaning and importance to the brand. These activities are overt, and both the company and the market audience understand this.

On another level, there is a series of decisions made by companies that ultimately convey an enormous amount of perception to the consumer about the brand. Because these are often influenced by factors other than simple marketing considerations, these decisions are sometimes not seen clearly as marketing decisions. The next section explores how these decisions absolutely affect consumer perceptions—and are, thus, absolutely marketing decisions.

Table 6.1. Defining Wine Marketing and Branding

Wine Marketing	Business activities that direct the flow of wine and related services from the winery to the consumer. (The business activities are usually defined as the Five Ps of Wine Marketing: Product, Price, Place, Promotion, and Packaging.)
Wine Branding	The perception of the wine product and its name in the consumer's mind.

THE FIVE PS OF WINE MARKETING

When marketing professionals talk about how they can affect the consumer's perception of a brand, the key factors in those perceptions are captured under four or five simple categories. Each of these categories plays a critical role in how the market perceives the brand. Every marketing campaign should be prepared to address not just some, but all, of these categories if it is to be successful.

1. **Product:** Clearly, how consumers respond to the product itself is critical to marketing success. We have all heard the maxim: "build a better mousetrap, and the world will beat a path to your door." It's not always true, but we have all heard it. It captures the essence of the role that product plays in consumer perception. If consumers really do see the mousetrap as being better, and they really do need a better mousetrap, then the direction of your marketing campaign is clear. It is always desirable to have a superior product.

2. **Price:** This is the most ambiguous of these categories, in terms of basic rules. When consumers think of a brand, price helps them decide about the quality of that brand. But what does that mean? Which is better: a lower price or a higher one? One would think that for two products of "equal" quality, consumers would prefer the lower priced one. But that is not always the case. In many industries, a higher price conveys luxury, desirability, and exclusivity. Consumers understand that the diamond that comes from Tiffany may have exactly the same certificate of quality as a diamond from a local small town jeweler—but that doesn't mean that lots of consumers will pay a lot more for that cute and distinctive little blue box. In the world of wine, almost all consumers believe that you get what you pay for—and that a more expensive bottle of wine is almost always better than a less expensive one. Choosing to price your wine

then becomes a true marketing decision, one that will ultimately determine how the consumer perceives the quality of your product, and the value of your brand.

3. **Placement:** In some industries, the concept of placement is quite confusing, but the wine industry gives us perhaps the best possible explanation of the importance of placement. In basic terms, placement refers to where and how the product is actually sold. In the most obvious example, wines that are placed on the very bottom shelf of a supermarket are almost always perceived to be of lower quality than those placed at eye level. In a similar vein, wines that are only available through fine wine shops and top white-tablecloth restaurants will have a different and more upscale image than wines that are sold in every convenience store. Where a company chooses to sell its wines will have a major impact of the consumer's perception of those wines.

4. **Promotion:** This is the one category that everyone understands—the nuts and bolts of advertising and public relations that surround a brand and draw consumer attention to it. These activities also create much of the image of the brand beyond the product itself. Sponsorships of major events, cooperative fundraising activities for high visibility charities, endorsement campaigns by celebrities, discounts and coupons, advertising in major magazines or local symphony programs, and joint promotions with internationally famous chefs all form part of the promotional activities of a winery brand. They all add to the perception of the brand in the mind of the consumer.

5. **Packaging:** In the world of wine, we face a particularly difficult challenge. While the goal is to create an international brand, the sales volumes and small budgets often limit the kinds of promotions that can be done. In most industries, packaging is simply considered anoth-

er part of the promotion category. But in the wine industry, often the only physical image the consumer will ever see is the label itself. In this case, the promotion of the brand is quite limited, and the label becomes the primary promotional element. For this reason, many in the wine industry add a fifth "P" to the four "P"s of classic marketing. Packaging is the be-all and end-all of wine marketing for some smaller wineries, and it has to stand alone.

THE IMPORTANCE OF BRANDING

Why is branding so important? Well-recognized brands make shopping easier for the consumer. No one wants to evaluate the advantages and disadvantages of several thousand different wine brands every time they go to the store. Many wine drinkers are willing to try new wines, but having gambled and won, they like to buy a sure thing the next time. It saves time and reduces risk.

Promoting a wine brand has advantages for wineries as well as consumers. A good brand reduces the winery's selling time and effort. Sometimes a winery's brand name is the only element in its marketing mix that a competitor cannot copy. Also, good brands can improve the winery's image, speeding acceptance of new products marketed under the same name. In other words, if the winery has a good reputation for its Chardonnay, consumers may be more willing to take a gamble on the winery's Sauvignon Blanc.

In the very competitive wine industry, it is critical that the winery differentiate its wine in some way from other wines that are competing for the same market. These differences may be rational and tangible—related to the quality of the wine—or more symbolic, emotional, and intangible—related to what the wine brand represents in the minds of the consumers. The wine brand name and all of the quality, price, and symbolic images the consumer recalls about a specific wine brand are what distinguishes a wine brand from its unbranded bulk counterpart.

Some wineries create competitive advantage with value—through providing decent quality and a low price. Other wineries create competitive advantage through producing the highest quality wine they can. Other wineries do not stop at creating a great wine; they use non-product-related images and stories to make the wine memorable to the consumer. In the wine industry, these intangible image associations may be one of the most effective methods for distinguishing one brand from all of the others.

Creating a successful wine brand entails blending all elements about the wine together in a unique and memorable way that appeals to the target customer. The wine must be of good quality, the brand name must be appealing and in tune with the consumer's perceptions of the wine, and the packaging, promotion, pricing, and all other elements must meet the consumer's tests of appropriateness, appeal, and differentiation.

Conditions Favorable for Successful Branding

According to Perreault, Cannon, and McCarthy (2008), it takes time to establish a respected brand. The following conditions are favorable to successful branding:

- The product is easy to identify by brand or trademark.
- The product quality is the best value for the price and the quality is easy to maintain.
- Dependable and widespread availability is possible. When customers start using a brand, they want to be able to continue using it.
- Demand is strong enough that the market price can be high enough to make the branding effort profitable.
- There are economies of scale. If the branding is re-

ally successful, costs should drop and profits should increase.
- Favorable shelf locations or display space in stores will help. (Wineries and their distributors must use aggressive salespeople to get favorable positions.)

KNOW THY CUSTOMER: WINE CONSUMER SEGMENTATION

The wise winery owner or brand manager will take some time to understand their target customer. Various New World countries have provided several ways to segment wine consumers. This chapter presents three of the more common methods.

Wine Consumer Segmentation by Consumption Rate

This consumer segmentation method is simple and useful, and developed by The Wine Market Council for the US market (2006). They have researched the different types of wine consumers based on consumer frequency, and have come up with the following classification scheme of three categories:

- **Core wine consumers:** 17.14% of the US adult population are classified as core consumers. These core wine consumers account for approximately 92% of the table wine volume consumed in the US. Fifteen percent of the core wine consumers drink wine daily, 48% drink wine a few times a week, and 37% drink wine weekly. Fifty-four percent of these wine consumers are female.
- **Marginal wine consumers:** 17% of the US adult population are classified as marginal consumers. This group consumes the remaining 8% of the table wine volume sold in the US. Marginal wine consumers drink wine less often than weekly but at least as often as every 2–3 months. Forty-six percent of these consumers regard

wine as a special occasion drink. However, when they do drink wine they drink about the same amount of wine as the core consumer, approximately 2.2 glasses per occasion. Fifty-seven percent of these consumers are female.

- **Nonadopters of wine:** 65.6% of the US adult population do not drink wine. They consume other alcoholic beverages such as beer and spirits, or do not drink alcohol at all.

The Wine Market Council found that an acceptable price range of wine for everyday use at home was US$6.50 to $19.00 and an acceptable price range in restaurants was US$12.00 to $29.50. There were no price sensitivities that differentiated the core wine consumer from the marginal consumer. For both types of consumers the most important factor when making a purchase choice was type or varietal of wine, followed by planned price range, brand, and country or place of origin.

Wine Consumer Segmentation by Motivation: Australian Perspective

Researchers in Australia have been the trailblazers in examining why consumers drink wine. They were the first to develop a consumer segmentation based on motivations for the Australian market. It can also be adapted to other countries. The researchers identified five types of wine consumers (Hall & Lockshin, 2000; Questar & Smart, 1998):

- **Premium Wine Drinkers** are those consumers who are highly knowledgeable about wine, and usually collect wines and have wine cellars. They especially enjoy rare wines, and gain satisfaction in obtaining an exclusive wine that others do not have.
- **Ritual-Oriented Wine Drinkers** are those consumers who enjoying drinking and learning about wine. They

take classes on wine, read books on it, and go to wine tasting regions on vacation. They also pay attention to wine critic ratings.
- **Image-Oriented Wine Drinkers** are those consumers who order wine at a restaurant or bar because they believe it will enhance their image as a sophisticated person. They will usually buy trusted brands because correct image is so important to them.
- **Basic Weekend Wine Drinkers** are those consumers who consume wine on weekends at barbeques and to socialize with friends. They have low levels of involvement, prefer to buy inexpensive wines, and do not want to spend time or effort learning about wine.
- **Social Wine Drinkers** are those consumers who usually only consume wine at social occasions, such as a dinner out, a party, wedding, or other celebrations.

Wine Consumer Segmentation by Shopping Behavior: US Perspective

A similar segmentation study was conducted by Constellation Wines in the US (2006) in which they examined shopping behaviors of wine consumers, and also asked them questions about motivations, preferences, and lifestyle issues. They found six types of wine consumer segments in the US, some of which can also be adapted to other countries. Figure 6.1 illustrates the six segments and the current percentage of consumers within each segment. Descriptions for each segment follow.

- **Enthusiasts:** These wine consumers are passionate about the entire wine experience. They research what they buy, and enjoy sharing their discoveries with friends and family. Currently they are 12% of the consumers in the US and provide 23% of the profits.
- **Image Seekers:** These wine consumers are both sophisticated and adventurous. They enjoy trying new wines and will often seek out those wines that achieved

MARKETING AND BRANDING WINE

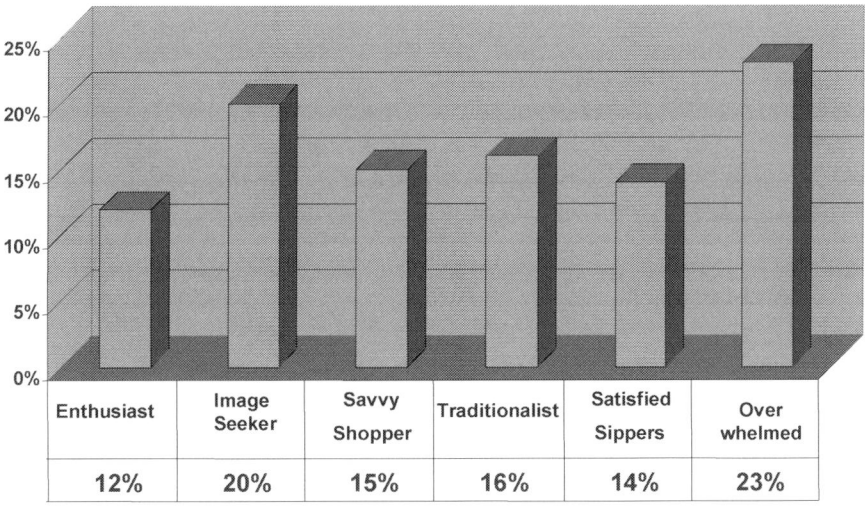

Source: Project Genome by Constellation, 2006   N = 3500 US Consumers

Figure 6.1. US wine consumer segmentation by shopping behavior.

high ratings by the wine critics. They represent 20% of US consumers and provide 25% of the profits.

- **Savvy Shoppers:** These consumers enjoy finding a good wine at a low price. They are value shoppers and believe that good wines need not cost a lot of money. They currently make up 15% of US consumers and provide 15% of the profits.
- **Traditionalists:** These consumers are brand loyal and look for consistency of flavor. They buy the tried and true and don't enjoy experimenting with new brands. They currently make up 16% of US consumers and provide 13% of the profits.
- **Satisfied Sipper:** These consumers don't really care where the wine came from in terms of country or region. They tend to prefer sweeter white wines and white zinfandel. They will often buy a glass of the house wine and aren't that brand conscious. They currently make up 14% of US consumers and provide 7% of the profits.

- **Overwhelmed:** These consumers find wine intimidating and confusing. When shopping they are overwhelmed by all of the wine brands on the shelf. The usually need help in selecting a wine. They currently make up 23% of US consumers and provide 11% of the profits.

WINE BRAND POSITIONING

Positioning the wine brand is at the core of a successful marketing strategy. A good wine brand positioning helps to guide marketing strategy by clarifying what a brand is all about and how it is unique from other wine brands. The strongest brands deepen their points of difference from other brands by convincing the consumer that they can provide greater benefit or value on the attributes that matter to the customer. For example, Volvo and Michelin have positioned themselves as products that provide safety and peace of mind, Intel has positioned itself as performance and compatibility, Coke as Americana and refreshment, Disney as fun, magic, and family entertainment, and BMW as styling and driving performance.

Deciding on a positioning requires determining: 1) who the target customer is, 2) who the main competitors are, 3) how the brand is similar to these competitors, and 4) how the brand is different from these competitors. In the case of Volvo, their automobiles appeal to people who most value safety and reliability and are willing to pay a little more to ensure that they get it. Volvo owners are less concerned with styling and driving performance. An automobile cannot be all things to all people and neither can a wine. Wine coolers can be fruity, fun, and refreshing, but a customer who pays $200 for a bottle of wine is most likely looking for other product benefits, such as outstanding quality and the prestige associated with owning an expensive bottle of wine.

Brand positioning involves understanding long-term pricing strategy. Pricing strategy, whether it is US$3.99 per bottle or US$100.00 per bottle, requires that the winery understand the investment they have in the land, bricks and mortar, global competition, costs of goods sold, quality of the wine, style of the wine, quantity produced (limited supply?), etc. Wine over US$30.00 retail transcends just being an important wine. It is a luxury good and usually needs a powerful story with an emotional connection.

The younger generation should not be overlooked as a target customer. The millennials, people born between 1977 and 1999 in the US, are a potential new market of over 76 million wine drinkers. Opportunities abound for wineries around the world to develop brands that appeal to this market segment. The positioning statement, logo, brand name, price range, and label design should be pretested on a group of these consumers before the winery commits to a course of action.

Once a brand has successfully positioned itself in the minds of the consumer, it is difficult to change consumer perceptions and that positioning. Therefore, the winery should carefully consider the positioning strategy. For example, there are several categories of wine: jug, popular premium, fighting varietals, classics, premium, ultra-premium, and luxury. Once a wine brand has been positioned as a jug wine, as in the case of Gallo, it takes years of good wine ratings and lots of advertising money to convince wine consumers that that winery can make an excellent premium wine, as in the case of Gallo of Sonoma.

The winery may not always have total control of the product's positioning in the minds of the consumers. Country of origin is an excellent example. Wine consumers may have the perception that all wines from Chile or Argentina are of inferior quality, when, of course, this is not the case. But convincing consumers to change their long-held beliefs is an uphill battle. In many cases, the French still believe that wines from the "New World" are inferior to French wines.

Being Consistent With Brand Image

Once a winery decides on a positioning strategy, in other words how that wine is special and unique, then the winery must be consistent with promoting that image in everything it does. If the wine is going to be positioned as a luxury wine, then everything the winery does to market that wine must reinforce the perception of luxury. This starts with an expensive looking label and bottle, carries over into all of the promotion pieces, and is reinforced by the marketing staff with their distributors and in the tasting rooms. For example, luxury wine advertising should only be placed in magazines, restaurants, and events that are considered "upscale."

Building a brand image takes time. The winery must beware of too much emphasis on short-term sales growth and current promotions without regard to the overall brand strategy and the image it is trying to build. Everyone that works for the winery must be aware of the brand's distinctive attributes and be ready to always represent the brand to distributors and consumers with consistency. For example, if the winery's strategy is to promote an image of "family" that goes all the way to the Spanish Land Grants in Sonoma County, then every piece of literature must consistently convey that image.

During the process of building an image for the wine brand, the winery may decide it is prudent to give the brand a "personality." This can be a very effective method for helping consumers differentiate the brand from the thousands of others available to them. A wine brand, like a person, can be characterized as being modern, old-fashioned, fun, exotic, or irreverent. The tasting room staff and tasting room decor, labels, advertising, and special events should reinforce this personality as part of the image building campaign.

Characteristics of a Good Brand Name

A carefully selected brand name can reinforce the positioning strategy and also contribute to the brand's image. It can help tell something important about the winery or its products. Following are some characteristics of a good brand name (Perrault & McCarthy, 2002):

- short and simple,
- easy to spell and read,
- easy to recognize and remember,
- easy to pronounce,
- can be pronounced in only one way,
- can be pronounced in all languages (for international markets),
- suggestive of product benefits,
- adaptable to packaging/labeling needs,
- no undesirable imagery,
- always timely (does not go out of date),
- adaptable to any advertising medium,
- legally available for use (not in use by another firm).

Building Brand Loyalty

Every business tries to develop brand-loyal customers: customers who not only like their brand, but will repeatedly purchase it instead of another brand. Airlines try to build brand loyalty through their frequent flyer programs. American Express has its Membership Rewards program, which gives cardholders points based on the amount they charge. The points can be redeemed for a variety of items, such as airline tickets, jewelry, and electronics. Safeway has its Savings Club in which members receive discounts on certain marked items in stores. Wineries build loyalty through newsletters, clubs and member-only events, support of local schools and charities, protecting the environment, members-only web sites, and quantity discounts.

Simply put, how likely is it that a customer will switch if the wine brand's price is changed, features are changed, or a competitor's products have perceived or actual superior features (their wine received a higher score than yours)? Strong brands have the highest level of brand loyalty.

In spite of easy switching costs, an onslaught of higher competitor ratings, new entrants into the market, and a vast array of lower price/value alternatives from New World countries such as Australia and Argentina, a winery tries to cultivate in its customers a nonspecific emotional attachment to the brand. Sometimes that emotional attachment comes from the special treatment the customers receive every time they stop by the winery or attend a special "members-only" preview of the new release. Other times, the loyalty comes from customer identification with the owners and the lifestyle they portray, such as Ravenswood consumers who identify with the "rebel" image of the winery and may actually tattoo themselves with the Ravenswood symbol. Another example is Bonny Doon Winery, which cultivates an image of fun with irreverent labels and unique wine club member parties that appeal to a loyal group of consumers.

Following are some tips for building an effective loyalty program (Keller, 2003):

1. **Know your customers.** Most loyalty programs involve the use of sophisticated databases and software to determine which customer segment to target with a given program.
2. **Change is good.** Wineries must constantly update the program to attract new customers and prevent other wineries from developing "me-too" programs.

3. **Listen to your best customers.** Suggestions and complaints from top customers must be carefully considered, because they can lead to improvements in the program.
4. **Engage people.** It is important to make customers want to join the program. This includes making the program easy to use and offering immediate rewards when customers sign up.
5. **Positive customer experiences.** Positive tasting room experiences build loyalty. Make them fun and exciting. Use every opportunity to help the customer feel a sense of camaraderie and belonging.

Research has found that loyalty-building programs in themselves are not enough for the customer. Just as in the case of the airlines, in which the customer demands on-time service, well-maintained planes, and a fair price, the wine buyer demands consistent quality and a reasonable price. With over a thousand brands of wine in the US alone, the customer has many options. Disappoint them once and they may never buy the wine brand again!

The Challenges of Taking a Brand International

Obviously, the challenges facing a company creating a national wine brand are considerable. In general, there is a lack of brand loyalty, and wine brands are usually fighting for shares of the market that would be unappealing to almost any other industry. At the international level, these challenges only become more difficult.

For most consumers, wine is an expression of personal taste and culture. In many markets, this means a primary interest in regional or local wines, with only rare explorations into the larger world of imported wines. In any market that produces wine, imported wines are going to have a smaller share of the market than local wines. And some markets, most notably those with slower economies and less expensive local products (Eastern Europe and Latin America), will always be a challenge.

The situation is exacerbated by the high cost of goods for most wine products, compared with the more traditional international branded products. There simply isn't the same kind of room in the budget for the traditional methods of international brand marketing, such as television advertising and major media promotions. Therefore, when attempting to establish an international brand, four major elements are critical:

1. **Top quality:** The first is a top-quality product. This means a clear category leader that establishes the company as being in the top tier of international wines. This is the flagship behind which the rest of the portfolio will sail into each market. It must be among the greatest wines of its region, and compare on some level with the greatest wines in the world. This wine not only establishes credibility with the key influencers in the media and trade, but also creates the demand among consumers for rare and wonderful wines.

2. **Brand story:** The second element is a strong story behind the brand: a way of capturing the imagination of the market and the attention of the media. This story must work in all of the target markets—and that alone may be a challenge. Because public relations campaigns offer more cost-effective ways to communicate to the various audiences than advertising, this story is critical. Personality is the key for long-term visibility and credibility.

3. **International relationships:** The third element is relationship building. This means not only establishing relationships with key customers in each market, but also attempting to build long-term partnerships with

there—and had been for more than 50 years. The issue, at the time, was that the winery was not that well known.

Hanzell's unique story is the "grand cru" farming standards that they use. Focusing only on the Burgundian varietals of Chardonnay and Pinot Noir, they use sustainable farming techniques, which are environmentally friendly to the special mountain vineyards of their Mayacamas, Sonoma Valley estate. With only 42 of their 200-acre estate planted, they use the same farming techniques as the top 2% of Grand Cru Burgundy. This includes stressed growing conditions and rigorous pruning to intensify the flavors of the wine, which results in very low yields—less than 3 tons per acre—and only 3000 cases produced per year. This creates a very rare, scarce, and exclusive wine that is sought after by wine connoisseurs around the world.

Wine marketing methods to support both brands include using only the finest materials for their embossed and engraved labels; expensive parchment letterhead and brochures; and bottles hand-polished and wrapped in tissue. Every detail of the grape-growing, winemaking, and wine-marketing process receives the utmost care and attention. Likewise, Hanzell Vineyards targets only the top 20% of wine writers, distributors, and retailers, who are invited to special educational events at the winery and other locations. During these meetings, the special story of Hanzell Vineyards is communicated in a relaxing and intimate environment to a very elite and influential group of buyers who will continue to "tell the story" to their customers. Thus, the brand grows successfully, with grace and exclusivity.

These two examples of wine brands illustrate the importance of understanding the uniqueness of the wine's story and then attempting to develop a complete brand strategy to communicate consistently the image and position of the wine. This takes much time, talent, dedication, and vision, but in the end it can pay off handsomely in customer loyalty and positive cash flow.

Summary

Creating a successful brand is a real challenge for wineries. Competition in the industry is intense. Before launching a brand wineries have to develop a long-term plan and be prepared to reinforce their brand image in everything they do. When consumers think of a wine brand they need to recall favorable images or mental associations about the brand. Creating strong, favorable, and unique associations takes time, money, and consistency. It requires that the winery understand its consumers. When consumers think of Coke they think "Americana and refreshment." Whey they think of Disney they think of "fun, magic, family entertainment." When they think of BMW they think of "styling and driving performance." When they think of a specific winery, what special image and perception comes to mind? That is the brand!

LEARNING OBJECTIVES:
- Describe the general process for distributing wine on a global basis
- Identify the historical and current context for wine distribution in the US
- Define the roles of wholesalers and brokers
- Explain how to find and work with distributors and brokers
- Identify the three distribution sales channels in the US
- Describe skills needed by distributor sales reps to sell wine successfully

CHAPTER 7

WINE DISTRIBUTION

Bruce Herman
Senior Vice President, Sales & Marketing, Foster's Wine Estates

Gary Long
On-Premise Sales Rep, Young's Market

The distribution of wine from winery to customer has been occurring for centuries, with the earliest records describing the transport of wine in ancient Canaan via amphorae—large terra-cotta pots with a pointed base. Wine has also been transported via goatskin bags, in barrels on old British sailing ships, and today in large ocean freighters that haul it in large case shipments or bulk. Refrigerated trucks and train compartments also transport wine between countries, states, or territories all over the globe.

The actual shipment of the wine is fairly easy. It is the complex regulatory requirements and identification of the proper distributor or broker that is more complex.

This chapter provides an overview of the process of wine distribution. It begins with a general description of the process for distributing wine on a global basis. Then it examines one of the most complex distribution systems in the world—that found in the US. Indeed, this system is so multifaceted that many in the wine industry say it is easier to ship wine from the US to England than it is to ship it between certain states. The historical premise for this system is described, as well as present-day distribution regulations and practices. This is followed by an explanation of the role of the wholesaler and broker, including information on how to identify one to sell your wine. Next is a list of tips on how to work successfully with wholesalers and brokers. The chapter ends with a description of "a day in a life" of a distributor sales representative.

WINE DISTRIBUTION ON A GLOBAL BASIS

The basic requirement to begin distributing wine on a global basis is a *clear strategy* and *commitment* to global wine marketing (see Chapter 12 for more detailed information on this process). This means that the top management team of the company must have identified a reason to export and/or import wine from other countries and have committed the personnel and financial resources to do so. Without this preliminary foundation, attempting to distribute wine globally can be fraught with frustration and disappointing results.

However, once this basic requirement is established, the next step is to *identify a market* that matches the company's wine style, image, and price point. This will require market research on the part of the company, and can also be the first place of contact with potential distributors. Some of the best ways to identify potential global distributors are to obtain recommendations from other producers who are already exporting, contact importers at trade shows and other wine events, and work with your local wine associations and agriculture trade offices. Embassies and consulates around the world also publish reports on various commodities, including wine and market share information.

In distributing wine globally, you have three *distribution choices*:

1. Identify an import **distributor** if you are interested in visiting the country to which you will export your wine and want to be involved in negotiation.
2. Use a **broker** if you do not have time to visit the country and prefers to have someone else represent the wines. This is the easiest method to start exporting.
3. **Sell** the wine directly by calling upon specific country distributors and retail establishments. This takes more time, but eliminates the margin made by the distributor or broker who is representing the brand.

The benefit of working with a global distributor or broker is their knowledge of the various markets. They will understand the legal structure of the distribution systems within the countries. In addition, they are knowledgeable about import tariffs, pricing, label requirements, and winemaking regulations. They also have established relationships with consumers, and know the specific cultural nuances of doing business in that country.

Another very important aspect of global distribution is *logistics*, or shipping wine safely so it is not spoiled by fluctuating temperatures. Most wine is shipped internationally by freight forwarders who use specially refrigerated trains, ship containers, and trucks. In addition, they take care that once the wine has arrived in the country that the proper shipping documents are handled and the wine is safely moved to a warehouse or other temperature-controlled environment. For this reason, having the appropriate insurance when shipping wine globally is very important.

A final consideration with global distribution is *payment*. Agreement on currency type, payment method (electronic, check, etc.), timing, etc., must all be negotiated and agreed to with distributors and brokers. Due do credit risks with certain customers, many experienced wine exporters will purchase insurance against nonpayment. Related to this is the issue of exchange rates. This can work for or against a winery. For example, several years ago when the US dollar was stronger than the Euro and Australian dollars, it was difficult for many US wineries to sell wine abroad because it was priced too high due to the exchange rate. At the same time, some European and Australian wine was being sold in the US at inexpensive prices, which cut into US wine sales. This situation has recently reversed, which allows some US wineries to sell their wine more cheaply abroad. However, chances are that the exchange rates will continue to fluctuate as they always do because of economic changes around the globe.

WINE DISTRIBUTION OUTSIDE THE UNITED STATES

Distribution complexity and methodology vary across each market around the world, mostly due to local laws, regulations, and business practices. While this chapter cannot possibly cover every distribution system in even the most populous countries, it is meant to provide some basic knowledge of systems, practices, and challenges that cross many cultures. Some countries allow beverage alcohol, namely wine, to be distributed directly from the winery/supplier to the retailer or restaurateur. Some countries require going through a "middle" tier of some sort. Some countries, as in the US and Canada, have laws that vary among the states or provinces inside the country. Beyond the actual delivery of the products, methods of payment and collection vary as well. The most important understanding, wherever one is distributing around the world, is to recognize the local laws, nuances, and business practices that provide the means to deliver the wine from the supplier to the consumer in the most efficient and effective manner.

WINE DISTRIBUTION IN THE UNITED STATES

From Colonial times until the passage of the 18th Amendment in 1919, beverage alcohol was sold and distributed in a free-market system. There were few rules and regulations about who could own retail establishments and how business was to be conducted between the retailer and the supplier. It was common to have suppliers owning their own bars, saloons, and taverns or giving incentives to retailers to carry their brands exclusively. Incentives to the retail trade may have come in the form of interest-free business loans and mortgages; equipment such as refrigerators, dispensing systems, glassware, and other supplies; or direct rebates to the establishment when they sold the supplier's brands to the exclusion of the competition. Lawlessness, alcohol abuse, and corruption were commonplace when the suppliers were selling directly to the retailers, with a preponderance of saloons tied to the suppliers of beverage alcohol. This abuse of beverage alcohol, public drunkenness, and the control of retail outlets serving the public by the large distilleries and breweries led to the American Temperance Union and the Anti-Saloon League, which saw all beverage alcohol consumption, moderate or excessive, as evil.

The economic challenges suffered by the retailers, partly caused by the exclusionary practices of suppliers, along with the social issues of abstinence resulted in the passage of the 18th Amendment, commonly known as Prohibition. Prohibition did not end the consumption or sale of alcoholic beverages. It forced the practice "underground" and offshore, leading to a black market for beverage alcohol, which was dominated by organized crime. Bootlegging or smuggling of alcoholic beverages became a large and very profitable business. All levels of government lost the ability to control, regulate, and collect tax revenue from this now "illegal" activity. During the Depression, Congress focused its attention away from the moral and religious issues of abstinence and began to consider the potential revenue gains if wines, spirits, and beers once again became legal to sell and consume. Ending Prohibition, in the eyes of Congress and a majority of the American people, could add much-needed revenue to local, state, and the federal government as well as much-needed jobs for the American worker. The "Great Experiment" became a great failure, which both the public and Congress changed on December 5, 1933 when the 21st Amendment to the Constitution was ratified, ending Prohibition and giving the states the authority to regulate the production, importation, distribution, retail sale, and consumption of beverage alcohol inside their borders.

Shortly after the 21st Amendment was ratified, Congress passed the Federal Alcohol Administration (FAA) Act, which set broad limits on the rules states might establish to regulate the sales, promotion, and merchandising of beverage alcohol within their borders. Congress was also concerned with the historic abuses of pre-Prohibition when suppliers owned retail outlets through a tied-house relationship. To insulate the retail tier from the supplier, many states adopted the three-tier system of distribution by placing an independent licensed wholesaler in between the retailer and the supplier. These states are known as open states, of which there are 31. Other states chose to become the wholesaler and retailer (or a hybrid of this structure) and these states are commonly known within the industry as monopoly or control states, of which there are 19.

The control or monopoly states not only regulate beverage alcohol distribution within their borders, similar to open states, but they also sell alcohol beverages wholesale and in many cases act as the retailer selling direct to consumers through state-owned and -operated retail outlets. The control states are: Alabama, Idaho, Iowa, Maine, Montgomery County Maryland, Michigan, Mississippi, Montana, New Hampshire, North Carolina, Ohio, Oregon, Pennsylvania, Utah, Vermont, Virginia, Washington, West Virginia, Wyoming.

The goal of control states is to promote responsibility and moderation in the consumption of beverage alcohol. In the regulated environment present in the US, control states are an alternative to the open or licensed states. They have a controlled distribution system owned and run by state employees that, in essence, substitutes the state for private ownership. Figure 7.1 illustrates the differences between the open and control state process.

The Role of the Wholesaler (Distributor)

Simply put, wholesalers (tier 2) purchase wines and spirits from suppliers (tier 1) and sell, deliver, and service the retail customer (tier 3). Wholesalers also collect state excise taxes and provide gallonage and sales reports to verify usage, consumption, and tax collection. In most states, wholesalers are allowed by law to extend credit to licensed accounts, provide frequent deliveries, provide full case, split case, or fill bottle requests to accommodate the needs of smaller retail customers and slower moving brands.

Wholesalers play an important role in the marketing and merchandising plans for the suppliers they represent. Wholesalers implement the suppliers' in-market point of

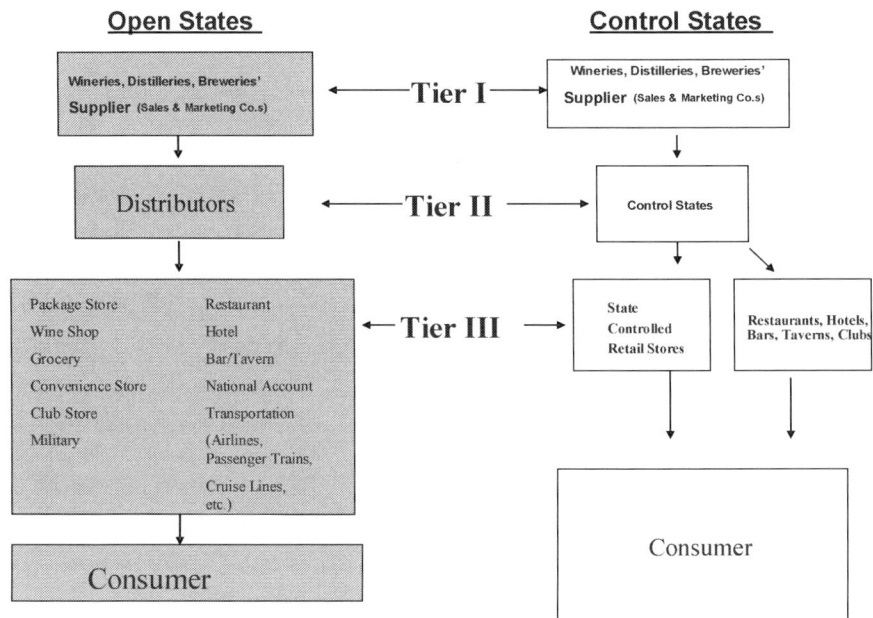

Figure 7.1. Distribution process in US open and control states.

sales programs by making product brochures, fact sheets, and posters available to the consumer through wine shops and package stores. They work with both suppliers and their retail customers to reset shelves, cold boxes, and place displays or wine racks in prime selling locations. Wholesalers train both on- and off-premise retail sales staffs on the use, taste, and food pairings of wines and spirits. They conduct sales seminars, tastings for wait staff, promote the latest cocktail recipes, print wine lists, table tents, and shelf talkers for both their retail customers and their suppliers.

Wholesalers also furnish the suppliers they represent with daily sales data, competitive market information, and category market expertise. Large wholesalers in major markets generally segment their sales force by trade channel (on-premise, off-premise, grocery, chain, club store, and national accounts); wholesalers in major metro-markets also provide specific ethnic market selling expertise about Hispanic, African-American, and Asian consumers. The information collected and given to suppliers, producers, and wineries by distributors, when pieced together throughout the country, gives them the ability to spot trends, target market their merchandising and selling activities, as well as anticipate problems, which gives these suppliers, producers, and wineries a competitive advantage for their wines and spirits.

THE ROLE OF THE BROKER

There are many different types of brokers who provide a service directly to wineries or wine suppliers. Brokers can act like wholesalers by providing the producer with a selling organization (sales people) that has a direct relationship with the retail tier. Brokers usually represent a number of different brands of wines and spirits, and focus their activities on selling their portfolio directly to restaurants, hotels, wine shops, package stores, or grocery chains in the marketplace. While the broker sells to the customer directly to the retail account, the producer usually bills the account directly instead of the broker.

Other brokerage organizations act more like sales and marketing companies, where they help to manage the relationship between the products they represent and the distributors chosen to sell these products within the open states. In this situation, the broker spends more time managing the distributor's sales force and trying to get as much selling time and attention focused on the broker's portfolio of wines and spirits.

Whether the broker sells direct to the retail trade or through a distributor, most brokers have one thing in common. Brokers do not take ownership of the wines they sell, but rather earn a commission from the producer once the wines are sold through to the distributor, the retail tier, or to the control state. This, you will note, is the fundamental difference between a broker who has a sale force that sells to the retail tier and a distributor. Distributors take ownership of the wines they represent by paying the producer or the sales and marketing company, usually within 30 days after an order has been processed by the producer. The distributor not only takes possession of the wine but also maintains the accounts receivables and assumes the risk of bad debts from the retail tier. Brokers, as stated earlier, generally do not take ownership of the wines they represent and do not take on the responsibility of the accounts receivables nor assume the risk of bad debts.

In large markets such as Metro New York and California, where there are many trade channels, ethnic markets, a large concentration of accounts, and a multitude of brands vying for distribution, brokers generally are present to fulfill a need not occupied by the traditional distributor. Brands on the periphery not represented by the first- or second-tier sales and marketing companies, as well as wineries representing their own brands, often utilize brokers to build their distribution and sales volume. Producers also utilize brokers because there is less competition within the broker's portfolio than within the distributor's selection of brands represented in the marketplace.

Generally, brokers do not represent major spirit suppliers or major wineries; therefore, smaller brands can gain a greater share of mind from the broker's selling organization than they can from a major market's distributor. In addition, some small restaurants and wine shops prefer to conduct business with brokers or small distributors because they believe that they are more important to the smaller broker or distributor and will therefore receive better service. Once brands grow in volume and begin to achieve momentum or greater critical mass in a market, brokers run the risk of losing brands as they move to a distributor. As brands grow, producers either decide to have them represented by their own sales force or represented by a larger sales and marketing company. In conclusion, brokers serve a need within the three-tier distribution system and generally represent smaller brands on their way up or more mature brands losing volume on the way down.

Finding a Distributor and Broker for Your Wine

First and foremost, it is important to have a vision for your brand or portfolio and understand who the target consumer is and what trade channels in each market you want to penetrate. Once you have a clear understanding and identity of who you are and where you want to go, it is much easier to choose a distributor or broker for your brand or portfolio of brands that align with your vision.

A good resource when starting out is to consult the Wine and Spirit Wholesalers Association (www.WSWA.org). They are a trade organization that represents most of the distributors throughout the country handling both wines and spirits. Another contact is the Wine Institute (www.wineinstitute.org), which is a trade organization that represents most of the wineries in the US. You will find a helpful resource section on their website and a link to other related organizations that may be used as a source for both distributor and broker leads.

The most effective, yet least efficient, way to identify potential distributors and brokers is to visit the marketplace and talk to both retail and restaurant customers whom you believe should be selling your bands. There is much information that can be gleaned from a market visit and talking to potential customers about who they like to buy wine from, who has the most knowledgeable selling organization, who services them the best, who has a compatible portfolio, and other key issues that would affect your decision-making process when deciding which distributor or broker you want to represent your brands.

Once you have identified a number of potential distributors or brokers, you will also need to have a clear vision of the competitive set and how you want the distributor or broker to approach the marketplace. Do you want them to be strong on-premise or off-premise? Is chain distribution a priority and do you need access to national accounts? Who in their current portfolio of brands may be your direct competitor and who may have the clout to demand more attention and therefore make the distributor or broker less effective on your wines? Remember that in today's consolidated marketplace competition cannot be eliminated, but how it is limited and managed could give you a greater chance for success. After you have sorted out these issues and interviewed as many distributors and brokers as possible, you will be in a good position to determine the best company to represent your brands.

Working Successfully With Distributors and Brokers

To work successfully with distributors and brokers, it is very important to *have a marketing and sales plan*. In addition, it is important to communicate with your distributor or broker early and often. Everyone is trying to get attention from their distributor or broker, and the sooner you can give them your plan, even if it isn't perfect, the better chance you will have of making certain that they don't forget about you and go on to selling someone else's wine. Your plan needs to include marketing as well as sales goals. For example, just telling the distributor your price and how many cases you want them to buy is not a plan—it's a sales allocation. Make certain that your distributor knows what your goals and objectives are for your brands not only for the current year, but also for the next 2 to 3 years.

Some questions to consider are:

- Do you want to be on wine lists, or poured by the glass?
- Does the distributor have copies of your latest and greatest wine reviews?
- Does the distributor know which items are highly allocated and which customers should be given the first shot at buying these sought-after wines?
- Have you effectively communicated which wines are going to be in short supply and which wines are going to be in long supply so that your distributor can manage them in a way that benefits your brands?

Review performance against the plan with your distributor or broker on a very regular basis. Nothing keeps your in-market agent's attention better than having brand review meetings every 60–90 days with management.

Stay focused not only on what your distributor or broker buys from you, but, more importantly, on what they sell to the trade. Although what you sell to the distributor or broker is what generates your revenue, what they sell through to the trade is the real indicator of how well your brand is doing in the marketplace. It is beneficial to not only know what they are depleting to the trade, but also to whom and at what price.

Distributors and brokers often spend too much time and money forcing out cases just to make a sales goal requested by the supplier. In the end, these efforts do not really help to build the brand for the future, but simply move inventory from one tier to the other.

You should have a *clear definition of your competitive set*; communicate it to your distributor and broker network so that they can use it to help guide selling and distribution activities. Also let your distributors and brokers know whether you want your wines to be sold in the chains or the club channel so that they can manage your distribution objectives efficiently and effectively.

Be consistent. Due to the size and complexity of the portfolios managed by distributors today, it is much more productive to present the goals and plans once and not change strategy in the near term. When directions or objectives change, the distributor can quickly lose focus or attention on the brand.

Don't limit your contact to just the wine press. You should include lifestyle editors, and travel and leisure writers, as well as food and restaurant critics. Don't overlook the radio talk shows and cable food shows to promote your wines. We live in an information age and there are hundreds of stations looking for things to talk about to their audience. Once you get press be sure that you have a way to communicate this to your distributor and to your accounts, directly, if at all possible.

Stay in touch with the wine trade. Winemakers and winery principles should make regular visits to the marketplace to meet and greet the people who sell the wines. Much can be learned and accomplished by getting to know your customer face to face. You must always remember that the truth is on the street and not in the boardroom, so stay close to your customers and work the market with your distributor.

Develop a way to *communicate what's going on at your winery* during harvest and at other times of the year directly to your retail accounts and to the distributor's sales force. The wine business is a people business, and the more you can communicate with your customers and distributor directly, the greater the share of mind you will achieve for your brands.

Set up a "trade only" section on your website. This makes it easy for sales people to download winery events, wine reviews, shelf talkers, sell sheets, and the like. It also is less expensive for you because you don't have to print, mail, and/or ship these valuable selling tools to you distributors. They can get them off your website and customize them for their accounts and their market.

Encourage visits from the trade to your property. Don't limit VIP visits only to owners or to buyers, but encourage clerks and waiters to visit your properties as well. They are usually the ones who have direct contacts with the consumer and can suggest your wines for trial.

A Day in the Life of a Distributor Sales Rep

A typical day for a distributor sales representative will vary greatly depending on what channel the representative is working in. In the US market, there are three major channels for distribution sales reps: 1) Off-premise chains; 2) On-premise; and 3) Off-premise fine wine shops & grocers (Figure 7.2).

Distributing Wine to Chain Accounts

It is most common for sales reps to start their career with a distributor in the chain division. In the US chains are

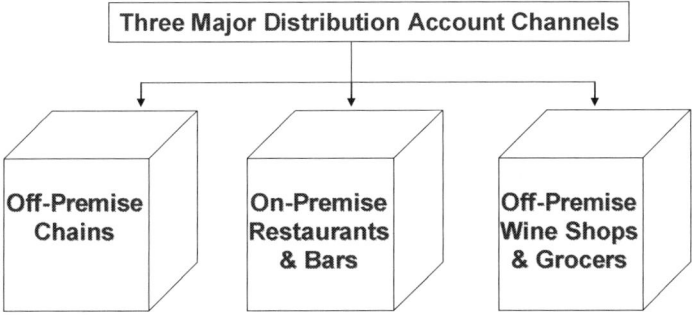

Figure 7.2. Distribution channels.

defined as grocery and retail stores that have several locations under the same "chain" name and sell wine to be consumed off the premises. Examples include Long's Drug Stores, SafeWay, and Cost Plus. With chains most programs and items are sold at the headquarter level by account executives. So it is the responsibility of the distributor sales rep to make sure that the programs and items are compliant with corporate mandates, as well as maximizing pull through at the store level.

The main drivers of sales in the chains are cold box placement, favorable shelf placement and facings, and displays. The cold box is often said to be some of the most expensive real estate in the store, due to the cost of its operation and limited space. Cold boxes typically feature white wines and, due to the limited space, frequently wines can be sold out. Therefore, it is the rep's responsibility to ensure that their products are fully stocked in the cold box by the store staff. Shelf placements are determined at the corporate level and each store has a shelf schematic based on the size of the store. Eye-level shelf placement is ideal and can increase sales by as much as four times over knee or ankle level shelves.

Some chains are more willing to give store-level autonomy to their managers and employees than others, so in certain stores reps can sell in displays that have a large impact on sales. This is especially important during OND (the October–November–December holiday period), when close to 40% of the annual wine sales occur in the retail environment. Pricing is also a major factor but is typically controlled by the negotiations of the account executives at the corporate level. It is the responsibility of the sales reps to audit the pricing to ensure it is correct, monitor shelf placement, influence display size and location in the store, and "merchandise" the products they represent. Merchandising could include restocking the shelf or cold box, dusting the bottles, and adding shelf talkers, coupons, or signage to the shelf or displays.

The General Market

After spending time working in the chains the typical career path for a representative is to move up to management within the chains or up to the general market. The general market is defined as the segment including the independent stores that do not have central corporate decision making. In the general market it is the responsibility of the rep to sell the products to a given account, and provide support and service to the account as needed. This can include working with the delivery drivers to correct any fulfillment errors that take place, merchandising the products in the store, and ensuring the correct level of inventories to avoid any out of stock situations. Additionally, sometimes the representative gets involved in the collection of money from accounts depending on the terms and laws in each market.

Distributing Wine to On-Premise Accounts

One channel in the general market is on-premise, where the product is consumed on the premises, such as restaurants, bars, or lounges. In other countries, this is also called on-trade. An on-premise rep will typically start their day later than other reps due to the fact that decision makers frequently work in the evenings and work a different shift from the standard day. It is often necessary to spend time out in the evenings in order to build strong relationships and encourage sales. In this role, in addition to handling orders and basic sales for the accounts, there is much more wine training and education given to the wait staff at the on-premise locations.

Sales in on-premise accounts are driven two ways: 1) by the glass promotions or as the house wine, or 2) placement on the wine list. The house wine is what is poured when a customer asks for a generic varietal, such as Chardonnay, instead of a specific brand. This is usually a less expensive wine. Most restaurants will also have a wine list. The structure of these lists can vary widely depending on the restaurant. Some restaurants prefer a smaller list with only one wine for each key varietal. Others will have lists that are many pages long and feature different regions, vintages, and styles. Sometimes a list can also be centered on a certain region or be focused on only European wines or American wines. Wines that are good sellers are typically put on by the glass because the decision maker feels that the wine will move fast enough that a whole bottle will be poured off in a reasonable amount of time. Wines by the bottle are typically the more expensive wines that are ordered with less frequency. In this channel the quality of a wine relative to the price charged is extremely important. While a wine may taste great, if it is overpriced it will not generate as many placements as a more value-driven brand of similar quality.

In order to optimize the samples a distributor sales rep receives they should preplan their day to taste several accounts on a given wine. Often wine lists are set for an extended period of time, but when changes are made it is always beneficial to have tasted an account on a number of wines to give them new options.

Distributing Wine to Off-Premise Wine Shop and Independent Grocer Accounts

The other channel is off-premise (also called off-trade in other countries), where the product is consumed outside the account, usually in a home. Examples of these types of accounts are independent grocers and fine wine shops. The off-premise channel requires a combination of skills from both on-premise and the chains. It requires the merchandising and placement focus of the chains, as well as the sampling and presentation skills of the on-premise. In the off-premise a rep can have a huge impact on sales by influencing shelf placement, cold box placement, and especially with displays. Key elements of displays are point of sale material and creativity.

Important Skills for Distributor Sales Reps

In all channels there are a few traits that can ensure success. The first is developing good relationships with your customers. The most important way to do this is building trust by following through on promises, helping to solve problems, and being truthful in your communication. Another key attribute of a successful sales representative is to preplan your day and methodically target your products to the accounts that would be the best venues for them. Lastly, seeing your accounts on a consistent and frequent basis is critical to building strong relationships.

CONCLUSION

In summary, establishing a successful process to distribute your wine is not necessarily easy, but with advance knowledge and perseverance, it will pay off. The wine industry is still basically a "relationship industry" all over the world. Therefore, it is necessary to build the relationships and learn the major regulations in each country, state, province, or territory to ensure your wine can be sold successfully. However, as this chapter states, there are associations that can help in this process, as well as brokers and consultants who can facilitate success.

LEARNING OBJECTIVES:
- Define professional wine sales
- Describe selling emphasis in three major US wine sales channels
- Explain decision criteria to create the annual wine sales plan
- Identify steps and issues in implementing and evaluating the wine sales plan
- Describe tips and challenges in selling wine internationally

CHAPTER 8

PROFESSIONAL WINE SALES

Armen Khachaturian
Director of Sales, Hanzell Winery

Elizabeth Rice
International Sales Manager, Delicato Family Vineyards

The old adage "if we build it, they will come" doesn't apply easily to the wine industry. Indeed, with thousands of wine labels in the global marketplace and new ones being launched every day from small wineries around the world, the task of marketing and selling these wines is extremely daunting. It is for this reason that professional wine selling skills are one of the most critical components of the global wine business.

Although marketing and selling go hand in hand, with marketing preparing the ground, it is the sales person who actually harvests the crop. In the wine industry, the successful winery sales person is usually on the road 60–80% of the time and is an expert at relationship building. People in other industries often envy the fact that they are involved in the glamorous world of wine, not realizing how long the hours can be and the extensive knowledge and experience they need to acquire to be successful.

In this very competitive global wine market, trying to convince a distributor, exporter, or retailer to carry your new, unknown wine brand that hasn't received any glowing reviews or medals is almost impossible. Therefore, this chapter provides an overview of the professional wine sales process and how to be successful. It begins with a definition of professional wines sales, and then provides an overview of the selling techniques needed to work in the three major channels of US wine sales. Next it explains the importance of creating an annual sales plan and how to implement it successfully, with methods to evaluate and revise as the market changes. Finally, this chapter presents challenges and tips in conducting international wine sales.

Defining Professional Wine Sales

There are many definitions of "sales" in the dictionary, but "professional selling" is different than traditional sales. With traditional sales we think of delivering or exchanging a product or service for money or its equivalent. This doesn't focus on the long-term relationship between seller and buyer. In the wine industry we want to promote consumer loyalty and long-term sales, and so relationships are important. Furthermore, much wine is sold through intermediaries, such as distributors, retailers, and exporters, so there is a further need to develop positive working relationships.

Therefore, the term professional wine selling can be defined as "the process of influencing a consumer to make a wine decision in their and your favor." With professional selling, it is critical to understand the needs of the buyer and do your best to offer them a solution. In some cases, it may not even be your product, but the long-term gain you create in focusing on their needs will most likely be repaid in the future. Professional wine sales is relationship-oriented selling.

Selling Wine in the Three Major Channels of the US Market

Wine is sold in the US utilizing three major channels (Figure 8.1). The first channel is through a distributor, also known as the three-tier system. The winery begins this process by selling their wine to a distributor at freight on board pricing (FOB), which is usually 50% of retail. The distributor then marks up the wine and sells it to restaurants (on-trade) and retailers (off-trade). They, in turn, also mark up the wine before selling it to the consumer. In order to maintain this process, the restaurants and retailers must deplete their inventory in order to continue to pur-

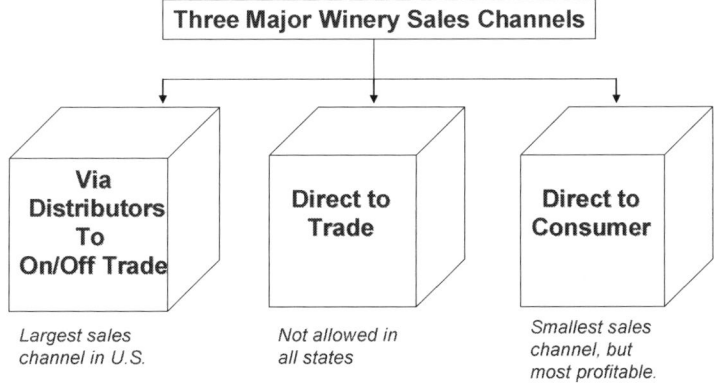

Figure 8.1. Three major sales channels for wineries.

chase your wine. If the wine does not deplete they will not reorder the wine from the distributor, who in turn will not reorder additional wine from the winery. There are many ways to avoid this situation, which are described in detail in the implementing of the sales plan section.

The second channel of selling wine is called direct to trade. By using this channel, you forgo a distributor and sell your wines directly to restaurants and retailers at wholesale pricing, which is usually 33% less than retail. This channel is currently available in Arizona, Connecticut, Massachusetts, Maryland, Oregon, Vermont, and Washington, but it is most commonly used in California. Using this channel requires a dedicated sales team to build, maintain, and nurture these individual relationships. One must also take into account the various aspects of warehousing, delivering, and invoicing the wine. If utilized properly, this channel can be financially more profitable than using a distributor.

The final channel of selling wine is called direct to consumer (which is described in more detail in Chapter 9). The most common methods of selling wine direct to consumers are through tasting rooms, wine clubs, newsletters, mailing lists, Internet sales, and special events. Although this channel does require a lot of time and resources, it is the most profitable because the wine is sold at full retail and it eliminates the middleman. In addition to the financial benefits, it is important to have a clear line of communication with your consumers and this channel allows for that to happen.

Creating the Annual Sales Plan

In order to be financially successful, you need to create a clear sales plan with specific goals. The goal may be as simple as selling the entire vintage of wine in 12 months. It may be as complex as selling 50% of the wine direct to consumers; 20% of the wine direct to trade in California and Connecticut; and 30% of the wine via distributors with 80% of the wine selling to restaurants and 20% to retailers. The process of creating a sales plan should be a collaborative effort between the CEO/President, CFO/Controller, Winemaker, and National and/or Regional Sales Manager(s).

Before beginning the process of creating a sales plan, you need to make sure that you have the proper distribution in place. If you don't have the proper distribution in place this would be the time to make those changes (see next section and Chapter 7 on selecting and working with distributors). You also need to know how many cases of each wine have been produced and are available to sell. Most importantly, the sales plan must keep in line with the company strategy that is already in place. With this information in hand, you can begin creating your sales plan.

Creating the sales plan begins by deciding what percentage of your wine you will be selling through each channel. Most established wineries strive to sell 50% of their wine direct to the consumer, but unless you already have a strong and loyal following or a really high rating from Robert Parker or the *Wine Spectator*, this can take years to accomplish. You must also include your plans for the tasting room, your wine release schedule for the wine club, and any special events that you plan to host. Next you must decide if you will be utilizing the direct to trade channel and/or the three-tier system. If you decide to use the direct to trade channel you will need to take into account the additional time commitments this channel requires.

Next, you will need to allocate the number of cases of wine by channel based on the quantity available. With regard to distributors, you will need to allocate by state, keeping in mind the individual needs for each market. Ultimately you want to create and/or maintain a pull in the marketplace so be careful to allocate the correct amount of wine to each market. You may also want to set aside a certain number of cases of each vintage to put into the wine library. A wine library is a term used when a winery warehouses and ages their wine for the purpose of future tasting events or to sell after a few years at a higher premium. Also don't forget that you will need a specific number of cases to be used as promotion during the year (i.e., tasting samples, special events, charity requests, wine judgings, etc.).

The final stage of the sales plan is focusing your efforts on specific depletion goals. You will need to decide what percentage of your business you would like to see on- and off-premise. There are pros and cons to both, but generally speaking, on-premise should help create brand awareness and exposure and off-premise should make the wine available to the consumer. The key to increasing sales via the direct to consumer channel is to lower the availability of your wine through off-premise channels, while increasing the awareness and value. This will help direct the consumer to the winery. Table 8.1 illustrates a checklist of some of the major decisions to include in your annual sales plan.

Tracking Sales

Once your annual sales plan has been created, you must create a system to help track sales. The most common way

Table 8.1. Checklist of Decisions for Annual Sales Plan

No. and/or Percent of Cases?	For Purpose/Channel	Depletion Goals by Date?
	Distributors for On-Trade	
	Distributors for Off-Trade	
	Direct to Trade	
	Direct to Consumer	
	Library Wines	
	Promotions/Tastings/Trade Shows	
	Charity Events	
	Other	

to track sales is by depletion reports sent by distributors. Most distributors also have the ability to furnish you with account-specific depletions. There are also a few companies like eSkye Solutions, Diver, and Divineware who collect this information from your distributors and deliver it to you via their proprietary software. The cost of these services may be significant; however, the time you save is invaluable. Utilizing these reports will ensure that you are on track to achieving your sales goals.

Implementing the Sales plan

Implementing your annual wine sales plan means working with the various channels you have identified in the planning session. The number of channels, as well as the quantity of wine your winery is selling, will also dictate the number of sales people needed. For smaller wineries at least one full-time sales professional is needed, as well as the assistance of the winemaker and owner who will be needed to call upon clients. If the winery is larger, the position of National Sales Director should be created, and several other sales professionals hired to assist. Some wineries also hire sales consultant or brokers.

Finding the Right Distributor

If you do not have an established distribution network or need to make a change to your existing distribution network, you must first choose the right distributor. To find out if you are the right fit for each other ask the following questions:

- Do they have wines from the same country or region as you?
- Do they have a lot of the same varietals as you?
- Do they have a lot of wines in the same price point as you?
- Will your wines fit into their portfolio?

- How many sales representatives do they have?
- What is their sampling policy?
- What is their market visit policy?
- Can they supply you with a monthly sales report?
- Are their warehouse and delivery trucks refrigerated?
- Do they pay on time?

Once you have found a distributor that is the right fit, you will begin working and building your relationship with them. Be cautious about jumping into the first distributor relationship you encounter. Also, be careful about not getting so desperate to place your wines that you will work with anyone. It is worthwhile to ask the right questions and build a solid, trusting relationship that will be a win–win for both of you.

Working With Distributors

In most cases distributors have hundreds of wineries in their portfolios. The only way to ensure that you will get the attention you need is by working with them to make sure that they are selling your wine to the appropriate accounts and helping you achieve your sales goals. This can be done by working the market with the distributor sales rep, participating in trade shows, presenting at sales meetings, planning key account luncheons, and communicating new press material as it becomes available.

Working the Market

Working the market is perhaps the most common form of support that a winery can provide to a distributor. This is when a winery representative travels to the respective market to spend a day visiting on- and off-premise accounts with a distributor sales representative. These visits allow the winery the opportunity to build or maintain a relationship with the individual sales representative, taste your wine and tell your story to sommeliers and wine buyers, and allow you to evaluate the marketplace. Before a market visit, you should contact the distributor sales representative to let them know what you plan on getting out of your day. You should also let them know if there are any specific accounts that you would like to see during your visit. During your visit you should take detailed notes about the accounts and make sure to immediately follow up on any questions, requests, or concerns that may come up. At the conclusion of your market visit, you should follow up with the sales representative to make sure that you tie up all the loose ends.

Trade Shows

Trade shows are another form of support that you can provide to your distributor. Trade shows are usually set up by the distributor to showcase their portfolio to restaurants and retailers. The distributor charges a nominal fee to the participating wineries to help offset their costs associated with the event. Each participating winery then sends their representatives to pour wine, tell their story, and continue to build relationships with specific accounts. It is important to write down the names and accounts of the individuals that stop by your table to help you follow up with your distributor. It would also be wise to write down any specific details that might help your distributor close the sale.

Distributor Sales Meetings

Distributor sales meetings are a great opportunity to get in front of the entire sales team to tell your story, taste your newest vintages, and communicate any new information about the winery. In order to present at a sales meeting, you will need to secure a presentation time well in advance through your distributor brand manager or contact person. You are usually given 20 minutes unless you ask for more in advance. Be sure to prepare for your presentation before you show up and be clear, precise, and to the point. Understand that your time is valuable and respect theirs as well.

Key Account Lunches

During your market visits you can also plan a key account luncheon. Sometimes it is difficult to see all the accounts that you need to see in one day's time. A great way to overcome this challenge is to invite a few key accounts to an intimate luncheon. This will give you the opportunity to build or maintain relationships with these accounts while supporting an existing customer. Once again, be sure to follow up with your distributor after the luncheon.

Public Relations

Press is an amazing selling tool only if it is used properly. The romance and glamour associated with the wine industry will ensure that you will receive attention from the press throughout the course of the year. Keep in mind that the press will feature other wineries as well. Therefore, it is crucial to communicate any new press immediately to your distributor.

In a world with so many wines, communication and follow up are perhaps the two most important ways to ensure success. If you only remember two things about working with distributors, communication and follow up are it.

Working with Restaurants and Retailers

Restaurants and retailers can have hundreds if not thousands of different wines on their wine lists and shelves. Therefore, it is important to work with them as partners to help them sell your wine. This can be achieved by providing retailers with in-store tastings and point of sale material and restaurants with wait staff trainings and wine dinners. Both can also benefit from visits to the winery to enhance their individual experience and receive personalized tours and tastings at your winery.

In-Store Tastings

A great way to increase sales in a retail store is by hosting an in-store tasting. In some cases the retail store will have a monthly calendar that goes out to their customers letting them know what will be tasted ahead of time. In order to participate in an in-store tasting, you must schedule this with the wine buyer or manager months in advance. In-store tastings allow the customer to try the wine before they buy it. It also provides an audience for you to tell your story and showcase your wines.

Point of Sale Material
Printed point of sale material (i.e., shelf talkers and neck tags) can also help increase sales of your wine without requiring you to physically be there. This type of point of sale material usually contains a rating of your wine by a regional or national critic. It provides information to the customer about the wine and gives the sense that the wine "must be good."

Wait Staff Training
A great way to help increase restaurant sales is by providing wait staff training. This can usually be arranged by the distributor sales representative before your market visit. Before a restaurant opens for service they hold what is called a staff line up or preshift meeting. This is the time when the Chef will go over the specials of the day and the management can let the wait staff know about any important details for the evening service. It is during this time that the entire service staff can give you their attention to listen to your story and taste through the wines. In addition to telling them about your wine you can also give them selling tips about your wine so that they will feel comfortable recommending your wine to their guests.

Wine Dinners
Wine dinners are another way to increase sales and raise interest in your winery. The planning process begins by choosing the appropriate restaurant. Then you need to choose the wines that will be made available for the dinner. The Chef then works with the winery to create the spe-

cific menu for the evening. The restaurant and winery can then work together to market the dinner to their respective mailing lists to ensure that the event will be a mutual success. There are many details that go into planning a successful wine dinner. The task may seem arduous but it is usually worth the time.

Trade Visits to Winery

Perhaps the best selling tool at a winery's disposal is the property itself. There is no better way to tell your story than by showing it. Make sure to invite sommeliers, wine buyers, and restaurant wait staff to visit you when they find themselves in wine country. When they do make it out to see you, make sure to take the time to walk through the vineyards and build and maintain your relationship. This will ensure that you stand out in their minds and when they are asked for a recommendation they will recommend your wines.

Maximizing Direct to Consumer

The direct to consumer channel is the most sought after method of selling wine because of its high profitability. Because the winery is selling directly to consumers at retail prices, they can profit the portion of proceeds usually taken by the distributors and retailers. Wineries can maximize and increase their sales through this channel by utilizing tasting rooms, wine clubs, newsletters, mailing lists, Internet sales, and special events.

Tasting rooms provide an amazing opportunity to maximize the direct to consumer sales because of their high foot traffic. They provide a great outlet to sell wine as well as to promote the various wine clubs, encourage mailing list sign ups, and publicize special events at the winery.

A properly written and laid out newsletter can be a great form of communication with your consumers and make them feel like they are an important part of the winery.

Keep in mind that newsletters can be in both electronic and print format. More detailed information on selling direct to consumers and utilizing the new Wine 2.0 relationship software platforms is available in the next chapter.

EVALUATION AND REVISION TO THE WINE SALES PLAN

Now that you have created and implemented the sales plan you need to evaluate it to make sure that your plan is working effectively. This should be done on a bimonthly basis followed up with annual reviews. The best way to evaluate the sales plan is by reviewing and comparing projected versus actual sales numbers in all three major channels. This will give you an accurate gauge as to whether or not you will be able to achieve your monthly goals.

Direct to consumer is by far the easiest channel to evaluate and revise. If, for example, you had projected selling 200 cases of wine direct to consumers in the month of January and have only sold 50 cases by January 15th, you will need to look and see why you are only at 25% of your sales goals. There are many possible factors for this. You could have simply overprojected your expected sales goal for the month; if so you will need to reappropriate those cases. Perhaps you have a consumer event planned for the end of the month where you project selling 150 cases. You might also have had a newsletter that went out the second week of January that will generate additional sales during the last 2 weeks of the month. Regardless of the reasons, the sooner you know whether or not you will make your goal the sooner you can make the appropriate changes.

Evaluating your sales through the three-tier system can be a bit more complicated. The major reason for this is that distributor depletions are not readily available to you. They are usually generated the first week of every month for the previous month's sales. In order to obtain your midmonth sales, you will need to call or email your brand manager and request this information (this also provides a great opportunity to keep in contact with your distributors and see if there is anything you can do to help them out). If the distributor midmonth depletions are below your projections you can discuss opportunities with your brand manager and see what can be done to increase sales.

At the end of the year, you need to evaluate the year's sales performance to see what changes will need to be made for the following year. This is also the time for annual distributor reviews. To prepare for the distributor review, you will need to compare their allocations to their depletions and look to see if the accounts that they have sold to are the appropriate accounts. You will also need to review the ratio of on- and off-premise sales and make sure they accurately reflect your sales plan. In addition to reviewing the previous year's sales with your distributor, you should also discuss your sales plan for the current year. This will ensure that there are no surprises for the year to come.

CHALLENGES IN SELLING WINE INTERNATIONALLY

Professional wine sales require a high level of relationship management, excellent communication, and efficient follow-up, but no where is this more true than in international wine sales. Though many of these issues are described in detail in Chapter 12, the following are challenges that specifically impact professional selling on an international basis:

Fierce Competition for Shelf Space

In the international area, a country loses its home field advantage and must compete with wines from all over the world. The competition gets fierce when it comes to pricing, quality, and space. The price of wines of similar qualities tends to be higher from the US than competing countries such as Argentina and Chile. Consumers all over the world are also now demanding better quality wines at lower price points, which some wine countries are overdelivering on. Furthermore, many wine-producing countries are given government subsidies to sell and promote their wines internationally.

In selling your wine to an importer or store you not only have to differentiate it from other US wines but also from wines from Australia, Chile, New Zealand, etc. Most of the major wine markets also go through trend cycles, and if California varietals are not in style, shelf space will be limited. On the bright side, the recent weakening of the US dollar has given suppliers a favorable exchange rate to sell their wines into key markets at a lower price point and still make necessary margins.

Taxes and Exchange Rates Impact Sale Price and Profits

Exporting brings a new meaning to taxes as a supplier must deal with import tariffs, extra protectionist tariffs, and higher sales tax. Taxes vary depending on country and international agreements. Many countries also add extra taxes onto imports of wine to protect their own local industry. On top of taxes, you also have to watch exchange rates. If a winery sells in US dollars this may impact the retail sale price of the wine if the exchange rates changes. If a winery sells in the country of import currency, exchange rates needs to be monitored as the fluctuations may negatively impact profits.

Label Requirements Impact Sales Logistics

Very few countries will accept a US label or UPC. Each country has its own regulations and requirements that a producer must abide by in order to import and distribute their wines. This becomes confusing, time consuming, and costly if you do not understand or abide by each country's rules. With new joint country trade agreements, label requirements are becoming more universal. For example, with the development of the European Union (EU) it was agreed that the member countries would accept one label requirement standard. In theory, it was a good opportunity for wineries around the world to sell wine easier into EU countries. But this is slowly changing as many countries are starting to add their own label requirements, such as France's pregnancy symbol or Germany's sulfite statements.

Be Patient About Payments

As in any aspect of business, getting paid is important. In international wine sales it takes a personal understanding of the foreign market and how businesses operate to determine when you can expect to receive payment. Typically, a 60–90-day payment lead-time is a sufficient amount of time for most importers and customers to pay. However, in some countries it is typical that the importer will not be paid by their customers for 120 days, therefore delaying the importer payment to the supplier. Late payments need to be evaluated on a case-by-case basis, as many times payment issues are a cultural challenge, not a personal oversight. Keep communication open and set goals and deadlines that are reasonable and possible.

Understand Distribution Restrictions

Importing into some countries is not as easy as finding an importing/distributor partner, printing a label, and shipping. Government boards (also known as monopolies in some markets) control importation and sales of alcohol in some countries. The government boards typically own the liquor stores, meaning they are not only the importer but also the retailer. The government boards usually will put out listing calls specifying country, style, and price points they desire to sell in their markets. A winery will then submit their wine into these calls and hope for the best. If not approved, the wine will not be sold in the market and the winery will have to wait for the next listing call or look for alternative routes, which usually involve significantly less volume. If a winery is lucky enough to get a listing, they are then guaranteed the listing for a certain time period and have an opportunity to sell more wine. Markets with government boards include Canada (excluding Alberta), Sweden, Norway, Finland, and Iceland. As a winery, you will also have to find a distributor or agent to then work with the government board and promote/sell your wines. Each government board is run differently, so gaining an understanding of how they operate is vital when attempting to import into these markets.

The Language of Wine Sales

Language is not as big of a challenge as one may think for communications with foreign countries. English is the predominant international business language, and where English is not spoken, translators are usually available. The one aspect where language is a challenge is in translation meaning. Be sure to research the meaning of brand names, images, colors, and any other communication materials before submitting into market. This will save money and embarrassment.

EXPORTING WINE FROM THE US TO INTERNATIONAL DESTINATIONS

Historically US wineries did not focus on exporting as a source of revenue because of the ability to sell all their wine in their own country, which offered fewer hassles and sometimes more profit. Exports were traditionally viewed as a one-off opportunity or too complicated to even attempt. There were a few pioneers though who focused early on exports and ventured beyond their borders trying to build a name for California wines, including Robert Mondavi, Fetzer, and Wente. In addition, Gallo was the first to put a major focus on exports by establishing a significant foreign sales and marketing force.

In today's global world, more and more wineries are finding it necessary to focus on exports. Reasons vary but mainly consist of: increased competition within the US from other countries such as Australia and Chile; opportunity to build the California image as a premium wine producer; more profit margins in many markets, especially with today's weak US dollar; and increased access to information and partners via the Internet. Wineries new to the export arena are finding exports to have unexpected challenges that are unlike anything they have experienced before.

Export Markets for US Wines

In the last decade, US exports have increased by 106% (Wine institute, 2006). Currently over 50% of US wine exports are shipped to Europe. When a US winery first decides to export, the UK is typically their first target because it is the largest importer of US wine in the world. Many wineries are met with a surprise, though, when they discover how price sensitive this market is. The average retail bottle price is £3 to £4 and wines over £10 represent only 1% of the wine market. These low retail prices are difficult for most US wineries to reach after adding compliant label printing, freight, import taxes, sales tax, and importer margin on top of their wine price. The UK is also one of the most competitive international market places for wine in the world, with every wine producing country vying for their chance to launch in this high-volume market.

The second largest import market for US wines is Canada. However, Canada has its own challenges with most of its province alcohol sales being regulated by government liquor boards. Competition is also fierce with every province boasting a diverse selection of wines from around the world. This leaves US wines with less shelf space to fight over versus what they would typically receive in their own country. After the UK and Canada, top US wine import markets include Japan, Germany, the Netherlands, and Denmark.

Selling Wine in Emerging International Markets

Another trend for exporters is to target emerging wine markets where wine consumption is growing and population is large. The most talked about emerging market is China. With a population of over 1.3 billion people, most wineries would only need to get 0.1% of the population to drink one bottle a year in order to sell out—at approximately 108,000 9-liter cases. Wine consumption is still very small at approximately 0.333 liter per capita, but China is already the 10th largest wine consumer in the world and is estimated to continue growing. Most wine consumed originates from local Chinese producers, but imports are growing quickly as consumer awareness has grown along with disposable income.

India is another emerging wine market with a large population of 1.129 billion people. Similar to China, wine consumption is still low at approximately 0.2 liter per capita but is estimated to grow with the rise in education and disposable income. Other emerging markets include Vietnam and Brazil. In all these countries, wine consumption is growing slowly, so focusing on these countries needs to viewed as a long-term investment.

SUCCESSFUL INTERNATIONAL SELLING TIPS

Once you have gained an understanding of the issues and challenges in selling wine internationally, it is useful to implement some of the following tips. It is important to recognize that specific selling techniques vary depending on your brands, distribution channel focus, and market strategy. However, these tips will assist in giving you a competitive edge if you take the time to understand and implement them.

1. **Be Prepared and Knowledgeable.** If you are in sales, management, winemaking, or marketing don't arrive to a country to work with your distributor and expect them to do all the work and presentations. You must be prepared with the right knowledge and tools for each market visit. Many times you will have to conduct wine seminars and dinners, so knowledge of wine, information on your products, and ability to speak in front of people are vital. Also, be sure to research the market to understand wine sales, distribution channels, and key customers.
2. **Time Management.** When it is 3 pm in California, it is 11 pm in London. Do not expect to have a conference call or receive information from your UK customers until the next day. Working internationally is not a 9-to-5 job. There will be conference calls at 4 am, weekend wine shows, and midnight flights across oceans. Time management skills are important to not only manage your own schedule but also be savvy of your international customer's time.
3. **Be Reliable and Efficient.** As with any job, being reliable and efficient is vital. Do not make promises you cannot keep and follow up after all meetings. So many business deals and relationships are lost because of lack of follow-up and unfulfilled promises.
4. **Build Personal Relationships.** A big key to successful wine selling is visiting your potential and current customers. The knowledge you gain from seeing the market and meeting the people involved in all transactions is invaluable. With today's technology, much international business can be done over the Internet and

telephone, but actually visiting the market and meeting the people will derive bigger results in the long run. In some countries, importers will not agree to any distribution agreements until they have met you in person. For the Brand Manager or Marketer, having first-hand knowledge of a market assists in implementing your brand strategies accordingly. It is also important to note that if a brand or promotion works in one country it does not mean it will work in another. Be sure to be flexible in promotions to suit the market trends and legal requirements.

5. **Be Open Minded.** Part of the fun of traveling in the wine industry is you get to have the opportunity to eat exotic foods, see historical sights, and experience local traditions. You must be open-minded to other people's way of life and accepting of local cultures. You are not required to try everything, but taking part in local foods and activities is a great way to build a good relationship with customers.

Conclusion

International wine sales is a very exciting aspect of the wine industry because it is always changing. There are emerging markets to research, fresh competition to investigate, consumer consumption trends to follow, and new regulations to figure out. Selling wine internationally is for those who like challenges, changes, and new opportunities. Traveling, learning a new language, studying about different cultures on top of understanding business fundamentals are all beginnings to a successful international wine career. Besides, in today's global world, even a basic understanding of international practices is important to be successful in any aspect of a business.

LEARNING OBJECTIVES:
- Discuss the pros and cons of direct wine sales
- Identify and describe the six direct wine sales channels
- Describe compliance issues regarding direct wine sales regulations
- Define Wine 2.0 and describe its growth potential
- Identify the four major social media components of Wine 2.0
- Explain how to use openness and engagement as a marketing strategy for Wine 2.0 issues

CHAPTER

DIRECT WINE SALES AND WINE 2.0

Janeen Olsen
Professor of Wine Marketing, Sonoma State University

Josh Hermsmeyer
BlogMaster, Pinot Blogger

Direct to consumer wine sales occur when wineries sell directly to the final consumer without the use of independent wholesalers and retailers. There are many methods by which wineries sell directly to consumers, including selling from a tasting room or cellar door, hosting parties, special events, and winemaker dinners where wine is available to purchase, having an wine club, and/or using telephone and mailing lists to solicit wine orders. Many wineries also create newsletters and email promotions targeting potential customers in hopes that they will make a purchase.

One of the largest growth areas in direct wine sales is via the Internet. This is happening around the world, and includes not only wine sold directly from wineries to consumers, but wine sold via the Internet from retailers, private wine clubs associations, and auction houses. One of the most exciting growth areas associated with wine sales is called Wine 2.0., and is about using the Internet to engage with wine consumers on their terms, in a time and manner of their choosing. This can include such social media components as blogs, vlogs, social networking sights, and other tools.

With this in mind, this chapter describes the advantages and challenges of direct wine sales and reviews the various sales channels. Next it provides an overview of the regulations for direct selling. Then it provides a definition of Wine 2.0 and its growing importance. This is followed by a description of the various social media tools, and then ends with a forecast of the growth expected in this area.

Advantages and Challenges of Direct Wine Sales

There are many advantages, as well as challenges, in selling directly to consumers that must be weighed against each other when investigating this sales option. The primary advantage to selling direct is that the winery is able to capture the margin that otherwise would go to the distributor or retailer. This margin typically ranges from 25% to 50%. For this reason, many wineries find the profits per bottle of wine are much higher when they sell direct to the final consumer. Another reason that wineries may sell direct to consumers and not use the services of distributors may be that the production volume is low and many distributors will not purchase small amounts. Smaller wineries often face this problem when trying to distribute wine and therefore are not able to sell to distributors. This is a very important means of reaching consumers as less than 17% of US wineries are represented in all states by distributors. Some small wineries sell 75% to 100% of their wine solely through direct sales (Walker, 2002). V. Sattui in the Napa Valley and Viansa in Sonoma sell 100% of their wine production direct to consumers. Even larger wineries may have smaller amounts of unusual varieties of wine that was not used in their blends, and direct sales allows the wineries to find a market for smaller production lots. They may also produce these unusual varieties or small production wines just to add value for their wine club membership, who are able to purchase wines not available to others.

With increased competition it is also becoming increasingly difficult for newer wineries to find room on retailers' shelves or restaurants' wine lists. It is also becoming more difficult to find major distributors who want to take on new wines. The additional problems within the distribution channel come from a lack of support on promoting products from the distributor's sales team who are overwhelmed with pressure to sell wines from the big producers. A new winery, whose marketing budget is relatively meager compared to the bigger existing wineries, is forced to rely on their direct to consumer sales strategy to overcome these challenges with distributors. These new ventures may find that initially the only method available to them to sell their wine is through direct sales. If they become extremely successful and their reputation grows, it may become possible for them to place their wine with distributors. Navarro initially was direct to consumer and only added distribution to ensure that their products were seen by gatekeepers in the A-Tier restaurant and retail accounts.

Even thought there are advantages in direct sales, these methods are not without their challenges. For example, even though the profits per bottle may be higher, there are still many additional costs associated with direct sales that must be covered. Money is required to build and operate a tasting room and to organize tours and special events. Internet and direct mail sales require the creation of promotional materials. Databases of potential customers must be created and constantly monitored. Packaging and shipping costs must be covered. Unless all of these activities are done well, direct sales may not turn out to be any more profitable than other methods of selling wine.

CHANNELS FOR DIRECT WINE SALES

Direct selling includes a variety of methods a winery can utilize (Figure 9.1). It is not uncommon to find that wineries use several of these methods together. The following describe the most common methods of selling directly to customers.

Tasting Rooms

Tasting rooms, or cellar door operations as they are called in some regions of the world, are a very popular method

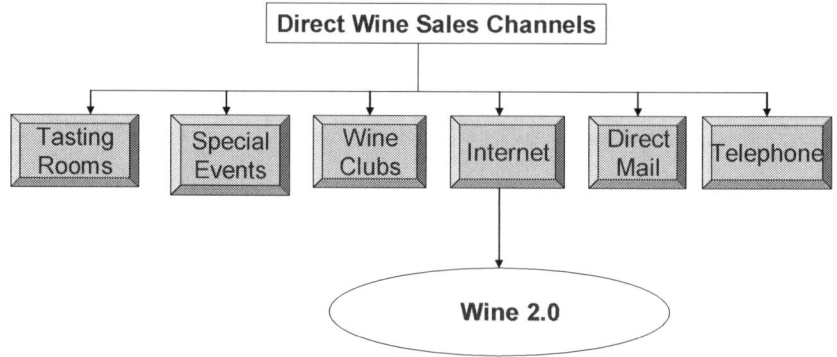

Figure 9.1. Direct wine sales channels.

of selling direct. In California there are over 600 wineries that operate some form of tasting room. Some tasting rooms are operated at the same location as the winery, while others may be located away from the winery, but in a popular tourist area. There is a trend for smaller wineries to come together and operate cooperative tasting rooms where products from several wineries are featured for sale. Both on-site and off-site tasting rooms allow the visitor to sample wines, learn about the viticulture and winemaking techniques, and purchase wine. Some tasting rooms that are located at the winery or near vineyards may operate tours for visitors as well. Many wineries are open to the public during set times of the day; however, there is a trend for more wineries to have their tastings by appointment only. By restricting the number of visitors at a time, the wineries hope to provide the guests a more personal and meaningful experience and create more loyal customers.

Many tasting rooms also offer other merchandise for sale besides wine. In some cases the merchandising efforts are minimal, perhaps focusing on wine openers, glasses, and other inexpensive wine-related gifts. In other cases, the merchandising effort may be substantial, selling products not related to wine, such as kitchenware, jewelry, art, and clothing. Many wineries report they earn a substantial portion of their revenues from non-wine sales. When winery merchandising is done well, products should be stocked and displayed in such a way as to support the overall brand image the winery is trying to achieve. Viansa and Beringer Winery are two examples of winery tasting rooms that do an excellent job of merchandising to promote their variety of brands while bringing in substantial revenue. In many areas, wineries are often associated with restaurants, providing another means to enhance overall revenues. However, local regulations and permit rules do not allow this option in many areas.

Special Events

Special events are another way that wineries can attract customers to the winery or to build brand awareness and create customer relationships. Branding is most effective in the wine industry when a winery can sponsor an event that will resonate with that individual when they are making their wine buying decisions. Special events often take place at the winery; however, many wineries also participate in trade fairs, festivals, and other events located outside of the property as a way to promote their wines. Regulations often restrict the sale of wine from such off-site events; nevertheless, wineries can use such events to familiarize the participants with the wine, build brand image and awareness, and collect contact information for later direct sales attempts. The types of events that wineries stage are quite broad as event planners are continually coming up with creative themes and ideas to bring in new customers. Mardi Gras parties, Italian festivals, movie nights, and parties featuring cuisines of all types provide popular themes for special events. The concert series and Robert Mondavi Winery and Wente Vineyards are popular with visitors.

Wine Clubs

Wine clubs take on different forms, but typically a person agrees to buy a set amount of wine on a regular basis. The customer provides credit card and shipping information and the winery automatically ships the wine. Wine club sales can be a great asset to wineries because they are a predictable, ongoing source of revenue. Both budgeting and production levels can be improved when wineries know they can count on a certain level of club sales each month or quarter. Wine clubs also allow a winery to stay in touch with customers who live to too far away to visit the tasting room on a regular basis. This is why many wineries consider wine clubs an excellent way to build brand loyalty. One of the biggest obstacles that wineries face is that, as of 2008, they are only allowed to ship to club members in 35 states. There are also several more states that wineries can ship wine to as long as the wine was purchased in person at the winery. Estimates suggest that 80% of all potential wine drinkers live in one of these states, creating a large market to serve (Tinney, 2007).

Signing up members for wine clubs occurs in several ways. Often visitors to tasting rooms are asked if they would like to become a member. It is not uncommon for tasting room employees to receive a commission for each new member they sign up. Usually an employee receives $5 to $25 per sign up plus incentives tied into monthly competitions. Another way that wine club members are signed is through an Internet site. This allows people who have never visited the winery to receive wine shipments as well. Wine club members are also signed up at consumer-based trade tastings, especially competitions where award winning wines are poured and consumers can sign up to receive these wines as part of their membership. The San Francisco Wine Competition tasting provides wineries such as Stryker and Stonegate a place for sign ups for wine club members.

As there are so many wine clubs for consumers to choose among, not only those offered by wineries but retail clubs as well, wine companies increasingly offer benefits to wine club members. Examples of benefits include discounts on additional wine and merchandise purchases, waving tasting fees at the tasting room, invitations to winemaker dinners, and other special events and parties (Coppla, 2003). Many wineries have created special tasting areas specifically for wine club members as a way to make them feel special when they visit the winery.

Newsletters and Direct Mail Campaigns

Newsletters and direct mail campaigns can also be used to solicit wine sales. Wineries can keep customers informed of what is taking place at the winery such as harvest and special events and also let customers know when new wines are released. Newsletters often contain educational articles about winemaking and viticulture as well as recipes and human interest stories. The purpose of newsletters and direct mail campaigns in not only to sell wine, but many wineries find it an excellent way to develop brand loyalty among customers. More and more wineries are using email as a means to distribute their newsletters to potential customers.

Telephone Sales

Telephone sales can be another means to reach customers. Although not as common, there are wineries that sell wine directly to customers over the phone. Many of these wineries have highly sought after wines that are allocated, so the calls are primarily made to qualified customers so that they can place orders when the wine becomes available. Haffner is such an example of a winery in Sonoma Country that sells most of its wine through telephone marketing.

Internet Sales

Internet sales of wine has grown dramatically in the last decade and is a useful means to create a strong brand image. Oftentimes the revenue is used to off-set website expenses when a website's primary purpose is to support brand loyalty and build customer awareness. There are still some barriers to consumers making online wine purchases, such as interstate shipping laws, lack of product knowledge, additional shipping and handling costs, the time delay in receiving the product, damaged wines shipped during unsafe weather conditions, and lack of online consultations at the time of making the wine purchase. We must remember that wine is still a mysterious product to most consumers! On the other hand, Internet availability helps consumers who already possess product awareness buy wines in areas where it is not readily available in retail outlets. Wines that have received good reviews or ratings are eagerly sought via the Internet. Small production or "cult" wines can sometimes be sought out using the Internet.

Compliance and Regulation Issues with Direct Wine Sales

Regulation of direct sales of wine to consumers is a very complex issue. There are currently 33 states that allow interstate shipping; however, the rules for licensing and shipping vary greatly among the states. For example, the amount that can be shipped to a consumer depends on the state in which they live. In Kentucky, it is two cases a year and only by wineries that make less than 50,000 gallons. In New York, the number of cases that can be shipped per year is 36 and there are no restrictions on the number of gallons the winery produces (Tinney, 2007). Wineries often rely on consumer fulfillment companies such as New Vine Logistics to manage their interstate shipment needs, such as storing inventories, managing the wineries' customer databases, and providing weather reports for shipping conditions. Some fulfillment companies are working within the three-tier system to provide shipping opportunities to more than the designated states. As shipping laws change, the wineries often are forced to wait for common couriers such as FedEx, UPS, and DHL to update their own shipping policies to reflect the current changes. This delay, sometimes months, can add to consumer aggravation within the states as eager customers try to get wine shipped to their state immediately upon hearing of regulatory changes.

Grassroots efforts such as Free the Grapes, working on behalf of the Wine Institute, are promoting direct to consumer sales by overturning restrictions to interstate shipping. Since the 2005 Supreme Court Decision that barred states from creating different shipping rules for out-of-state wineries than for in-state wineries, the trend has definitely been to make direct to consumer wine sales more feasible, but wineries must monitor compliance with shipping laws carefully as the laws are constantly changing and the penalties for breaking the rules can be severe.

THE ADVENT OF WINE 2.0

The term "Wine 2.0" attempts to evoke the movement from one generation of ideas to the next. Proponents of Wine 2.0 are often either technically savvy or are online businesses themselves, and their efforts are directed mainly at the segment of wine consumers recently dubbed millennials (wine drinkers aged 21–30). It would, however, be a mistake to conclude that only technically adept wineries can benefit from understanding and incorporating the concepts behind Wine 2.0 into their marketing plans. The tools and concepts outlined in this section can be used successfully by wineries, brand owners, and retailers at every level of technical expertise.

At its core *Wine 2.0 is simply about using the Internet to engage with wine consumers on their terms, in a time and manner of their choosing.* Wine 2.0 derives its name from the equally nebulous term "Web 2.0." While Web 2.0 is itself difficult to define and is in constant flux, in general it refers to an evolution of the Web away from a model centered on information storage and retrieval and toward a model of the Web as a social platform for interacting and communicating with others.

In "Web 1.0" information on a website wasn't updated often and users weren't given the ability to easily add comments or to submit reviews. By and large, content was created to be passively consumed rather than interacted with.

In contrast, Web 2.0 leverages user-generated content to add context to information and encourages social interaction by providing tools to share and communicate with others. By collectively contributing reviews and sharing commentary not only on products, but also on the businesses that produce them, consumers are able to become highly informed about the goods that interest them extremely quickly and in surprising depth.

This fundamental change has created both opportunities and challenges for businesses that have long been accustomed to passive consumers. Today's buyers don't trust press releases and are wary of experts. They rely on word of mouth and recommendations from trusted sources above all else when deciding where to spend their money, and they want to build meaningful and authentic relationships with the people behind the brands they are passionate about.

The wine industry is uniquely positioned to take advantage of this evolution in consumer behavior, and those wineries, brands, and retailers that are taking the first steps toward doing so are at the forefront of what is called Wine 2.0.

What Is Wine 2.0?

Wine 2.0 is an acknowledgement among some wineries and wine retailers that market forces (user-generated content, easy and quick access to peer reviews, instant publishing of commentary via blogs and message boards) will inevitably change how wine is bought and sold.

How exactly wineries should respond to this new environment is still an open question. What is clear is that increasingly consumers are turning to the Internet and Google to research the goods they buy, including wine. As has always been the case, consumers also like to talk about the wine they drink and their experiences with the wineries they visit. Unfortunately for wineries and brand owners, not all of these discussions are positive, brand-building conversations. Indeed many will be decidedly negative.

Compounding this problem is the fact that what consumers choose to say about a wine brand can have a great deal of influence on the market. Blog posts in particular show up prominently in Google search results, making bloggers disproportionally influential to prospective wine buyers researching their next purchase.

Openness and Engagement as a Marketing Strategy

Wineries can't hope to stop these conversations from happening, nor can they hope to control them. It would be a PR catastrophe for winery employees to attack or try to marginalize everyday consumers who've gone public with their opinions. The very best wineries can do is to join in the conversations, engage directly with the market, and use the conversations as an opportunity to forge an image of openness and to build positive word of mouth.

While it might seem a daunting task, simply acknowledging bad reviews via comments on websites, blogs, and message boards is often quite effective in diffusing awkward situations. People like to feel their opinion is important and that they have been heard. If you provide them with that kind of personal validation you have the opportunity to turn them into a source of positive PR instead of a blemish on your brand's online presence.

The particular tactics a winery uses to implement a strategy of openness and engagement are too numerous to list, but effective online marketing has so far included starting winery blogs, frequenting and commenting on the sites of influential bloggers, and being a helpful and prolific poster on popular wine message boards.

Components of Social Media

There are currently four major components of social media for Wine 2.0. These are: 1) Blogs, 2) Vlogs, 3) Message

DIRECT WINE SALES AND WINE 2.0

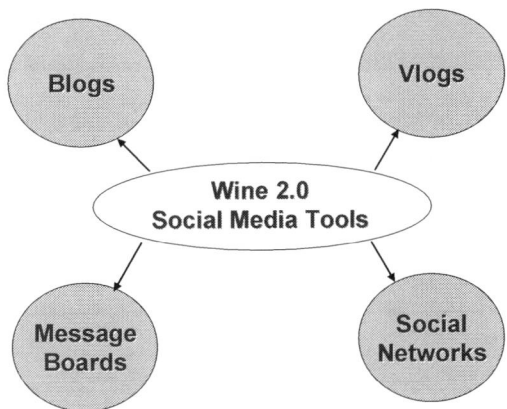

Figure 9.2. Social media components.

Boards, and 4) Social Networks (Figure 9.2). Each of these is described in more detail in the following paragraphs.

Blogs are one of the most visible forms of social media. They are cheap to create and require no technical knowledge to operate. Anyone can create a blog and start posting their wine reviews for people to read (and for Google to index) in a single afternoon. These pages are cached and easily searchable by any consumer interested in your brand.

It is very important to treat bloggers with respect and to always respond to criticism in a professional manner. Every interaction is public and recorded by Google, so future sales could depend on how well you are able to address the needs of dissatisfied customers who have taken their grievances public. Again, oftentimes issues can be resolved by simply acknowledging that errors were made or by offering to replace what you suspect is a corked bottle.

Blogs are run on a vast array of platforms and software packages. The most popular are Blogger (www.blogger.com), Wordpress (www.wordpress.com), and MovableType (www.movabletype.com).

Online video is one of the most influential forms of social media. Wineries have only recently begun to experiment with marketing via video. There remains great potential for well-crafted video content to "go viral" (spread via social networks like a virus) and thus create a substantial increase in brand awareness. The key is to create remarkable content—literally, something worth remarking about. Currently the leader in online video is YouTube (www.youtube.com) by a sizable margin.

Message boards include popular sites such as eRobertParker.com and egullet.com. Both are well read, have amassed large communities, and are frequented by extremely passionate wine consumers. Cultivating a positive image on wine message boards can be an effective way to market your wine brand and to speak directly to a rabid consumer base. Unfortunately, it is also an easy way to get sucked into "flame wars" (heated confrontations that often include personal attacks) due to the anonymity and concomitant lack of civility inherent in online communication.

Social networks are websites that allow users to create online identities that can be customized to reflect their personal preferences. Both Facebook (www.facebook.com) and MySpace (www.myspace.com) are good examples of social networking sites. Both have features that allow users to denote other users as friends and thereby share their profiles with them. Groups can also be formed based around shared interests such as wine.

Currently wineries are having a difficult time determining how best to take advantage of this communications channel. It may well be that the best way to leverage social networks and the vast amounts of data contained in them will be in interpreting, rather than influencing, consumer preferences.

Growth in the Wine 2.0 Space

Growth in the Web 2.0/Wine 2.0 space is accelerating. MySpace was sold for $580 million in July 2005 to News Corp. In October of 2006 YouTube was purchased for $1.65 billion by Google. A year later, in October 2007, Microsoft purchased a 1.6% stake in Facebook for $240 million, putting its value at somewhere around $15 billion. Social networking sites MySpace and Facebook are among the top 10 most frequently visited sites on the Internet. At the time of writing, Facebook was adding over 200,000 new users a day. CNET News reports that two new blogs are created every second.

Selling wine online is appealing to wineries for many reasons, not the least of which are the high margins inherent in direct to consumer sales. Tasting rooms have long been a key profit center for large and small wineries alike. Used intelligently, blogs and social networks can help wineries drive more traffic to these profit centers.

Influential customers with blogs offer opportunities for wineries to forge personal connections easily and for little cost. Leaving comments on blogs and using the various tools outlined above to reach out to current and prospective customers helps cultivate strong relationships and spread positive word of mouth.

Moreover, by making their preferences explicit, users of blogs and social networks such as Facebook and MySpace are tremendous sources for market research. Being able to keep a finger on the pulse of the market, in real time, is extremely valuable.

Polling and other formal market research continues to show that millennials are willing to spend large amounts on wine and that they are constantly searching for new experiences. By reaching out to this important market segment using the tools that they are comfortable with and in ways they find appealing, wineries can ensure that they are well-positioned to profit from this new generation of wine drinkers.

Conclusion

In summary, it is clear that as the wine industry has become more and more competitive; many wineries are looking for new methods of selling wine to consumers. Selling direct to consumers has received more emphasis in marketing plans and many wineries have made direct sales their primary method of distribution. The advantages of direct sales ensure that future growth of this marketing channel. But it won't be an easy road as regulations and cost structures make direct sales of wine a challenging venture.

LEARNING OBJECTIVES:
- Define wine tourism
- List the benefits of wine tourism for wineries
- Describe motivations of wine tourists
- Identify the experiential components of wine tourism
- Describe the wine tourism development process
- Gain insights from two wine tourism case studies: Sandalford Winery in Western Australia and the Okanagan Valley in Canada

CHAPTER 10

WINE TOURISM

Donald Getz
Wine Tourism Professor, University of Calgary, Canada

Jack Carlsen
Professor of Tourism Studies, Curtin University, Western Australia

Liz Thach
Wine Business Professor, Sonoma State University

Wine has become a powerful motivator for short- and long-distance travel. Although it is linked to gastronomy and more general cultural interests, there can be no doubt that the allure of wine and "wine country" has been increasing globally. Both the geographic spread of new wine regions and new wine consumers will continue to fuel demand for this form of special interest travel.

Wineries are the core attraction in wine tourism. Even though many are not built or managed as attractions, there is increasing recognition that wine tourism works to the benefit of most wineries and they are adapting to this market. More and more wineries are being built as architectural landmarks and tourist attractions, as hospitality and function centers, and even as self-contained vacation and conference estates.

This chapter provides an introduction and overview of wine tourism. It begins with a definition and the benefits of wine tourism, and then moves on to describe the motivations of wine tourists and the experiential components they desire. Next there is an overview of the wine tourism development process and the many partners involved in a successful wine tourism campaign, as well as innovative wine tourism programs. The chapter concludes with two case studies of successful wine tourism destinations: the Okanagan Valley in Canada and Sandaford Winery in Western Australia.

Defining Wine Tourism

"Wine tourism" has several meanings, depending on whether the focus is on the wine tourist, the destinations that employ wine tourism as a competitive strategy, or the wineries and wine industry that use their appeal to generate direct sales and other marketing-related benefits (Carlsen & Charters, 2006). One of the most common definitions of wine tourism focusing on the consumer is: "visitation to vineyards, wineries, wine festivals, and wine shows for which wine tasting and/or experiencing the attributes of a wine region are the prime motivating factors for visitors" (Hall & Macionis, 1998 p. 197). However from a wine tourism region and winery perspective, the definition can be "a strategy by which destinations develop and market wine-related attractions and imagery, and a marketing opportunity for wineries to educate visitors and sell their products directly to consumers" (Getz, 2000 p. 4). All of these perspectives are interrelated and need to be examined through the lens of the wine tourism system.

Benefits of Wine Tourism

The potential benefits of wine tourism to wineries, communities, and destinations have been identified in many studies (Table 10.1). Dodd and Bigotte (1997) said the majority of American wineries rely primarily on tourism for their survival, reflecting both their inability to market widely and the fact that profit margins are highest at the source. More specifically, wineries can reach new customers through tastings, developing loyal customers, testing new products, and creating additional revenue streams through retailing and food. Research has proven that one of the best ways to create and keep a loyal consumer is having them visit the winery location—the "home" of the wine. Increasingly, wineries are being designed as landmark tourist attractions, while older wineries have made substantial investments in visitor centers and services. Sandalford in Perth, Western Australia is profiled later as an example of a successful tourist-oriented winery.

The classic wine regions of Europe will always hold great touristic appeal based on their combination of wine, gastronomy, scenery, and history, whereas New World wine

Table 10.1. Benefits of Wine Tourism for Wineries

Develop new customers
Increase wine sales
Test new wine products
Higher margins through direct sales
Foster brand loyalty
New links with wine trade
New partnerships within the community

regions have had to "invent" themselves as destinations. While Napa and Sonoma in California have achieved iconic status—that is, they represent the essence of "wine country" and are must-visit appellations—numerous other wine regions have to fight for competitive advantages.

Australia is one country that has taken wine tourism very seriously, including formulation of strategic plans at the national, state, and regional levels (Carlsen & Dowling, 2001), and establishment of the position "wine tourism planner" within the Winemakers Association. In Canada, not particularly noted for wine exports other than ice wine, both Niagara in the east and the Okanagan Valley in the west have become well-established wine destinations. The Okanagan, and its town of Oliver (self-proclaimed "wine capital of Canada") are profiled in this chapter.

The economic benefits of wine tourism have been documented in many settings (Carlsen, 2004; Carlsen & Charters, 2006; Getz, 2000; Hall, Sharples, Cambourne, & Marcionis, 2000; Mitchell & Hall, 2006), but it cannot be assumed that benefits will always outweigh costs. Furthermore, issues surrounding the sustainability of wine tourism from economic, ecological, and social–cultural perspectives are of increasing concern. These issues are discussed generically, and are illustrated in the Okanagan case.

Motivations of Wine Tourists

From the demand side, wine tourism is a form of special interest travel based on the desire to visit wine-producing regions, or in which travelers are induced to visit wine-producing regions and wineries while traveling for other reasons (Getz, 2000). Consumers typically want a bundle of benefits from their wine tourism experiences (Charters & Ali-Knight, 2000; Williams & Kelly, 2001). It tends to be a subset of cultural tourism in which history, scenery, food, wine, and unique cultural opportunities constitute the overall appeal. Getz and Brown (2006) modeled the ideal wine tourism experience based on consumer research, identifying core wine, cultural, and destination elements. A general surge in "experiential" travel is behind much of the growth in wine, gastronomic, and cultural tourism, as today's more sophisticated traveler wants to do unique things, learn about matters that interest them, and co-create memorable experiences.

Researchers have revealed significant differences among wine consumers and wine tourists. Brown, Havitz, and Getz (2007) used ego-involvement theory to show that some Calgary, Canada wine consumers viewed wine as being central to their lifestyle (they tended to be older males) and this translated into a high level of travel for wine, including specific links between wine preferences (e.g., for French and Italian) and destination preferences (Brown & Getz, 2005). Lesser involved wine consumers (they tended to be women) also traveled for wine, but held somewhat different motives and preferences. Many couples traveled to wine regions and wineries together.

Hall and Macionis (1998) classified wine tourist into three groups: 1) wine lovers, 2) wine interested, and 3) curious tourists. Charters and Ali-Knight (2002) added another group whom they called the "hangers on." Highly involved wine lovers are more likely to buy wine; they want to learn about, and taste it. They have an interest in the links between wine and food. They also want to expand their knowledge about the wine making process. Those with lesser involvement are typically more interested in the social aspects of dining and drinking.

Dodd (1995) concluded that the wine tourists in Texas were highly educated and had higher household incomes than other tourists. Wine tourists do not usually travel on the cheap package deals (Dodd, 1995) and the best target

Table 10.2. Top 10 Motivations of Wine Tourists

1. To taste wine
2. To gain wine knowledge
3. To experience the wine setting (e.g., meet the winemaker; tour cellars and vineyards)
4. To be in a rural setting—beauty of vineyards; learn about farming—agritourism
5. To match food and wine—culinary tourism
6. To have fun—wine festivals and events
7. To enjoy wine culture—romance and elegance
8. To appreciate the architecture and art
9. To learn about the "green" aspects—ecotourism
10. To enjoy the health aspects of wine

market for the industry will include affluent, mature, and senior couples (Lang Research Inc, 2001).

In a review of the wine tourism literature, Thach (2007) identified 10 major reasons wine tourists around the world are motivated to visit wine regions (Table 10.2). Though some differed by country – such as health reasons being more of a motivation for European wine tourists - the top three reasons were universal.

THE EXPERIENTIAL APPEAL OF WINE REGIONS

Roberts and Sparks (2006) provided insights into the factors that enhance the wine tourism experience. Through focus groups, they identified eight enhancement factors that provide context to the wine tourism experience:

- Authenticity of Experience
- Value for Money
- Service Interactions
- Setting and Surroundings
- Product Offerings
- Information Dissemination
- Personal Growth—Learning Experiences
- Indulgence—Lifestyle

Charters and Ali-Knight (2000) believed that wine tourist expectations are likely to vary from region to region; no one set of critical success factors will apply everywhere. Williams (2001a, 2001b) studied the evolution of wine region imagery as reflected in the advertising pages of *Wine Spectator* magazine. He concluded that imagery shifted through the decade of the 1990s from an emphasis on wine production and related facilities to more aesthetic and experiential dimensions. The imagery of wine country as a rural paradise has been conveyed to wine consumers, in which leisure, cuisine, scenery, and outdoor activities are bountiful. Bruwer (2003) thought the appeal of wine regions to be based on difference of place, and these differences must be branded. Both natural and cultural features are important, but attractiveness is also related to distance (real and perceived) to markets.

Wine tourism and wine exports should be mutually reinforcing. Sharples (2002), for example, suggested that the reputation (i.e., how well known it is for quality) and export of Chilean wines fuels wine tourism to that country. Consumers who have experienced a wine-producing region might be more likely to become loyal customers and to spread a positive word about the wines. What is unknown is the dynamics of this interaction and the resulting pattern of travel preferences and choices. Chaney (2002) argued that many consumers simplify their wine choices by picking them on the basis of country of origin, and noted that many retailers display wines by country and region of origin.

The notion of secular "pilgrimage" is also relevant. In social settings such as wine clubs and wine tastings, or among wine-loving friends, there is a high probability that word of mouth, as well as formal information about wine regions, will be shared. An element of status might very well be associated with visits to famous or even out-of-the-way wine regions. Many wine-related websites provide evidence that consumers think in terms of pilgrimage, and marketers use this concept in their wine tourism promotions. For example, Lee Foster (2003) expressed it this way: "Every traveler with an interest in wine and food owes himself or herself, at some point in life, a pilgrimage to Bordeaux. I will always remember my own journey to this gustatory shrine."

What sets a pilgrimage apart from other special interest travel is that very specific sites hold deep meaning for the visitors. There will be a search for authenticity, often manifested in seeing the actual grapes, physical plant, and personnel that produce favored wines. It might also be argued that famous wine regions, like Bordeaux, are pilgrimage destinations even for those who prefer wines from elsewhere. If true, this would suggest that the Bordeaux "brand" holds "equity" for the destination as well as its wine producers.

THE WINE TOURISM DEVELOPMENT PROCESS

Key components of wine tourism development include wineries, wine routes, tours, guides and interpretation, visitor centers and museums, packaging, special events, links with gastronomy, and themed accommodation and retailing. Developing and marketing wine tourism is a joint responsibility of destination marketing organizations, many governmental agencies, wineries, and wine industry associations.

Because of the large number of players involved in developing a wine tourism region, a high level of cooperation among stakeholders is critical. Table 10.3 provides a list of some of the partners that need to be involved. These include not only the obvious partners of hotels, restaurants, and retail stores, but support agencies like police services and medical facilities. In addition, partnership with local

Table 10.3. Wine Tourism Development Partners

Restaurants	Other tourism attractions
Hotels	Golf/spa/tennis venues
Tour companies	Transportation
Retail shops	Winery/Grape associations
Other wineries/vineyards	Tourism boards
Police and medical	Chamber of Commerce
Environmental agencies	Employment agencies
Government agencies	Colleges and Universities
Other regulatory agencies	Financial institutions
Neighbors	Other
Distributors	

government and environmental groups for permits, road agencies, and other infrastructure support is needed. Finally, neighbors should be involved to make sure they are aware of such issues as increased traffic and potential noise pollution.

Though most wine communities understand the benefits wine tourism can bring in terms of new jobs and money flowing through their shops, there are still concerns with environmental and infrastructure issues. That is why it is important to get them involved in the wine tourism development process up front. Even local universities, employment agencies, and other tourist attractions such as golf courses, and spa and tennis resorts should be consulted and invited to participate in the planning process. Finally, and obviously, financial support must be obtained, and a marketing campaign designed that includes not only brochures and a website, but signage, event planning, and evaluation.

Thach (2007) outlines a five-phase wine tourism development process (Figure 10.1), which begins with an assessment phase in which local wineries and grape growers in the region meet to conduct a needs assessment to see if there is enough interest and support to begin the develop-

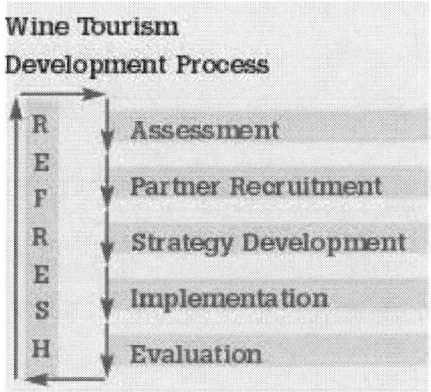

Figure 10.1. Wine tourism development process. Reprinted with permission from *Wine Business Monthly*, by Thach (2007).

ment process. If the answer is yes, the next phase is partner recruitment in which a steering committee is formed with representation from major partners such as hotels, restaurants, transportation services, financial partners, local government, and other tourism groups.

Once the steering committee is formed, the team embarks on a wine tourism strategy development process that emphasizes the unique beauty and specific competencies of their region. During this process, they create a vision statement for the region, as well as long- and short-term goals. The fourth phase is implementation of the strategy, which includes much hard work and substantial financial resources. It is generally a multi-year-long plan with prioritized goals, a clear action plan and accountability, and monthly meetings of the steering committee to assess progress and make any necessary revisions. Ideally a full-time director position for the wine tourism effort is funded and appointed.

The last phase is evaluation, which should include ongoing monthly, quarterly, and annual evaluation of progress against goals and key metrics, such as number of visitors, total winery revenues, total community revenues attributed to tourism, etc. Evaluation also includes a year-end assessment and needed revisions. Finally, the complete strategy should be reviewed and "refreshed" every 3 to 5 years to ensure the wine region is staying innovative and attracting not only new consumers but luring back old consumers time and again through exiting and entertaining programs. Table 10.4 provides a list of both standard and new types of wine tourism programs.

Marketing the Wine Tourism Destination

Marketing of wine destinations is done by tourism organizations as well as individual wineries and wine companies. The wine industry often associates their brands with specific appellations and promotes visits to their wineries. As the number of potential wine destinations has been steadily increasing, both owing to the spread of wine production

Table 10.4. List of Wine Tourism Programs

Wine trails (routes) and signage
Special winery events (e.g., music and wine, chocolates and wine, grape stomping, weddings, corporate events, etc.)
Passport program (check off each winery you visit and receive gift/certificate)
Wine tours (special tours of wineries and tastings)
Self-guided tours of wineries
Wine cities (e.g., a town that has wine tasting rooms and other wine-related activities)
Wine education centers
Experiential programs (blending wine, picking grapes, etc.)
Interactive winery websites and kiosks
Wine partnerships (e.g., golf and wine; spa and wine)
Wine ecotours (environmental wine programs)
Wine culinary tours/seminars (wine and cooking trips and seminars)
Wine adventure tours (jeep tours of vineyards; kayaking on river/ocean near winery)
Wine teambuilding events (companies bringing employees for wine blending, harvest, etc.)

and the growing importance assigned to wine tourism, it is becoming more difficult to find competitive advantages.

Awareness of the wine-producing region can be increased through repeated purchase of wines (which might have pertinent information on the label), discussions with other wine consumers, formal wine tastings, background reading on wines, or comarketing by wineries and wine destinations. However, the vast majority of wines are sold without much in the way of destination information or imagery on the labels.

Marketing efforts might not always be a determining factor in attracting wine tourists. Research by Dodd (1995) concerning visitors to Texas wineries found that word-of-mouth recommendations were the most important source of information used, followed by previous exposure to the winery's labels and other sources. Brochures were most important to visitors living more than 30 miles from the visited winery.

Mitchell (2006) provided some analysis of reasons for purchase (and nonpurchase) by wine tourists after visiting wineries. He found that it is the experiential elements of the visit, such as service and social aspects, that are most influential in postvisit purchase behavior. He also pointed out the importance of dissatisfiers that, if absent, can compromise the experience but in themselves do little to enhance it. In contrast to previous studies, Mitchell found a high likelihood of brand loyalty among New Zealand winery visitors, especially if they visited more frequently, which results in increased postvisit purchases.

O'Mahony, Hall, Lockshin, Jago, and Brown (2006) found that high involvement correlates with higher postvisit purchases of wine, but the relationship is a complex one. Without overreliance on market segmentation (which can be misleading when based on a single factor such as demographics) it is evident from this research that there is a continuum of wine involvement from low to high that may or may not change after a winery visit.

The Issue of Too Much Wine Tourism: Sustainability and the Environment

Researchers are increasingly concerned about the impacts and sustainability of wine tourism (Poitras & Getz, 2006). For example, Skinner (2000) argued that as wine regions become increasingly involved with, or even dependent upon, wine tourism, the need to sustain tourism as an economic resource is essential. Carlsen and Ali-Knight (2004) examined the case of Napa Valley, California from the perspective of managing wine tourism through demarketing. In Napa, perhaps the world's most developed wine tourism destination, a number of strategies have been employed to limit development, modify demand for winery visits, encourage high-yield wine tourism, and promote alternative attractions and regions. The authors concluded that mass tourism and wine tourism are largely incompatible, and that marketing has to seek approval and involvement with the community.

Williams and Dossa (2003) argued that "conserving the natural resource base in wine regions is a product development function that requires the collaboration and sound planning of many partners" (p. 26). More particularly, sustainable wine tourism will depend on identification and management of unique issues pertaining to the resources used, specific forms of wine tourism development, and the specific impacts caused by wine-related tourism. Cambourne, Macionis, Hall, and Sharples (2000) noted that in the context of tourism development in rural areas, changes to the "winescape" (i.e., physical, as in land devoted to vineyards; social, such as overcrowding at wineries; and cultural, such as commercialization) can have a significant impact on the wine tourism potential of a destination. Development occurs alongside the "need to retain or attract people in rural areas, maintain aspects of 'traditional' rural lifestyles and agricultural production, and conserve aspects of the rural landscape" (Hall et al., 2000, p. 11).

Making certain that tourism and wine developments directly benefit residents will always be a challenge, so pertinent decision-making criteria have to be enforced when proposed developments are reviewed.

CASE STUDIES IN WINE TOURISM

The rest of this chapter is devoted to two cases studies on wine tourism. The first is a short case on regional wine tourism development in the Okanagan Valley of British Columbia, Canada. It highlights most of the phases of the wine tourism development process. The second case focuses on a specific winery, the Sandalford Winery in Western Australia, which highlights the variety of successful programs the winery has implemented as well as its innovative partnerships.

The Case of the Okanagan Valley, British Columbia, Canada

The Okanagan Valley of Canada has historically been considered a "peaches and beaches" destination, primarily attracting families looking for an affordable, summertime vacation. A beautiful area with slopping mountains cradling a long, beautiful lake, complete with its own mythical "Loch Ness monster—Ogopogo," the valley was known for its abundant crops of apples, peaches, and other fruit to feed Canada. In the 1930s, the first winery was built and slowly others were developed. Next came a series of golf courses and seasonal tourists. The Okanagan Valley was a perfect location for wine tourism development and expansion (Figure 10.2).

Figure 10.2. Lake Okanagan wine region. Photo by Don Getz.

The Wine Tourism Assessment Phase

Research on Okanagan wineries was conducted by Getz and Brown who studied winery goals, visitor-oriented services and facilities, and future plans. The Okanagan is not an all-year destination, and this is clearly reflected in the statistics. Visitor-oriented employment is small and highly seasonal. Only a few of the larger wineries offer a complete visitor experience, making them "destination wineries." Nevertheless, the trend is toward more facilities and services to both attract the tourist and maximize on-site spending.

The major challenge is to attract dedicated wine tourists, especially outside the short summer season. The valley has over 5,000 acres planted in grapes, and by 2007 some 80 wineries were listed; most are in the "farm" and "estate" categories and are small by international standards. Land available for vineyards is possibly nearing capacity, without replacing soft fruit orchards, and the availability of water for irrigation is another potential limiting factor on expansion.

Wine tourism in the Okanagan faces other challenges. The local population is relatively small, although Kelowna has grown quickly and has surpassed a population of 100,000. Major cities are a long drive away, prohibiting day trips (Calgary is about an 8-hour drive away and Vancouver at least 4 hours). Air access from those two big cities to Kelowna is good, but it is minimal in the south of the valley. In winter the roads into the region are often subject to poor driving conditions as several mountain passes must be crossed from east or west.

Wine Tourism Strategy Development

The survey of Okanagan wineries revealed a strong desire to capitalize on cellar door sales, and this was definitely leading to expansion. What the owners and managers wanted most, to help realize their goals, was the development of more and particularly better quality accommodation and food services in the region, special events, and more effective marketing to individual tourists. The wineries also recognized the need to promote quality wines to make Okanagan wine tourism more competitive.

Implementation Phase: Creation of a Wine Capital

Oliver, a small town (population 4,000) in the south Okanagan, declared itself to be The Wine Capital of Canada. This branding initiative was formed through the Oliver and District Community Economic Development Society, and is now a major economic development platform for the sub-area. The claim of "Wine Capital of Canada" was based on the following facts:

- Oliver is home to 13 wineries—more than any other community in the country.

- Approximately 3,000 acres of quality wine grapes are grown in Oliver (20% of total Canadian production).
- Wines produced from Oliver grapes receive over one-half of all Canadian wine awards.

Led by the Oliver and District Community Economic Development Society, the Wine Capital of Canada Masterplan was unveiled as a strategy for economic rejuvenation, followed by a Wine Village Accord that was adopted by Town Council. Next, a very innovative proposal, called Agriculture Resort Area, was put forward with the aim of encouraging wine tourism while preserving the rural character and agricultural lands of the Town. Specifically, this concept includes provisions for country inns, a heritage development, and a wine village core.

Evaluation Phase

Oliver's innovations and determination have achieved some concrete successes. Its main street is being renewed to reflect the wine theme. A major private investment established the Wine Country Welcome Centre (with restaurant, wine store, and planned retail shops) in the old main street fire hall. Recently, it was announced that a development company was prepared to invest in new resort-style development within built-up Oliver.

Part of the evaluation phase also included gathering stakeholder opinions and potential solutions on sustainability issues. Recommendations included the necessity for environmental impact assessments on all major new projects, and particular attention to the accumulating impacts of tourism on economy, society, and the environment. Preserving "ruralness" and the agricultural land base were identified as priorities, and these are being taken seriously by Oliver. However, it remains uncertain how wine and traditional fruit crops will evolve, given that wine appears to be more lucrative. Finally, Oliver and the Okanagan illustrate the need to pursue high-yield, wine tourist niches, not mass tourism and all its negative impacts.

The Case of Sandalford Winery, Swan Valley, Western Australia

Sandalford Winery, founded in 1840, is one of Australia's oldest and largest privately own wineries. It lies in its original position on the banks of the Swan River, upstream several kilometers from the center of Perth, the state capital of Western Australia. In 1991, Peter and Debra Prendiville acquired the winery and started what they call a "real tourism venture." Today, Sandalford wines are not only recognized in retail outlets domestically, but also in hotels and restaurants and in 36 countries on five continents.

The estate seeks to offer a world-class experience though a careful blending of innovation and authenticity. Visitors are encouraged to learn the principles of wine making, wine tasting, and wine appreciation. Additionally, events held at Sandalford raise the image of the Swan Valley appellation. Promotion of the Swan Valley's wines, food, and other "products" like local arts and music, attracts local, national and international audiences.

Marketing Strategy and Target Consumers
The winery's marketing strategy follows that of the Swan Valley Tourism Committee and Brand Western Australia. Sandalford caters to three specific target markets, namely the corporate, tourists, and locals. The corporate sector includes a wide spectrum, ranging from large, nation-wide or international companies to smaller, local organizations. The corporate market is high in yield, and demand has been rapidly growing. This sector is also steady during the times when other market segments are low.

The tourism market includes both national and international groups, fully independent tourists (FIT), and groups brought by travel agents, inbound and wholesalers. Sandalford promotes in advertising and guidebooks, particularly with the aim to attract the FIT market. The tourism market is not as highly profitable and is seasonal, whereas the local market, including weddings and other functions, is the most consistent.

Winery Tourism Programs
"Sandalford Cellar Door and Emporium" is the most profitable part of the operation, generating high demand during the peak summer season. In addition to tasting and purchasing wine, visitors can enjoy wine appreciation sessions. The Emporium offers a wide range of luxury, corporate, bridal, or special occasion gifts items as well as promotional products branded with Sandalford's logo, such as T-shirts, wine glasses, carry bags, and wine decanters. Especially useful for tourists, Sandalford offers free delivery on all case sales throughout Australia, with a worldwide delivery service also offered to the international visitors. The winery created partnerships with Japanese tourism organizations offering visitors to Sandalford discounted rates for selected products, plus home delivery in Japan.

Sandalford invested in developing *guided tours*, called "The Sandalford Experience," that provide visitors an overview of wine production. Suspended catwalks wind through the various wine making facilities, allowing visitors first-hand experience including all the sights and smells of an operating winery, while using interactive multilingual headsets. Visitors have a chance to learn the history of the estate, by watching a video in the winery's 54-seat theaterette as well as wine tasting during the 90-minute tour. They are also provided with take-home wine education booklets, and aroma sets for wine identification.

Tours change seasonally to reflect what is taking place within the winery. During the summer months, guides take visitors to the vineyards where guests may taste the grapes straight from the vine. During the vintage season, visitors get the opportunity to sample fresh-pressed grape juice. During cooler months, to supplement the experience, Sandalford's visitors have a chance to sample 10-year-old "museum" wines alongside the latest releases. All this generates year-round interest and repeat visits to the winery.

Another available experience is *wine blending.* Under the guidance of experienced staff, visitors can blend their own wines and either take them home or enjoy them over a meal in the Sandalford Restaurant. Sandalford's senior sommelier guides participants through the educational process of blending between different varieties, vintages, and regions before the actual blending occurs (Figure 10.3).

"Wine Maker for a Day" aims to attract corporate and incentive markets, and is recommended as a team-building activity. A take-home bottle of blended wine is provided with personalized wine labels that include corporate lo-

Figure 10.3. Blending seminar at Sandaford Winery. Photo courtesy of Duncan Turner.

gos. For individuals and groups, the labels may include special messages (for instance "Happy Birthday," "To my lovely wife").

"Miss Sandalford" is Perth's premium *luxury boat cruise*, featuring an Orient Express style, polished teak flooring, and lounge seating. During this very popular cruise from downtown Perth, tourists experience the history and scenery of the Swan Valley while sipping fine wines and snacking on gourmet cheeses. Additionally, the boat is a unique venue for weddings, meetings, or private charters. Because of its success and uniqueness, the cruise is featured on Qantas in-flight videos on both domestic and international flights to Perth.

The award-winning *restaurant* at Sandalford offers its guests fresh local cuisine and educates them about wine–food connections. Harmonized food and wine menus, and seasonal choices (e.g., sampling of mulled wine during winter months) make the restaurant one of Swan Valley's most prestigious fine dinning choices.

Sandalford's 12 *function areas* offer a range of distinct atmospheres, while making the most of vineyard and garden views. Their landscaped grounds are perfect for the large-scale summer concert series, encompassing a man-made lake, a natural amphitheater, and installed sound and light system. Audience capacity reaches up to 12,000 guests per evening. Sandalford is host to many concerts, attracting both domestic and international guests. This series of summer concerts, plus participation in festivals such as "Taste in the Valley" and "Spring in the Valley," play an important role in promoting the whole wine region.

Partnerships

Sandalford winery is actively engaged in the tourism industry, both as a member in formal organizations and as participant in festivals, the Swan Valley Branding Campaign,

Spring in the Valley Committee, and The Swan Valley Food and Wine Trail. It is located at the beginning of the Swan Valley tourism trail, which makes it a visitor service center. Winery staff are trained to provide general tourism information for the Valley. Additionally, Sandalford's staff are aware of other products and events in the region and encourage tourists to use other facilities.

Sandalford Winery is committed to the promotion and development of tourism in Western Australia. It actively supports the meeting, incentives, conferences, and exhibition market by offering regular staff training and client inspections. Sandalford is a member of the Perth Convention Bureau, Tourism Western Australian, and the Tourism Council of Western Australia (TCWA). It is an accredited tourism business supported by TCWA, which assures customers that all aspects of the business are of a superior standard.

A significant partnership has been developed in the Middle East market. Sandalford Wines are included on the house wine list in The Burj al Arab, known as the world's only "seven-star" hotel, as well as the Emirates Palace Hotel in the United Arab Emirates (UAE). Sandalford wines are also served on all Emirates Airlines international flights in business and first class. Events and concerts organized by Sandalford are promoted to international audiences in partnership with Emirates Airlines, and the Western Australia state's event tourism division, Eventscorp WA. The partnership with Eventscorp is significant, as the organization is the events division of Tourism Western Australia and makes a significant contribution to Western Australia through event and business tourism each year.

Continued Wine Tourism Efforts: The "Refresh" Phase
Sandalford Winery believes that to attract the largest number of visitors, constant improvements are needed. All operations are studied by management to improve its position in the wine tourism market. Works in progress (as of

2007) include a boutique accommodation retreat to be located on the property. This is to be a competitor with other world-class, luxurious day spas. The complex will consist of 16 chalets, each one situated by private, man-made lake, swimming pool, tennis courts, entertaining rooms, and various massage and therapy rooms. The whole spa will be orientated around wine-inspired products.

Conclusion

In conclusion, wine tourism is simultaneously a form of consumer behavior, a strategy by which destinations develop and market wine-related attractions and imagery, and a marketing opportunity for wineries to educate visitors and sell their products directly to consumers. Wine tourism reinforces wine exports, as educated consumers who have visited a wine-producing region are more likely to become loyal customers and to spread a positive word about the wines.

Wine tourism is a growing phenomenon that only helps to increase wine awareness and consumption in all the wine-growing regions around the world. It has been proven that it is economically beneficial to the communities in which they are located, the people in the area, and the wineries themselves. The experience of visiting a wine region, regardless of whether one is a serious wine collector or a casual visitor, is one that increases the tourist's appreciation for wine and its role in society, a benefit that leads to sustainable growth for both the specific wine region and the industry as a whole.

Special thanks to Karolina Anna Koziara for her work on the Sandalford Winery case, and to Lisa Poitras for her work on the Okanagan and Oliver case.

LEARNING OBJECTIVES:
- Define wine media and four major roles
- Explain the importance of wine media
- Identify the three major wine scoring systems
- Describe wine public relations and the two major responsibilities
- Explain the relationship between the wine reporter and the public relations professional
- Identify the five communications vehicles used for public relations
- Describe future issues for wine media and public relations

CHAPTER 11

WINE MEDIA AND PUBLIC RELATIONS

Megghen Driscol
Public Relations and Communications Consultant

Tim Matz
President/Managing Director Jackson Wine Estates International

Tor Kenward
VP, Public Relations, Beringer-Blass

The wine industry is remarkably fragmented, and with the thousands upon thousands of choices available to consumers today, making an educated choice can be extremely daunting. Research shows us that a consumer will trade from wine to another adult beverage if they become intimidated or overwhelmed by the choices available. To make this process easier and safer, today's consumer has come to rely upon third-party endorsements as a way of selecting their wines or validating their choice.

Most people wouldn't dream of purchasing a computer system without doing the proper research to find the best quality or the greatest value for money. However, when it comes to wine, consumers may not need to do any research because the wine critics already have.

This chapter describes the very important role of the wine media, as well as their relationship with the public relations staff of a wine organization. It begins with a focus on wine media by describing the role of wine media, why it is so important, and the impact of wine scoring systems. The second half of the chapter describes the role of wine public relations, including definitions, relationships with wine media professionals, and the communications vehicles used for public relations. The chapter ends with a brief overview of some of the future issues for wine media and public relations.

Defining the Role of Wine Media

Wine media may be roughly defined as *the group of writers, journalists, and reporters who evaluate and communicate the benefits of wine to the public.* The role media plays in the wine industry is paramount to the current success and future growth of individual companies and brands. While it provides individual challenges and opportunities for wineries, brands, and journalists, it has been a critical contributor to the overall growth and expansion of wine consumption in all regions of the world. As with the influence of the press throughout our daily lives, media coverage of the wine industry has similar effects.

There are four major roles media plays within the wine industry:

1. As consumer advocate where the writers or journalists feel they have a core obligation to protect the consum-

er from overhype, misrepresentation, and too much irrelevant information.
2. As a third-party, unbiased endorser of brands where there is vested interest to be as objective as possible.
3. Introduce and educate consumers to new brands, styles, varietals.
4. Provide newsworthy interesting and fact-based information for the consumer to make a more informed purchase choice.

The role media plays within the wine industry is one of the most important parts of the marketing mix (when communicating brand information) within the wine industry, but it extends beyond that as it is also the means to deliver and communicate messages about the industry, companies, and social responsibility issues relating to alcohol. On the marketing side, the role of media mostly pertains to brands, products, and companies communicating messages of imagery, and rational or emotive benefits.

The Importance of Wine Media

Media possesses an enormous influence over the wine-consuming public, using magazines, journals, newsletters, radio, television, and the Internet to deliver timely news, key messages, and quality cues about wineries, brands, varietals, and products. There are a handful of key influential writers and publications that have national influence, but there are also many markets where the local wine critic has tremendous influence and a strong following. Almost every local paper has a weekly or monthly wine review, either by syndication or a lifestyle critic who happens to cover wine as a part of their beat. Both the national media and the local critic are very important.

As this book is focused on New World wine regions, it is notable that the role of media is more relevant in the US than perhaps any other large wine-consuming country in the world. This is due to perhaps four reasons:

1. Historically the US is not a wine-consuming nation; therefore, third-party endorsements are heavily relied upon.
2. Due to the relative newness of our industry, the average consumer's palate is not as educated, and hence relies on knowledge gathered through the media.
3. It is the nature of the American consumer to seek guidance via critics and experts, no different than when evaluating movies to view or electronics to purchase.
4. The wine business is so fragmented and there are so many choices (which change slightly each year) that the overwhelmed consumer depends on recommendations of the "experts."

Role of the Wine Journalist

The role of the wine journalist is to act as a nonbiased, third-party critic evaluating wine from the objective characteristics as well as the subjective differences within each wine. They will give guidance, direction, facts, and opinions about different wines. While knowing the facts about wine is consistent among wine critics, it is important to understand wine critics are individuals with different opinions. Every wine critic has a unique palate; therefore, their preferences for wine styles will vary. Most wine journalists are somewhat or extensively trained in how to go about tasting wine, and hence know how to maximize their use of the senses. While they will likely agree on the facts such as grape variety or residual sugar, it is not uncommon for wine critics to judge a specific wine completely differently. The subjectivity in tasting comes into play because each individual has unique, varying degrees of taste, likes, and dislikes. For example, one person might enjoy floral elements and pick up that characteristic more readily than spice. Additionally, a person may have been exposed to a broader range or even more exotic tasting foods in their younger years, and have developed tastes much more defined than someone with a more limited diet.

Many wine journalists, with their trained palates, tend to classify wines in a "box" such as Rhône style or Bordeaux style or Beaujolais style. After the taste profile is put into a box, subjectivity comes into play because varying taste buds will gravitate to different flavors. An example of subjectivity is when judging or tasting a Syrah/Shiraz wine from the same grape type—each will taste very differently depending upon many factors including country of origin, microclimate, vinification techniques, or oak maturation—the options and outcomes can be staggering. For example, an Australian Shiraz may have bold, fruit-driven characteristics with hints of white pepper versus a classic French Rhône Syrah, which will likely be more earthy and savory in style. Both can be fabulous wines, just very different.

Gatekeepers

A gatekeeper is a term used within the wine industry that defines a person or group of people who have extraordinary influence impressing their opinions within the wine community. For example, a gatekeeper in the retail segment would be a key buyer of a major chain. As it pertains to the media, a gatekeeper is someone whose opinions, views, and expertise are respected, accepted, and credible within the wine industry. A gatekeeper can be a respected wine writer, but is not limited to just journalists.

There are many experts who do not have a media outlet to express their views; however, they can be as influential and important as a local or national reporter. These gatekeepers are often called upon by the media to provide perspective, content, and personal recommendations either in print, on television/radio, and the Internet. This group is made up of high-profile Master Sommeliers, Masters of Wine, key retailers, and restaurateurs. For example, having the Master Sommelier at the Ritz Carlton in New York promoting your wine in his/her restaurant or on a panel discussion can add tremendous value to your brand's reputation and credibility.

WINE SCORES AND RATING SYSTEMS

While scores or ratings are perhaps one of the most controversial subjects within the media, they are also one of the most influential measures when promoting wine. Wine scores provide a quantifiable rating system for the industry to differentiate wines, therefore allowing the trade and consumer to make informed decisions. Due to the influx of so many types of media and the speed of delivery to the reader or listener, wine messages have become like so many other messages—the latest or most memorable sound bite. For example, if a consumer sees that a specific wine has received a 90-point score or a five-star rating, he/she will feel that the wine is a safe, acceptable, or even preferable choice. Table 11.1 outlines the three major types of wine scoring systems used by the media.

Table 11.1. Three Major Wine Scoring Systems

System Name	Scoring Method	Examples
20 Point UC Davis System	Rates wines on a scale of 1 to 20, with 20 being highest	13 to 20 = average to outstanding quality
Star or X System	Rates wines on a scale of 1 to 5 stars or "x"s, with 5 stars being highest	*Wine X Magazine*, *San Francisco Chronicle*, *Restaurant Wine Magazine*
100 Point System	Rates wines on a 50 to 100 point scale, with 100 being highest; used most frequently	*Wine Spectator*, Robert Parker, *Wine & Spirits Magazine*

Of course, getting a good review in a national publication is sure to result in additional sales and brand recognition, but a very targeted message in a small market can do wonders in that specific city. For example, getting a "best buy" rating in *Consumer Reports* magazine is extremely powerful on a national level, just as getting a positive wine review from a respected critic in a local newspaper will have tremendous impact in that local market.

DEFINING WINE PUBLIC RELATIONS

Wine public relations can be roughly defined as "developing and promoting positive winery news and brand information via advertising, journalism, and special events for the public." According to Posert and Fransen (2004), professionals in wine public relations are experts in the art of "spinning" a message within the media to promote a particular wine business.

Obviously, within the wine industry, one-on-one communication with consumers, distributors, and wine journalists is beneficial; however, this doesn't always have the scope of brand communication that is desired. Therefore, good wine public relations ensures that the message reaches much larger audiences, via newspapers, magazines, journals, radio, television, and other media channels. By promoting news and information via a third party, it is often perceived that there is more credibility in the message (Academy of Wine Communication, 2003).

After consistent messaging from a wine public relations department, eventually it is possible that a reporter, critic, or journalist will actually use the brand communication points within their articles. However, this is a very unpredictable process—it can happen overnight, or it can take years. An example of this happened rather quickly with the Charles Shaw brand, also known affectionately as

"Two Buck Chuck." This brand was released in early 2003 and sold exclusively through a grocery outlet called Trader Joe's in the Western US. Through the endorsement of this trusted grocery store, thousands of consumers purchased the wine at a price of $1.99. With such a low price, and a taste profile that appealed to consumers, Charles Shaw immediately became an overnight success. Cases of wine flew off the shelves, and the success of the brand was so phenomenal that it was covered in newspapers and magazines across the country, as well as on the national evening news on a major TV station.

Penfolds from Australia is another example of a wine brand that saw incredible growth from a well-thought-out PR plan. While it took much longer to achieve fame, it has perhaps been more effective in the long term. This brand was virtually unknown in the US, even though it had been sold in Australia for more than 150 years. When it received *Wine Spectator* magazine's "Wine of the Year Award" in the mid-1990s, demand for the wine was so great that stores had to allocate the amount they sold to customers because they couldn't keep it on the shelves. Both of these wine brands are good examples of excellent public relations campaigns promoting positive brand messages and stories.

Public Relations Role

The role of wine public relations is to work in partnership with top management, sales, and marketing to deliver consistent messages on company/corporate affairs and on the wine brands within the portfolio. Within an organization, wine public relations fulfill two key responsibilities:

Responsibility #1: The primary point of contact and communication liaison to anyone outside the organization addressing issues in corporate affairs, government relations, and crisis management. They are responsible for delivering a consistent, clear, and concise message to all the appropriate parties, ensuring that what should and needs to be said is accomplished.

On occasion public relations professionals have to address very sensitive issues, problems, or challenges presented in the marketplace. Examples could include a specific wine quality issue that penetrates the public perception, or a highly allocated wine that has been recently written about and is in great demand. In these cases, as with any others, it is imperative that the public relations professional communicate the facts accurately so the message is not misperceived in anyway. Facing these types of issues is perhaps the most challenging responsibility within public relations, and can either make or break a company's reputation.

Responsibility #2: The communication liaison on all the brands to the outside media, trade, and consumer. This responsibility is to express the facts, attributes, and imagery that best convey the message the brand wants to communicate. This is where public relations work closely with sales and marketing to ensure consistency on how the brand is marketed and communicated across all tiers.

Because public relations professionals have two very different roles and responsibilities, it varies on where this group reports, sometime into marketing and sometimes into executive management. Either way, the sensitivity of the information requires that public relations staff is closely linked to the marketing team as well as senior management in an organization.

Relationship Between the Wine Reporter and the Public Relations Professional

Public relations professionals work very closely with wine writers, reporters, and journalists. These are, without a doubt, the most important relationships a public relations person can have, and precisely why a seasoned professional is so desirable to savvy organizations. The influence of this core group on brands and companies is tremendous and requires time, energy, and professionalism by both parties.

The wine reporter relies upon the public relations professional just as much as the PR professional relies upon the reporter. This is always a fine balancing act between both parties because each has individual goals, but both need the other to perform their responsibilities.

Most reporters are constantly under deadline. If you can provide a credible and interesting hook or angle, the reporter will use it. The key is in establishing a relationship with that person and having first-hand knowledge of what they write about and who their audience is. The easiest way to alienate a reporter is to pitch an irrelevant or uninteresting idea or provide a story idea that would have no meaning to their readership or viewership.

Communication Vehicles

There are five major vehicles used by wine public relations and media professionals. Each of these is described in the following paragraphs.

Consumer Publications

Consumer publications consist of any medium that is targeted at the ultimate wine purchaser. These could include newspapers, magazines, newsletters, or Internet/websites. Wine-specific publications provide a precisely targeted forum to reach the wine consumers. The largest are *Wine Spectator*, *Wine Enthusiast*, *Decanter*, and *Wine & Spirits*, which have tremendous influence with both the trade and consumer. Other consumer publications such as *Ladies' Home Journal*, *Gourmet*, *Bon Appetit*, and *Esquire*, while not as targeted, can reach a much larger audience base, resulting in increased brand awareness and sales.

Wine critics who educate consumers by describing and promoting wines from specific wineries primarily use consumer publications as a medium. Often, wine companies will also use consumer publications to advertise their individual brands. Advertisements are most often used to promote brand attributes and activities as well as educate consumers on the various differences of wines, regions, and grapes from around the world. Advertising is also an effective forum to promote accolades and scores from other publications, enabling the company to reach a much larger audience.

The newspaper, one of the consumer publications, is primarily used as a means to promote wines by individual wine companies (mostly through advertising) and provide a vehicle for wine writers to critique wines (mostly through editorials or columns). This is a consumer-driven medium that targets the wine purchaser in a given market. In the eyes of the consumer, a third-party endorsement will always carry more weight over a paid advertisement.

Trade Publications

Trade publications are specifically targeted at suppliers, wineries, restaurateurs, retailers, and distributors. Some examples of trade publications include *Santé, Impact, Market Watch,* and *Beverage Dynamics,* to name just a few. Trade publications are a popular medium used by wineries to inform the trade about their brands and their company. They are also a means for wine writers to provide descriptions, knowledge, and opinions on various wines and wineries. Additionally, almost every state has a local beverage publication such as the *Pennsylvania Beverage Journal, Massachusetts Beverage Journal,* and *California Beverage Journal.*

Radio

Radio is a medium that is less utilized in the wine industry relative to many other industries, perhaps due to cost and reach to the specific targeted audience. When used, it is primarily a vehicle for paid advertising in the traditional sense. However, a seasoned PR professional will always look for opportunities in radio to provide a platform for an interview between a show host and a wine expert, whether it is a wine critic, winemaker, winery representative, or industry advocate. Local radio shows are a terrific way to generate publicity, although they are used much less often than traditional print publications.

Television

Television, though less used than written publications, is a medium employed within the wine industry in two ways. First, it provides a platform for traditional advertising—although this is not a common practice for wineries, mostly due to cost, reach, and general societal pressures of limiting wine exposure to specific targeted age groups. Secondly, television is a medium used to educate and promote wines via specific wine-related shows. In order to get television coverage one must have a compelling story to tell. It must be noteworthy, informative, and/or entertaining. While it is not always easy to get coverage, the results can be exceptional. Larger, better known companies will often spend a percentage of their public relations budget on product placement. While it is impossible to say how effective this practice is, the feeling is that if a consumer sees brand X on their favorite television show, it is considered to be an endorsement of that brand by a celebrity or famous personality.

Internet and Blogs

Almost all wine companies (including individual brands) and publications now have websites, which are geared toward the consumer. As discussed in the Wine 2.0 chapter, the Internet has become a formidable medium to communicate information about the wine industry. It is most effective in terms of cost efficiency, and speed to market. Due to the sophisticated abilities to segment specific groups or individuals, messages on the Internet can be written and delivered in customized ways to ensure effective, targeted communication. This is true whether via email or a website; hence, communicating can be conducted passively or actively.

Blogs have become increasingly more important and are proving to be a very powerful tool in public relations. Many blogs provide commentary or news on a particular subject such as food, politics, and, of course, wine. A typical blog combines text, images, links to other blogs, Web pages, and media related to its topic. The ability for readers to leave comments in an interactive format is an important part of many blogs and has opened the opportunity for average consumers to become wine critics themselves. The most essential element of a public relations plan is consistency of message and with the flood of new blogs, this discipline is more important then ever. The PR professional now needs to monitor and respond to blogs.

The Future of Wine Media and Public Relations

The media and public relations will always play a critical role in the wine industry. With the influx of new brands coming into the marketplace, living in a society where information overload exists, and consumers more hurried now than ever before, the roles of public relations and the media will continue to be platforms that create brand awareness. PR helps to communicate product attributes and qualities to consumers so they can make an informed choice.

In the overall marketing mix, public relations will continue to be one of the most cost-effective ways of obtaining and retaining brand recognition and loyalty—something that can differentiate a brand from the clutter, increase sales, and provide long-term visibility.

The media, whether it be via newspaper, radio, television, or the Internet, will continue to be a critical vehicle used to communicate consistent, clear, and concise messages to the targeted audience.

LEARNING OBJECTIVES:
- Identify the six major trends driving the global wine market
- Define direct and indirect wine exporting
- Describe success strategies for global wine marketing
- List the six major issues to consider for export readiness
- Describe market choice considerations for exporting wine
- Explain issues to consider when branding wine for export

CHAPTER 12

GLOBAL MARKETING AND EXPORTING

Larry Lockshin
Professor of Wine Marketing, University of South Australia

Tony Spawton
Professor of Wine Marketing, University of South Australia

The term globalization is familiar to all working in the modern business world. No longer are markets mainly domestic and local; businesses of all types are competing with others from around the world. The wine sector, too, is now part of this globalization process. In some respects wine has been one of the most localized of products. Until the middle of the 20th century, much of the world's wine was purchased and consumed within 50 kilometers of its production. At the same time wine has been one of the first agricultural goods widely traded between countries and regions. Records as far back as ancient Egypt identified superior wine-growing regions, and by Roman times, individual vineyards had reputations for quality and were in high demand.

Nearer to the modern day, wines from Bordeaux have been highly demanded in England, as have been ports from Portugal and sherries from Spain. Over the same period of the last 150 years or so, new wine-producing countries, called the New World, have developed production and some level of exports to their Old World parents. This was part of the colonial mentality of creating agricultural sectors to export back to the home country. Historically, this trade was often at the mercy of political struggles in the Old World. For example, Britain affected wine growing and exporting from Australia both positively and negatively as they raised and lowered tariffs depending on whether they were at war or peace with France, Spain, and Portugal. The decision for Britain to join the EEC, as it was then, meant that wine exports ceased for almost two decades.

This chapter provides an overview of the key trends in global wine marketing. Then some of the success strategies for entering the global market are discussed. From the strategies, direct and indirect exporting are defined, and examples of how some of the best companies are organized for marketing outside their domestic sphere are provided. The chapter continues with a detailed list of specific tactics for exporting, including some views on organizing for export, an export readiness checklist, market choice considerations, distribution systems, competitor analysis, advertising and promotion, and pricing. The chapter concludes with sections on branding for export, communication and promotion issues, and the export value chain and logistics.

Key Trends in the Global Wine Sector

Why is the wine industry becoming more international? First, trade liberalization and the mutual recognition of enological practice have *reduced tariff and nontariff barriers to wine marketing*. The European Union agreements with bilateral partners, like Australia and South Africa, have reduced the various technical nontariff barriers, such as fill level monitoring in bottles, nonacceptance of certain wine production practices, and the like. Nevertheless, nontariff barriers of differential labeling requirements depending on market eventuated in the agreement that the mandatory statement (alcohol by volume, region, packer etc.) were agreed only for a small number of countries and a new demand that the statement "sulphides added" should be in a range of languages with the spelling of "sulphides" varying with the language. Similarly, the requirement for traceability, ingredient labeling, and limitations on entry into the distribution systems of markets continues to differ across markets. Add to this the growing drive to reduce alcohol abuse by various means, which includes the curtailment of advertising and promotional tactics, restrictions on the usage of regional names and traditional winemaking expressions still operate to frustrate the global trade in wine. China's entry into the World Trade Organization (WTO) has resulted in the reduction of import duties on wine, which not only will allow the prices for imported wine to drop, but also will make domestic producers more competitive as they now have less of a price advantage. China has also developed a standard for grape wine, which will allow acceptance of Chinese wine in the world market from 2008. All of this is occurring while total volumes exported around the world grow by less than 1% each year. Trade is growing, but actual volumes have not grown much at all as the make-up of who is importing and exporting continues to change (Table 12.1).

Table 12.1. Six Major Trends Driving Globalization of the Wine Market

Reduced tariff and nontariff barriers to wine marketing
Economic development and wine tourism
Increased wine production around the world
Changes in wine consumption (decreases in Europe and increases elsewhere)
More wine sold in supermarkets and discount stores
HORECA—growth of hotel, restaurant and café culture

Another major trend has been the *economic development* of the world. Higher standards of living are associated with higher wine consumption. Although the relationship is not a direct one, as living standards and wages rise, so too does the demand for bottled wine. Wine sold in bulk, as in "fill your own containers," is typical in traditional wine-drinking countries; bottled wine is the main import and export in the New World and the newly developed countries, such as in Asia. The growth of food safety regulations and the drive for "product integrity" by importing markets with the increasing strategy of "mis en bouteille" (i.e., bottled at source) will accelerate the trend to packaged wine globally. This will be balanced by increasing emphasis on lowering greenhouse gas emissions by shipping bulk and then bottling in the final market. This type of "branded" bulk wine is different from traditional bulk wine exporting. This will result in changes in the structure of the supply chain and the transfer of bottling from the producing regions to the one where the product is sold. The implication of this will be a depletion of the "value-added component" (packaging and transport support industries) in the producing regions with their transfer to the regions where the wine is sold.

Economic development and the heuristic and hedonic lifestyle associated with wine have fostered tourism and multicultural interests. A generation ago, people were happy to consume the traditional foods and beverages of their parents. If their parents migrated to a new country, such as the US or Australia, often their traditional foods were looked down upon and their children craved the "normal" foods and beverages of their new country. Now people are much more interested in preserving their traditional foods and beverages. Italian, Thai, Indian, Malaysian, Greek, Spanish, French, and Chilean foods are all available in restaurants and even in many supermarkets. Tourism and television have opened the eyes and the mouths of consumers in developed countries, and wine has been part of that awakening. Wine consumption has markedly increased in places like the UK, Scandinavia, Canada, Australia, Holland, and now Asia and even India and China as a result.

At the same time, *wine production has spread widely.* Wines are available from over 70 countries in the world, although only a few constitute the major wine-exporting countries. France still leads the world in total wine exports (Table 12.2), followed by the major Old World countries. But the New World has emerged as a major competitor, especially

Table 12.2. Top Exporters of Wine in the World, 2003

Country	Wine Exports (ml)	% of World
France	1,514	20.8
Italy	1,291	17.7
Spain	1,235	16.9
Australia	536	7.3
Chile	402	5.5
USA	329	4.5
Portugal	311	4.3
Germany	277	3.8
Moldavia	275	3.8
South Africa	239	3.3
Argentina	185	2.5
World	7,295	

Source: OIV: Situation Report and Statistics for the World Vitinicultural Sector in 2003.

in key export markets, like the UK and the US, and in new markets, like Asia.

The next global trend has been the well-documented *decreases in consumption* in the Franco-Latin countries and the accompanying *increases* in the Anglo-Saxon and Asian ones. Overall global consumption has not increased very much in the past 10–15 years, only about 2% overall, but regrettably is expected to fall over the next decade unless the decline in consumption in the Franco-Latin countries is arrested and new markets are opened up in Northeast Asia and elsewhere. Although better quality wine is being made around the world (a higher percentage is being sold at the premium, $US5.00 and above price range), total global consumption has hardly changed. There are very recent hints from the US that substantial consumption increases are beginning, but the ability for wine marketers to convert these potential increases to sales has been slow due to a lack of commitment to wine marketing; as a result there is an increasing surplus of premium and ultra-premium wines in most New World- and Old World-producing countries.

Another global trend has been the very fast increase in the percentage of *wine sold through supermarkets and discount stores*. Across the world, even in such traditional places like Italy, over 60% or more of total wine sales are now ending up as fast moving consumer goods (FMCG or packaged goods). The positive side to this is that it increases the availability of wine to more people, who would not normally go into a wine specialty store to shop. Oversupply has produced good-quality bargain wines, although recent statistics show that people already drinking wine are buying most of these, but some are using these inexpensive wines to extend the number of drinking occasions. The downside, and a very large one, is that the packaged goods system favors larger companies with consistent quality, strong logistics, brand management, and enough capital to engage in regular high/low supermarket promotions. As even the specialty stores become corporate chains, the number of SKUs (units on the shelf) has shrunk from 1500 or so to 800 in the specialty chains, and to 350–500 in the supermarkets. This again reduces the chance for smaller producers to be on the shelf.

The other side to this trend is that availability should create more wine drinkers. Some of these drinkers will develop into wine-involved or interested consumers, who want to experience the diversity of the global wine offer. These consumers and the existing involved buyers will make sure that independent specialty stores remain and the better managed ones will prosper. Even in supermarkets, where there is high demand, specialty wine shops within the supermarket are being developed and trialed. As long as there is demand for interesting and unique wines, the market will respond and smaller wineries will have outlets, though only the better managed ones will survive.

The final trend is that *HORECA (hotel, restaurant, café)* is still strong in many traditional wine-drinking countries and it is growing in many New World ones as well. Sales have been lost to the overall social trend of quicker eating and eating at home in some Old World countries. But the major change has been the disappearance of the generation of café-based wine drinkers. This has reduced overall consumption, but these people mainly drank the lowest quality level of wine and are being replaced by choosier drinkers. One of the big issues here is the lack of good data, due to the large number of small privately owned firms in this sector. New World countries seem to be embracing the café society and, as noted above, international cuisine. This opens the door for smaller wineries to find sales outlets in the HORECA channel in the faster growing wine markets.

One of the key differences between the wine trade and other consumer products is the relative fragmentation of production. Even with alcoholic beverage companies buying into the wine trade (Diageo, Pernod Ricard, Allied Domecq, for example), the sector is much more fragmented than comparable beer, spirits, and other consumer beverages (Figure 12.1). This has two major implications for wine producers. First, there is a lot more consolidation that is likely to occur. As the retail channels consolidate in major wine-consuming countries, there is a greater impetus for suppliers to become larger in order to deal with them. Second, on the other side of the ledger is the opportunity for smaller wineries to stake a claim in the global market. The next section looks at some successful strategies for global marketing among large and small wineries.

Success Strategies for Global Marketing

Technically, there are no global wine companies as there are global soft drink companies. Wine is still produced and labeled by the country of origin. Some of the largest wine companies are multinational, in that they make and sell wine from more than one country. However, they do not

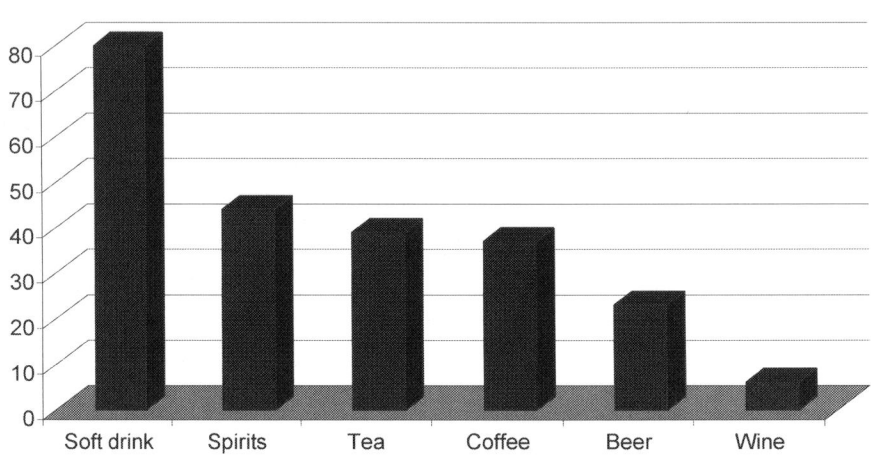

Figure 12.1. Global market shares of the major beverage categories (Rabobank, 1999; reprinted with permission).

use the same brand name for wines from different countries, in contrast to the strategies of Coca-Cola or Pepsi-Cola. One exception so far is the new range offered by Lindemans (Foster's of Australia) with wines from Australia, South Africa, and Chile, each labeled appropriately, but using the Lindemans brand. Of course, major retailers, like Tesco's in the UK, sell wines from many countries under their home brand label. Most wine companies are exporters: making wine in a single country and exporting that product to other markets. Developing a strategy for exporting marketing depends mainly on the size of the wine company.

The larger wine companies have more options for developing international markets. Direct exporting is the main one, and it will be covered below, because it suits all sizes of wineries. Large companies can invest directly in other countries and produce wine there for either the domestic market or to export. BRL Hardy, now owned by Constellation Brands of the US, bought an estate in southern France, La Baume, and then expanded production by buying grapes from around the region. The resulting wines are sold in France and exported to the UK and back to Australia.

Another method of internationalization is to buy an existing winery in another country. Constellation Brand's purchase of BRL Hardy in Australia, Foster's Brewery's purchase of Beringer in the US to merge with its Mildara Blass wine business, and Pernod Ricard's (France) purchase of Orlando Wynham in Australia are all examples of this strategy. It is usually driven by gaining immediate access to a producing winery and the accompanying economies of scale in distribution and selling and the circumvention of nontariff barriers designed to discourage exporters entering a market. The other issue for exporters such as Australia or any country for that matter with a currency value less than that of the import market (Euro, Pound, or US dollar) is the "in-market cost" (the cost of distribution incentives and advertised promotions) and retention from margins earned in that market are paramount, rather than using a "repatriation and pay" method from the home market. This has driven (where possible) distributor and agent ownership of joint venture arrangements to facilitate this. Contract production and marketing joint ventures are growing in importance, with BRL Hardy/Stellenbosch Wines an example. BRL Hardy has similar ventures with wineries in Sicily; in a similar manner Mondavi and Antoniori have joint ventures in Chile.

All of these strategies focus on the need to gain sufficient critical mass to support distribution volume expectations and the maintenance of hard-won retail selling space. The Beringer buyout by Foster's, for example, allowed an immediate increase in distribution for Australian wines into Beringer's home market by the existing sales force, while at the same time the Beringer wines could take advantage of Mildara Blass's larger market position in the UK. Retail consolidation means that large supermarket and discount buyers are aiming to reduce the costs by purchasing the widest range of wine needed from the fewest suppliers, thus reducing the purchasing and logistics costs. These multinational producers can bring a sales book (the list of wines available from a single sales person) listing wines from multiple countries, with multiple brands at a range of price points. The joint ventures use the same strategy but also have the strategic choice of whether to use the well-known existing brand name or to develop a new one. Often a new brand name is chosen, because the existing name has such a close tie to a different country, but the company name may include the joint partners as a kind of sub-brand to identify and build on existing brand equity, such as the brand "Yellow Tail" developed specifically for the US market through a joint venture agreement between Casella Wines in Australia and Deutsch and Sons in the US.

Smaller wine companies, typically of less than 1 million cases and most often much smaller than that, are forced to use direct or indirect exporting methods to targeted countries as their main means of internationalization, as they do not have the critical mass of volume and capital required to undertake the globalization strategies of the larger corporations and are therefore committed to export as their prime market entry strategy. The larger companies choose this route as a stage of their market development to gain early market share and to establish themselves as a part of a new category within wine. As the required efficiencies of competing in the FMGC market intensify, the alternative strategies detailed above have become more necessary and more attractive to maintain market share once it has been established and to provide the logistics and service such as just-in-time (JIT) deliveries and stocking requirements the supermarkets demand from all suppliers.

DEFINITIONS AND EXAMPLES OF EXPORTING

The rest of this chapter will focus on direct exporting, but first a short paragraph on *indirect export*. This was the favored method of export when negotiants, merchants, and brokers bought wine at the winery and undertook the total export process on behalf of the producer. These indirect facilitators are usually country of production focused, but have the advantage of both expertise and resources to provide effective export support and logistics consolidation to gain the better full container (FCL) shipping rates than the premium lesser (LCL) rate. The advantage to the winery is that their product is bought under the trading terms of a local market transaction and paid in the local currency. The main disadvantage is that the wine could be sold in any market or any segment of that market without any reference to the exporting winery. This method is still a recommended method for very small wineries to enter the market that have neither export experience or the skilled staff to deal with the complex export marketing process and the maze of nontrade barriers that need to be taken into consideration along the way (Table 12.3).

Now we examine *direct exporting*. The first ingredient to successful wine exporting is a strong management desire and commitment to the process. Without this, most export strategies fail. Exporting requires a large commitment in resources and time to be successful. Potential wine exporters should ask themselves what they have to gain by exporting, given that every part of the process is more expensive than the same thing in the domestic market. Gaining market information, travel, packaging, shipping, warehousing, insurance, managing the sales force, and reacting to competitors all cost more when it is done at a great distance. Typically wineries choose to export because they believe there is little room for expansion in their domestic market or that prices and margins will be higher overseas. These assumptions should be tested before the final decision is made. Some wineries make the decision to export based on some desire for prestige or for travel to distant markets, rather than on more objective criteria of sales and profit growth. Countries with either small or shrinking domestic markets are the home of most of the larger exporting wineries in the world; Australia, New Zealand, Chile, and South Africa are countries with limited domestic markets; France, Italy, Spain, Portu-

Table 12.3. Definitions of Indirect and Direct Wine Exporting

Indirect exporting	Use of a facilitator to export your wine for you (e.g., negotiant, broker, association). Good method for small wineries just getting started.
Direct exporting	Establishing an export strategy and structure within your company to export your wine. This requires strong commitment from top management, financial resources, and a clear brand clarity.

gal, and Argentina are countries with large, but shrinking, domestic bases.

Once the decision to export is made, the key next step is to decide the target country for exporting. Wineries new to exporting are advised to focus on a single market at first to learn how to do it and to have the requisite management time available. The easiest markets to export to are typically those that have already been targeted by other wine companies in your region or country or already have a well-developed wine-consuming culture. This means there is already some market recognition for your country or region of origin; distributors and retailers are aware of the styles and need less cajoling and information to make a decision. Of course, long-standing export markets may have less opportunity than some newer ones, but this must be balanced against the cost of informing a new market not only about your brand, but also about your country or region, which is a long and expensive process.

Examples of Three Australian Exporters

Here are three brief examples of Australian companies that successfully took this path.

Banrock Station Winery

The first is the very large BRL Hardy and their brand, Banrock Station. This brand was developed from a large planting of grapes in Australia's Riverland region, the warm and dry inland area responsible for the majority of Australia's commercial wines. Initially Banrock Station was developed for the Australian market as a slightly higher priced commercial (sold in bag-in-box or casks) wine with an environmental positioning. The old station (large farm) was completely rejuvenated by opening up the original wetlands, which had been drained, changing irrigation practices to limit water usage, integrated pest control, and building environmentally aware buildings. The wine was sold in brown labeled bottles and casks, the same wine in each format. Initially seeds to plant trees were attached to each package to reinforce the environmental message.

The venture was so successful in Australia that it was taken almost in its entirety to the UK market. The same positioning was used, which, just like in Australia, was unique for a large and relatively inexpensive wine. As sales grew, the company invested in a state-of-the-art environmental cellar door, with passive heating and cooling, water recycling, and fantastic views over the rejuvenated lagoons and wetlands surrounding the vineyards along the Murray River. Even though the cellar door is more than a 2-hour drive from Adelaide, it is one of the most visited cellar doors in Australia, with many international visitors choosing to camp in the parkland developed out of the old sheep station. Here, the export strategy was almost exactly the same as the domestic one, because the positioning was relevant to both countries. The size of the vineyards meant that the domestic Australian market was not large enough to absorb all the sales, so export to Australia's (and BRL Hardy's) largest overseas market was the best way to expand sales. BRL Hardy took advantage of the fact that they already had a sales force and good distribution in the UK to launch this brand with good promotion and shelf space. Now, 12 years after its Australian launch, the brand sells over 7 million cases, mainly in Australia and the UK, but also in the US and other export markets.

Bartonvale Winery

Peter A. Smith, a retired Australian aerospace engineer, developed a different type of export strategy for a brand new small winery called Bartonvale. He started the vineyards with the idea of growing a top-level Shiraz in Barossa Valley and exporting it at very high prices to the lucrative

US market. He had no plans to sell the wines into the domestic Australian markets, where high levels of competition would limit his chance of gaining cult status and high prices. He chose vineyard sites that were near existing high value locations and used cuttings from vineyards already supplying cult-level wines, like Penfolds Grange and Peter Lehman's Stonewell Vineyards. He hired Rolf Binder of Veritas Winery to make his wines, because Rolf already had several of his own cult wines selling in Australia and the US. The idea was to make exclusively Shiraz and Shiraz blends in a style that critics like Robert Parker would rate highly.

He looked for an importer distributor that had experience selling these types of wines to high-end specialty wine stores and exclusive restaurants. Peter Smith had a four-page list of questions to ask prospective distributors about their existing portfolios: how much time and effort they spent marketing new wines, how much he would have to bear of the promotion costs, how they developed an executed marketing plans with other small wineries, etc. His chosen distributor, Dan Philips of the Grateful Palate in California, already had a portfolio of Barossa Valley Shiraz, including some cult wines, like Torbreck and Three Rivers. He was able to go into small exclusive retailers with a shelf of these cult wines and had access to many top restaurants.

Once the wines were made and the quality fit the plan, small amounts (all that was available) were sent over to the US at retail prices of about $20 to $30. These were the lower end wines of the range, though still with relatively high prices. This helped establish awareness and prepare the market for the more expensive offerings. Within 2 more years, the higher priced wines showed their potential by receiving Robert Parker scores from 92 to 97 points, and the strategy was moving exactly to plan.

Kingston Estate Winery

Wineries often think that they need to have a premium product image to export. Take the case of Kingston Estate in the Riverland of South Australia. Located in a traditional "bulk wine" and therefore perceived low-quality region, Kingston Estate decided to build its reputation in the commercial and commercial premium export markets. Using bulk wine sales as leverage it gained access to the BOB (Buyers own Brand) of the Scandinavian monopolies, the growing "by the glass" dispensing of the café market in the UK, and the commercial wine segment in New Zealand, by leveraging its own branded product as the "house wine" in the process. This successful marketing strategy was so effective that this became the basis for the launch of the Kingston Estate brand into the domestic market in Australia. This dual marketing strategy of using export as leverage for domestic marketing success has made Kingston Estate one of the leading wineries in Australia.

Each of these cases, though different, illustrates key parts of a successful marketing strategy.

1. First, a particular market and set of distribution channels should be chosen that match the company's ability to produce and market the wine.
2. Second, the wine must be of the quality that fits its intended price and position, whether this is a low-priced commercial wine or a high-priced cult wine.
3. Finally, with the strategy and wine in place, the company must execute it properly.

All aspects of the product and packaging must fit its intended profile; the distributor should have access to the intended outlets and be able to deliver the marketing strategy developed with the winery, whether this is mass distribution through supermarkets or exclusive distribution in selected shops and restaurants. Finally, good follow through over time is necessary to maintain the momentum in the market by managing the sales force or distributor, reacting to competitors, and enhancing the brand's reputation and awareness. We now discuss the organizational structures and the basics of branding necessary for successful exporting.

Tactics for Exporting

As noted above, successful exporters first must commit to the long-term process of investing and maintaining their export activities. Whether the company is large or small, someone must be responsible for the export activities. In a large company, an export manager- or vice president-level position is usually dedicated to these activities. Smaller companies may just add to the duties of the marketing manager or general business manager and, as exports grow, the companies usually will hire an assistant to deal with the paperwork and logistics of exporting, while the manager deals with the actual marketing and selling. In one sense, exporting is no different than selling on the domestic market. You must manage all the same activities, but now at a greater distance, with a few extra difficulties thrown in, such as language differences, regulation differences that may affect packaging and even the wine processing, and cultural differences in how wine is sold.

Organizing for Export

In organizing for export, it is important that responsibilities be clearly defined. Someone must take overall responsibility for the activity, but in most cases there is too much to do for a single person, who also has domestic marketing responsibilities. In this case, it is useful for a second person to take charge of information collection about the potential target market and begin investigating potential distribution partners. Often there is an export advisory service maintained by the government or even the trade association. These entities help provide information about target markets, regulations, and sometimes even potential partners. They can help arrange meetings in the chosen market with key informants and advice of upcoming trade events where more information and meetings with prospective distributors can occur. Some countries provide this service for free or at a low charge, while others add "user pay" fees depending on the amount of service or information required. These services are usually very cost-effective for smaller wineries, with little information or experience in the potential target country. There are often consulting companies that offer similar services, but charge more and advertise a more customized product. Wineries should ask for references and investigate whether these customized services are worthwhile before jumping in.

Larger companies can have a quite complex organizational structure for multinational marketing. Typically, there is an overall marketing manager or vice president who is responsible for all marketing and sales activities in all markets. When a large domestic market is involved, there can be a national sales manager, who may or may not report to the marketing manager. Many companies choose to have a matrix-type organization, with some crossover responsibilities for different aspects of the export market. For example, larger companies usually appoint specific market or regional managers, such as the European marketing manager, or the Asia-Pacific marketing manager. These people may have single country or even subcountry regional managers in place for large brands. Each of these managers is responsible for overseeing the distribution system, setting targets for different channels or even specific retail chains, providing sales promotional support for the sales team, setting prices or price ranges, managing advertising, and working with the brand managers.

The matrix comes into effect when companies also have specific brand managers responsible for their large brands. The brand manager works with the regional managers, both domestic and export, to maintain specific brand positioning in all markets. They often have responsibility for the overall packaging, advertising, and promotions to make sure all fits within the brand's positioning strategy. They must ensure reliable supply to all markets and adjudicate when production does not meet demand, or develop strategies with the individual market managers to increase sales, when production increases.

Orlando Wyndham has global marketing managers for its Jacob's Creek and Wyndham Estate brands. They also have regional managers for Europe, North America, Asia-Pacific, and Africa-Middle East. Within each of these regions there are country managers for the largest markets, like the UK, the US, Canada or for regions, like Scandinavia. There are also specific country brand managers, like the brand manager for Jacob's Creek in the UK. This brand manager would report directly to his or her country manager, but the brand manager would also be working with the Jacob's Creek global brand manager to make sure that all activities fit within the overall positioning of Jacob's Creek. For example, one of the major retailers may ask for a price reduction in Jacob's Creek and be willing to place a very large order. The proposed price might be outside the range that the local brand manager has authority to provide. The global brand manager for Jacob's Creek would be involved in the decision as to whether this low price is approved or a different bargaining strategy is proposed to the retailer. The country manager would be involved, but would not have authority to reduce prices beyond a certain set point. Their role would be to provide advice on negotiation to keep the price within the expected range and then to oversee any logistics necessary to import and warehouse an unplanned for increase in Jacob's Creek for this promotion.

At the same time they would be providing the same services to the other brand managers in the UK.

A smaller company must make similar decisions, but with fewer brands and fewer outlets these decisions might be handled through discussions between the overall marketing manager or business manager in the home country and his or her agent in the export market. The same planning must occur, but with usually one brand it is not necessary to have a complex management structure.

The next steps in organizing the company for export involve a set of activities summarized below in an "export checklist." It is important that individuals within the company or those acting as part of the distribution channel are aware of their roles and responsibilities. They should have definite timelines established for completion of their tasks, so that all necessary activities are performed in the right sequence. It is no use having a label ready and printed if the wine is not ready for bottling, or having a distributor actively seeking placements when the wine has not been approved legally for import into the country.

EXPORT READINESS CHECKLIST

Wineries should have a process for export market operations, almost a checklist of activities to ensure that nothing is missed or can go awry. The export market for wine is highly regulated by virtue of wine being an alcohol and therefore subject to alcohol-based regulation as well as that of food safety and product liability. The salability of wine has always been in question as wine is a perishable, fragile, low-value, high-transportation-risk product.

To enter the export market wineries should have detailed knowledge of the following factors.

1. *The external capabilities of your winery*: the quantities of wine and the styles of wine you have available for

sale and whether they match the expectations of the market segment you intend to export to. Be thorough and assess these not just for a single shipment but for a strategy of long-term development.

2. What are *cash resources and credit line facilities*? In direct export payment terms tend to be long, 120 days plus. Market development costs are high to negotiate contractual arrangements and establish a logistics network to ensure you can get your product to market in a saleable form.

3. Ensure you are acquainted with *tariff and nontariff barriers* and their operation and requirements. Mutual acceptance of enological practice may be becoming common, but be very aware of exemptions of various processes or additives by various markets. Your winemaker association should keep a register of these market requirements that you as the exporter need to be conversant with.

4. Labeling is used as a nontariff barrier. *Label approval* is often mandatory. Ensure the claims you make are acceptable and that the label layout complies with the mandatory requirements of the importing country. Ensure you do not breach copyright and intellectual property law, and brand law. Some markets require health warnings, standard drinks statements, ingredient details, winemaking process information, and traceability coding as the basis for possible product withdrawal.

5. Understand the *styles of wine* that are selling in your proposed market. This will require some market research as most markets and cultures have differing taste expectations. Taste is the key determinant to product choice in all markets.

6. Have a system to help you systematically *choose your export market of choice* and whether you are able to both compete in and service that market. Remember that export marketing is much more difficult and com-

petitive than just selling in your home market. If you are unsuccessful in selling wine in New York or Houston, then perhaps you should be not be considering export anyway.

Market Choice Considerations

If you are ready to proceed, consider level of consumption for market choice. Be very careful of using just per capita consumption. It is deceptive as it includes the total population not just the alcohol-consuming population. You need access to consumption studies that specifically tell you:

- who is drinking wine (gender, age, and income profiles);
- when do they drink it (as an aperitif or as an accessory to food);
- where is the main location of consumption (home or outside the home);
- frequency of consumption;
- the occasion of consumption for less frequent consumers;
- where do they buy their wine (supermarket or specialist outlet);
- what purchase cues do they use and their relative importance (country of origin, variety, region, brand name);
- price paid based on usage occasion (everyday vs. special occasion pricing).

This information should form the basis of both the strategy and positioning you take in the export market, and how you design your marketing mix to be a part of the consumers' repertoire of wine brands.

Distribution Systems

The key issue is that you are aware of the permitted levels of marketing and market access available to exporters. The choice of agent/distributor is critical in this context. The company should be accredited and respected to be able to deal with the resellers in the market. The greatest challenge for any exporter is to gain a compatible match of agent/distributor. All arrangements with resellers or agent distributors should be by *contract to supply arrangements*, be performance based, and have *termination and dispute clauses* and their jurisdiction nominated. Distribution systems can be classified as follows:

1. **Regulated open market** (e.g., UK). The characteristics of these markets are those of FMGC marketing, direct negotiation with the retailer, freedom to market directly to the consumer, multiple outlet selling and availability, societal control on consumption and abuse (drink-drive, underage drinking, curtailment of advertising). Resellers are concerned with margin and stock turn.
2. **Central/state government controlled** (Canada and Scandinavia). Controlled distribution, curtailed availability, central purchasing and on-selling to resellers, limited direct sales to the consumer (Canada only), products need to win a "listing to be sold," advertising and merchandising regulated in government-owned stores, price promotions common, product trial and tastings on-premise only.
3. Mixed systems of the above.

Advertising and Promotion Opportunities

Advertising and promotion opportunities should be made available to you from your winemakers' associations, such as:

- which wine shows and expos;
- the role of the wine press;
- the ability to advertise and the media available to capture the target market attention;
- tastings and tasting promotional opportunities.

Competitor Analysis

Portions of your competitor analysis should also be made available to you from your winemakers' associations. Questions to consider are as follows:

- Is there a domestic industry and how strong is it and what defensive strategies are being adopted to limit imports?
- How parochial are the consumers and are they prepared to switch preferences?
- How many other wineries from your country are in the market and at what is the status of these wines in meeting consumer category requirements (expressed as share of category)?
- What is the market positioning and availability of wines from your country across the "usage" spectrum?
- What are competitor and local marketing expenditures on promotion and brand building?
- What are relative cost efficiencies and margin generation for resellers?

Pricing Levels

The price levels operating in the market are an essential analysis. The cost escalation of export is a significant factor in determining where wineries compete. You need to go into export to make a profit in the longer term, but you also need to price your products to recover your additional cost of exporting. These costs will include additional packaging costs, compliance certification, label modifications and special print runs, reinforced shippers, containerization, shipping, insurance, duties, and/or excise.

The first consideration is the general price range in the export market, as the greater the price range in each category the better the opportunity for cost recovery. For example, the price range of £3.99 to £9.99 for commercial premium wines in the UK market allows a range of positions where profit can be generated, whereas in the ultra-premium price range of £9.99 to £15.99 there is price congestion resulting in brand limitation at reseller level. From a winemaking point of view, the winemaker needs to aim at *overdelivery of intrinsic quality* to justify the imported wine cost-derived price premium. Failure to deliver the intrinsic value proposition will lead to market failure or the need for loss-making price discounting in order to drive sales.

BRANDING FOR EXPORT

The simplest thing, of course, is to maintain existing positioning and branding activities in all markets. This makes the job of the marketing or brand manager much easier. True global marketing companies strive to maintain a single brand image in all their markets. This is easier said than done. A pair of Levi jeans, for example, has different positioning in its home US market than it does in many international markets, where it is seen as more exclusive and at a higher relative price position than at home. This is not a bad strategy, and some wine companies specifically pursue higher prices and positions in export markets than in their home markets. Marketers should be aware, however, that as price rises sales generally decrease, so a large selling brand in the home market may not achieve equivalent export sales if it is positioned at too high a price point. This goes back to the initial reasons for export, and the overall brand positioning must fit within these parameters. The initial branding decision, therefore, is whether to maintain the existing brand position, allowing for country differences, or develop a new brand or brand extension for the export market(s).

Most companies decide it is easier and better to maintain their existing brand and make adjustments for different export markets. We will discuss this branding strategy first, and then briefly discuss developing a new brand for export, because many of the same processes are used, but larger changes are made (Table 12.4).

Product or Wine Style

The first decision is whether the wine style itself is acceptable in the target market. This can be done by taking samples and tasting them with knowledgeable wine channel members in the target market: importers, retailers, or restaurateurs, who sell wines of the same type and price point. Sometimes it is necessary or advisable to change the blends or style for a specific market, especially if the market has large enough potential to warrant separate blends. Many large multimarket Australian wines, for example, have slightly fruiter blends for the UK market and slightly more oak in their US wines. Smaller wineries do not make any adjustments in wine styles for export markets. The cost in tank space, materials, and inventory does not warrant this investment. Sometimes, wineries will remove wine earlier from barrel or leave some in barrel longer as a reserve wine for a specific market, because these activities are not as costly as having separate blends for each market. Obviously making any changes to wine styles must be driven by strong knowledge of the tastes of the target market.

Packaging

Certainly the packaging is often changed for different markets. There are often specific label requirements for information to appear in certain font sizes and positions (front or back label), which are different from the home market.

Table 12.4. Major Considerations When Branding for Export

Style of wine
Packaging
Product line considerations
Pricing
Communication and promotions

Care should be taken in making too extensive or unnecessary changes. If possible the front label should be the same for all markets. This may mean changing the domestic label slightly, but the overall cost of having only one front label may be much less than having different ones for each market, which can not only affect label costs, but bottling and inventory costs as well. Back labels usually must be changed for export markets. Again, the fewer different labels the better, because of the reduced inventory costs and increased flexibility in allocating or reallocating wines to different markets. Sometimes a split back label is most cost-effective, with a generic top back label and a different label with required information below. The rest of the package should be as much the same as possible for the above-mentioned reasons.

There are no specific recipes for a good or attractive label. Label designers often speak of designing above the price point, so that a $10 bottle of wine has a $15 to $20 look to it. Wine producers are advised to use well-recommended label designers with experience in wine label design. When considering a design that will be used for domestic and export, it is important to maintain simplicity. Cultural differences between countries can make very complex labels unattractive in some places. Even brand or company names (if it is the brand) should be considered in the light of multiple languages. What may be a well-known landmark or place name domestically can be nearly impossible to remember or pronounce in another country. Most successful brand names are simple one or two relatively short words. The label design and the brand name should clearly link to the brand position. For example, the Banrock Station label proclaims, "Good earth, Good wine," clearly communicating its sound environmental position along with the focus on the value of the product. Initially the label was made in a recycled style brown. Now after 12 years on the market, the label has been brightened, because the positioning is more towards the wine quality, but still a drawing of the river and trees is visible to connect the brand's position. As recommended above, it is very useful and inexpensive to test label concepts with key distributors, retailers, and restaurateurs in the export market before launching.

Product Line Considerations

One of the key questions in branding for export is the size and composition of the product line: How many different varieties? Different price points? Package sizes? These decisions mainly focus on the company's strategy for the overall size of the brand and where it is to be sold. The decision will be quite different for a brand aiming for 1 million cases or more than for one attempting to sell 10,000 cases. Too often wineries with relatively small brands develop and try to sell far too many product variants, which not only confuse the market, but this adds unnecessary expenses in production, packaging, inventory, and selling costs.

We will look at each of these decisions in turn. First, we discuss the range of wines by style or variety. Wineries should consider that knowledge of their country or region will be much less in export markets than in the home market. Consumers and even trade buyers may have only general associations of varieties and regions: Australian Shiraz, California Cabernet or Chardonnay, New Zealand Sauvignon Blanc. This makes the task of marketing a broad range of grape varieties more difficult, especially for the smaller producers. At the higher price points, buyers will have some knowledge of region and variety: Barossa Shiraz, Napa Cabernet, Marlborough Sauvignon Blanc. This actually makes it more difficult to sell other varieties from the same region, such as Cabernet from Barossa or Pinot Noir from Marlborough. The best strategy is usually to go into these higher price point markets with one or maybe two varietals and establish the brand name, especially among the trade and high involvement buyers. Then the line can be expanded as awareness of the quality of the brand grows. The winery may use different well-known regions for its varietals, but the key link here is to the variety–regional–quality association.

We do know from market research that consumers choose more by grape variety than by country or region of origin in the higher volume outlets. So, at lower price points, in supermarkets and discount outlets, selling by variety with a brand identification works well. Here the strategy is to gain shelf space and grow brand awareness. Successful wineries often go into the market with two to four varietals simultaneously to maximize shelf space and reduce overall logistics costs. Both the trade and the winery benefit from this strategy of multiple varieties of the same brand. As brand sales and awareness grow, more varietals can be added to the mix. Jacob's Creek is a good example of a wine that originally entered the UK market with its Shiraz Cabernet and a Chardonnay, while in its home Australian market it had a wider line of offerings. Now the line consists of a range of blends, such as Shiraz Cabernet, Cabernet Merlot, Shiraz Grenache, Chardonnay Sémillon, and a range of single varietals, such as Shiraz, Cabernet, Merlot, Chardonnay, Riesling, and the newly released and successful sparkling Pinot and Chardonnay. All of these wines are branded Jacob's Creek and sell within a relatively close price range, with the blends slightly cheaper than the single varietals. Growing the range like this allows the consumer more choice within the same brand and at the same time acts as a very positive promotion with the large amount of shelf space.

There is some debate as to whether a different brand name should be used in export markets than in the domestic market. We believe that there is more leverage from maintaining the single brand across markets than from trying to start a new one. Sometimes there are issues that make this strategy impossible and a new brand has to be developed for the export market. The local name may be prohibited in the export market or too difficult to spell or pronounce. In this case, the new brand should still make a connection to the home market through its name and positioning. Wine remains one of the few multinational products where the country and region of origin are highly important.

Pricing

Another decision is whether or not to extend the brand to different price points or to launch new brands in those price points. Larger wine companies usually have different brands for different price points. This is especially true for high-volume brands selling at the lower, $5 to $10 US price point. It is often difficult to convince the trade and consumer that the same brand that sells for $7 has a higher quality wine for $15. There are some exceptions to this rule, but generally it is easier to create a new brand at a higher price point than to try and extend the range upwards. This is due to two reasons. First, consumers wonder what the difference is if they taste the high priced wine and it is not to their liking (usually it will have more concentration and oak treatment). Second, the segment that buys wines at $7 is not the same segment that buys at $15, so the promotion of the higher priced wine to the existing consumers is not very effective. You have to promote to the higher price buying consumers anyway, so a new brand is not much more costly to launch.

One of the major issues in deciding price points for a line with different quality and price levels is the distance between the suggested prices. If the distance is too small, then when the upper line goes on promotion, it actually intrudes upon the price point of the lower line. This can quickly denigrate the consumer perception of the lower priced wines. Managers should keep in mind the ongoing retailer strategy of lowering price points for wine, both on and off promotion. A wine that starts with a list price, for example, at $15.99, may soon find itself selling on promotion for $13.99 or even less. Soon that becomes the new "regular" price and the wine is now within $2 to $3 of the next wine in the same line below it. When a new promotional price occurs, the wine may end up at the same price as the lower quality wine in the same line. Managers must first set suggested retail prices that allow for discounting and, second, strongly resist resetting the suggested price after long discounting periods.

This leads to the general discussion of pricing wines for export. As we state above, there usually is little difference in the overall positioning of wines for export versus the same wines in the domestic markets. However, the actual price points often do differ between markets. This is typically due to duties and taxes, which often differ substantially. These taxes and duties affect all wines equally, so the relative prices of wines are not affected. But also, wineries need to consider the extra channel members and their mark-ups, which affect the final retail price. Usually there is an importer, who assists in receiving the wine and clearing it through customs. This importer can also be a distributor, or may sell the wine to distributors for resale to retailers. We will discuss channel arrangements in the next section. Wineries should be aware that there often are extra margins in export and maintaining equivalent prices to the home market can take some planning. Sometimes wineries either have to accept a lower margin or consider moving into a slightly higher price point to maintain existing margins.

Wine companies should investigate each potential export market's price points before they make the final decision to enter that market. Depending on the market, there are various publicly available reports, often in local trade journals, that give an idea of the price points in the market. Again, wineries can easily discuss these issues with potential distributors or importers, who know the market well. It is also useful to visit the market and note the brands and regions, which you know from your home market and see where they position themselves as to price. Given the well-developed international wine market, a useful strategy is to decide whether to be above, below, or equal to an existing brand or brands in the chosen market. Of course, price has a direct relationship to the overall positioning of the wine and its potential volume of sales. New brands in the market looking for high-volume sales usually enter at a price point below that of wines with similar quality. Kumala from South Africa is a good example. The wine quality matches many of the mainstream Australian and American brands in the UK market, but it is priced below those from Jacob's Creek, Hardy, and Gallo. In only a few years, it has moved solidly into the top 10 wine brands in the UK.

Smaller wineries will not be looking to sell wine into these lower price points, but the lessons are applicable at any price point. Look for wines of similar quality and prestige and decide where you want to position relative to them. Palliser Estates, from Martinborough, New Zealand, one of the ultra-premium wines from that country, for years has used relative price as part of their marketing strategy. Their aim is to be at or very near the highest priced wine in their category, which is clearly part of their overall positioning. At the same time, wine quality, packaging, and communication must meet the same positioning objectives. As international markets become more sophisticated, there is little room left for wines that do not deliver the total package of quality, look, communication, and prestige at the appropriate price. The problem is that these developed markets have little room for new brands without pushing out another brand. Therefore, the new brand must bring something special to the market. Often this is easiest done with a slightly reduced price to existing wines. This of course reduces revenue and itself has a detrimental effect on the wine's position in the market.

Communication and Promotion

Price must be accompanied by relevant communication or promotion strategies appropriate for the market. Larger brands, selling hundreds of thousands or millions of cases, obviously have larger budgets and can afford more mass appeals. These can range from price promotions accompanied by cooperative advertising and shelf talkers to billboards to sponsorships. Very few wines have the budgets to use mass media, such as television, for promotion, although it can be argued that this medium reaches more consumers per dollar spent than any other. Mass media only works in countries that have a relatively high proportion of wine drinkers. Otherwise such broad exposure is wasted on many nondrinkers. Another issue with mass media is laws prohibiting or restricting the advertising of alcoholic beverages. France and many of the Scandinavian countries do not allow advertising for wine or other alcoholic beverages. Other restrictions exist in many countries. Wine exporters can usually get a guide to these laws from their home government's foreign trade office or the branch in the destination country.

The first rule of communication strategy follows on from pricing and positioning: it must be appropriate to the overall position of the wine. Like packaging, there are no hard and fast rules. In fact, a good communication strategy often breaks through consumer inattentiveness by being unique and creative. That said, we will first look at some of the standard practices and then move to some examples of more creative promotions, including sponsorships (Table 12.5).

Table 12.5. Promotion Strategies in Wine Export Markets

Price promotion campaigns
Tastes of wine in store
Restaurant promotions of wine
Country promotion with association support
Press releases
Sponsorships

Retail stores often demand that wineries participate in various *price promotion campaigns*, especially for different holidays. The fact is that this is when many infrequent consumers come into the market and regular consumers buy more wine. So it is important to participate. The key issue with price promotions is to work to keep the discounts reasonable and within the image range of the brand. Sometimes it is possible to provide other, value-adding offers, without dropping the price of key wines, but the opportunity differs by country. Supermarkets in the UK, for example, still use price promotions almost entirely as their means of promoting wine for various holidays and rarely give wineries other opportunities to promote their wine. In other countries, wineries can promote and add value to their offer without reducing price by bundling their wines with another product, such as buy three bottles of this wine and receive a free set of wine glasses, corkscrew, apron, or even another type of wine made by the same winery. These value-adding, nonprice promotions are better for attracting attention without reducing the perceived price point of the wine.

The ultimate, though expensive, promotion is to *give tastes of the wine*. There is nothing better than experience to introduce and sell a consumer on your wine. These types of promotions should be managed carefully due to their expense. Whether it is in a specialty wine store or restaurant, be sure that as many as possible of those tasting your wine are potential buyers. There is no sense in offering tasting of a $30 bottle of wine in a venue that caters mainly to $5 buyers.

It is very positive to *work with restaurants* in export countries, where the foods might differ from the home country, to develop promotions matching food with the wine. This may be a free glass of wine with a specific dish, or a tasting of wines with multiple dishes for a fixed price, or just a wine by the glass promotion. Back labels that extol the virtues of the wine with roast lamb, for example, do not add much value in countries that hardly consume any lamb. Food-based promotions with local cuisine can overcome this and at the same time build brand recognition with the trade and consumers.

Another excellent promotional scheme is to *work with wineries from the home region or country as a group in specific export markets*. We noted above that consumers choose wines by variety at lower price points, but region becomes much more important at higher price points. Food-based promotions by a group of wineries, either at a single or at a group of restaurants, can help build awareness and understanding of specific regional characteristics, which is essential to gaining prestige and higher price points. Individual wineries are not often able to do this on their own. The argument can be made that smaller wineries will do much better developing events for their region than they will by spending the same amount of money on their own brand in export markets. At its ultimate, these regional promotions can include wines and wineries that may be unable to send representatives, and have to rely on their local distributors, but still can add value to the overall promotion and their own brands. The Australian wineries often work in this way. Wineries usually pay differential participation prices for these promotions based on their size. Wineries can choose to participate directly or pay a fee and provide wine for their local agent to pour. The wineries that participate directly obviously gain more, but even those that cannot afford to help round out the overall promotion and help develop the awareness of the whole region.

Press releases and *sponsorships* are standard means for wineries to gain awareness and increase sales. The rules and techniques really don't differ much from the ones used in domestic markets. It is just more difficult to source information as to whom to send the press releases to and how to focus them. It often pays to hire a local agency that can manage this for you. Be sure to choose one that is well established in the wine area and has a good reputation for press releases that create activity on behalf of the brand. In essence you are buying local knowledge, about whom to send to for different types of releases and how to write those releases in the proper language for the country and culture. It is very rare for a domestic company to understand and be able to promote a winery in a different country.

Sponsorships follow much the same rules. It is important to understand the kinds of consumers you are seeking to influence, before deciding or developing any sponsorships. The people drinking your wine may differ dramatically in different countries, even if the wine is sold at relatively the same price points. Again, it often pays to work closely with your distributor, if they have the expertise, or with a good PR company that understands the wine market. Sponsorships should deliver awareness and, in the best cases, tasting or purchasing opportunities with your target market.

THE EXPORT VALUE CHAIN AND LOGISTICS

The contraction of market opportunities in the future will mean a heightening of competition between wine industry organizations. This competition will be undertaken at all levels of the value chain: country versus country, region versus region, brand versus brand, and retailer versus retailer. Realizing this inevitability, there has been a series of acquisitions, mergers, joint ventures, and strategic alliances as the industry organizes itself to improve the efficiency of value delivery, and by doing so protect and/or gain market share.

The wine product is becoming more homogeneous with a *consistent quality now a common expectation* as a key performance indicator (KPI) of wine salability at all levels of the distributive chain, so the natural variability due to climate and grape vine disease are being negated by better techniques of vineyard management and the use of quality-enacting technologies in the winemaking process. On a global basis, there is a growing homogeneity of offer as the grape varieties used are concentrated into:

- White wines: Chardonnay, Sauvignon Blanc, Chenin Blanc, Gewürztraminer, Riesling, Sémillion
- Red wines: Cabernet Sauvignon, Pinot Noir, Syrah (or Shiraz), Merlot

New varieties are being trialed, but their consumer impact is not expected until the longer term.

The result is that the way that companies compete will change from a product focus to a service focus facilitated by *database marketing, relationship management*, and *networking cooperation* between the various members of the supply chain. Before going on to discuss these we need to spend some time on the transactional marketing activity of export marketing.

Wine is a product where there is still a "high salability risk" on the part of the distributor, retailer, and consumer. The practice of claiming climatic variations is no longer accepted and a general level of quality is now expected as the norm. In addition, poor practice such as secondary fermentation, wine "in bottle" variations, unbalanced wines, oxidation in young wines, and crumbling corks are no longer tolerated at any level. The failure of the winemaker to provide saleable product results in a penalty being charged for product returns and replacements by the retailers.

The improvement of packaging practice has been significant toward providing salable products, with "clean room" conditions in bottling halls or at least for the "filling heads" now common practice. Closure has become a key issue due to widespread product failure of the commonly used cork stopper, resulting in a switch to the "Stelvin closure" (screw cap) or to new high technology cork closure substitutes on many commercial and premium wines in some markets. Their use will continue to grow.

The salability integrity of the wine product is fundamental in a FMGC market as well as a superpremium market. The distributive and consumer salability expectations are the same irrespective of the positioning. Wine is a low-value, perishable, and fragile export. The protocols for shipment for export need to be thorough to ensure salability and to maintain market value. Precautions that need to be taken to ensure passage include:

1. Product damage can result due to heat/cold/vibration/excessive hydraulic movement in the bottle. Wine containers should be stored at 20°C, and below the waterline of the container ship.
2. Vibration causes a wine bottle to rotate 10 times per 600 miles of transit so special export shippers and dividers are necessary, or the use of "sleeves" for ultra-premium bottles. Without these precautions the risk of label, bottle rotation, and capsule scuffing is inevitable.
3. In transit condensation inside the container needs to be guarded against as collateral damage to shipping cartons can be substantial, causing collapse and resulting in breakage and other damage.

4. Breakages due to rough handling can also cause collateral damage to the remainder of the shipment.
5. The container needs to be secure from risk of pilferage.

Similar conditions apply for road and rail transport and interstate shipment. *Remember the wine bottle is one of the only packaged products (with bottled water) where the product is displayed as a part of the dining or entertaining ritual* on most usage occasions, so the maintenance of a high level of package presentation is essential.

CONCLUSION: FUTURE PREDICTIONS FOR THE GLOBAL WINE MARKET

The global market for wine will come under increasing pressure as overproducing countries endeavor to overcome and dispose of excess wine beyond domestic requirements. This position will accelerate the commoditization of wine with the major supermarket retailers aiming to maximize their margins through a global sourcing network direct from producers via a contract packaging arrangement leading to the inevitable rise in BOB wine marketing.

The narrowing of the global varietal range will facilitate this, reducing both retailer buyer and consumer purchaser risk. This trade already exists at the distributor level of the market, and the retailers will enter the market directly using brokers and packers to source and supply on an ongoing basis. This entry will add to the price pressures on producer margins with the gravitation of retail pricing increasingly downwards, pulling brands into the commercial premium sector of the market and where only brands with a significant critical mass of say, in excess of a million cases, will survive.

The lack of difference and price-justified quality differential in the super/ultra categories will lead to a widespread contraction of winery numbers, via acquisitions, mergers, or just plain "going out of business" with only those that have developed a differentiated marketing plan offering niche attribute characteristics, such as unique variety or grape source (e.g., Torbreck in the Barossa Valley, which markets a unique Shiraz sourced exclusively from 100-year-old vines) or unique terroir (e.g., Priorato in Spain with its interpretation of Grenache, the Marlborough region in New Zealand with its unique characteristics of Sauvignon Blanc), or where the wine industry is interlocked into a associated industry such as the tourism-focused Niagara Peninsula, Upper New York State and increasingly parts of the Napa Valley.

The luxury segment will continue to prosper, driven by perceived scarcity, prestige, and status consumption (e.g., champagne, cognac and the Grande Cru of Bordeaux and Burgundy). As fortified wine consumption contracts they will be eventually joined by both ports and sherry, their commercial level consumption having long been eroded by the long-term trend towards table wine consumption.

Commoditization and global sourcing of wine will eventually lead to a general lowering of nontariff barriers as contract production and packing become the norm, with a move to favor lesser regulated and thus lower cost (subsidies aside) producer sources such as Eastern Europe, South America, and South Africa. Wineries need to develop marketing strategies within the constraints as noted above. However, the potential to grow international brands at the commercial and popular premium level as well as at the ultra-premium and luxury levels still exist for those willing to invest in the long term.

LEARNING OBJECTIVES:
- Define NPV
- Describe financial decisions in vineyard and winery projects
- Explain how differences in real estate pricing impacts wine economics
- Describe the pros and cons of equity verses debt financing
- Identify merger and acquisition trends
- List major publicly traded wine companies
- Identify the two ways wine finance is different from other industries

CHAPTER 13

Financial Aspects of Wine

Robert Eyler
Professor of Economics, Sonoma State University

Tony Correia
Accredited Rural Appraiser, Correia-Xavier, Inc.

Financing a winery, or projects like custom crush, a vineyard purchase and development, or importation, all have one thing in common: the investor should only seek to engage in the project if expected to be profitable. Predicting the outcome of any project correctly uses a lot of diverse information.

This chapter provides simple ways and models for wineries to forecast the profitability of projects. One important point to start: the wine business is not financially different than other businesses. In a winery, there is a production facility, which chooses inputs and processes to optimize profit. Once the product is made, the wholesale price is determined somewhat by the market and somewhat by the firm; however, the bottle price dictates much to the winery budget.

This chapter has three sections. First, a brief refresher on finance basics provides a foundation for this chapter's ideas and conclusions. Next, applications of these financial fundamentals show examples for different parts of a wine project. Finally, some ancillary topics on financing choice and other issues round out our look at finance in the wine industry.

Financial Foundations

Finance is applied economics, related to making decisions concerning the purchase of both real and financial assets. Real assets, such as grapes, machines, labor, etc., dominate production and day-to-day business at most wineries, while financial assets, such as trademarks, loans, ownership, trusts, etc., drive financing real asset purchases. For the most part, we focus on real assets here; we must remember that financial assets provide the funding for these purchases, regardless of their form. Wineries and vineyards are affected by housing, land, and general financial markets fluctuations. We view the winery and vineyard owners as trying to maximize profit with respect to constraints they face. The decisions at each step of the winemaking process, from rootstock to bottles on store shelves, must be made with a consistent focus. We use a net present value idea throughout this chapter to analyze these choices.

Net Present Value

Net present value (NPV) is a financial concept, used generally for decisions concerning plant and equipment acquisition, or gaining ownership in a firm through venture capital. The same rules apply in the case of financial assets, such as goodwill when buying a brand or label from another entity. The value today of future profits, less the initial cost to acquire the asset, is the NPV in terms of future cash flows. A simple formula for this is the following:

$$\text{NPV} = \sum_{i=0}^{n} \frac{\text{ECF}_i}{(1+R)^i} - I_0$$

where ECF is expected cash flows (cash net revenues from the project), R is interest rate or rate of discount, I_0 is initial cash outlay for purchase of capital for time period $i=1,\ldots n$, where period n is the last period of an asset's usable life or when the asset is fully depreciated.

This formula acts like a formula for profit. If NPV>0, expected profits are greater than zero, and the investment should be made. If NPV<0, expected profits are negative, and the investment should not be made. If NPV=0, flip a coin. (The interest rate at which a project experiences a zero NPV is known as the internal rate of return or IRR. This rate should be compared to other rates to determine if the project is viable at all.) For similar investment choices, the investment with the highest NPV is chosen. If no choices have an NPV>0, a new array of choices should be sought, including additional search costs in the new figures, or the project should be abandoned.

Financial Decisions in Vineyard Projects

Wine is an agricultural good and must be treated as such fundamentally. However, it has large differences that separate it from most other agricultural products. The good's price on an open market for many firms is found by subjective judgments concerning wine quality rather than scientific means. There is a 3 to 5-year wait for the first crop after planting. It is also an alcohol-based product, which subjects it to different, additional commerce laws than many other agricultural goods. From rootstock choice to trellising to trucking the harvest out of the vineyard, the choices made in all areas of the vineyard dictate how much the grapes are worth. For a vineyard project, the owner or entity leasing the property should assemble all the costs, estimate the potential revenues (starting in the first year of viable yield), and make a spreadsheet. The costs are by far the easiest of the cash flows to assemble and conceive in the initial stages. The revenue and the rate of interest to use in discounting future cash flows in the vineyard are more difficult.

Example 1: Vineyard Development

Anyone who has started a vineyard, or budded new rootstock onto old vines, knows there are a number of decisions in a vineyard project. Costs include stakes, trellising, irrigation, labor, equipment or equipment rentals, fencing, permits, etc. There is also an implicit cost of digging up land that was used or could be used for other purposes. This *opportunity cost* is not a part of many expense tallies for vineyard development, but even if the land was not being used, it has an alternative use! Once these explicit and implicit costs are assembled, the owner can assess what is a realistic return on this investment by estimating revenue per ton, assuming a certain output per acre planted. Table 13.1 shows a brief summary of costs in the vineyard segment.

The price per ton or revenue is the most volatile data in the project. ECF represents the expected cash flows for

Table 13.1. Vineyard Cost Categories

Rootstock choice	Vineyard management fees
Budding	Frost protection
Trellis	Labor
Stakes	Harvesting
Irrigation	Trucking
Pest control	Permits
Weeding	Ancillary legal costs
Pruning	Vineyard acquisition costs
Fencing	Contingency fund
Leafing	Insurance
Taxes	Debt service and other land costs

each subsequent period (starting with the current period to its endpoint, whether that is a week, month, or year). It is prudent to run different scenarios under different prices, assuming you can identify most of your costs and then run sensitivity analyses on a spreadsheet by changing the price per ton as revenue. Figure 13.1 shows the grape prices in certain California Districts.

Once you have this set up, the present value analysis described above is easily performed, allowing a decision between varietal, irrigation type, trellising, etc. These decisions are typically driven by the physical characteristics of the site and the winemakers' preference. Another choice, if you want to produce and sell wine, is to buy grapes or juice on the open market, but the financial assessment of that choice is basically the same. The basic equation below summarizes the calculation of vineyard expected cash flows (including trucking, unloading, and any costs associated with the grapes per the contract; bulk juice is available directly from large growers or through B-to-B websites, such that the search cost should be included):

$$\text{Vineyard Revenue} - (\text{Production Costs} + \text{Acquisition}) = \text{Vineyard ECF}$$

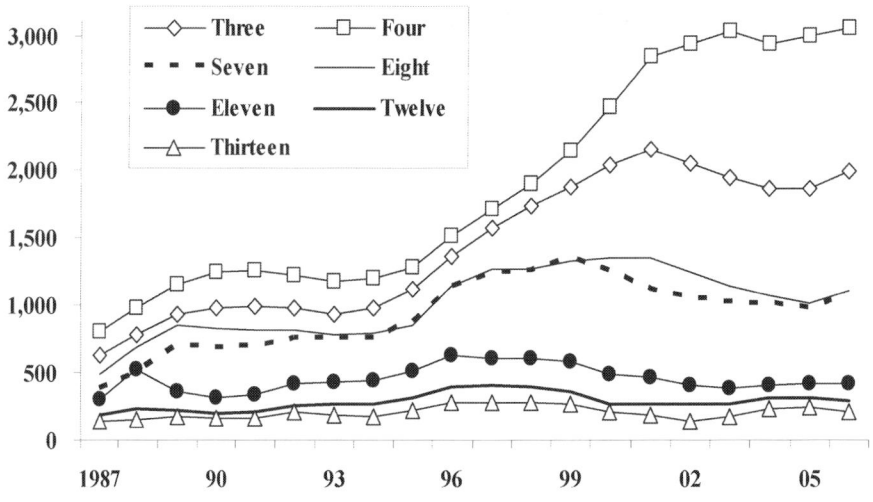

Figure 13.1. Average prices, grapes/ton, selected California districts. The locations that represent the districts shown are in the Grape Crush Report from www.nass.gov. 3: Sonoma and Marin Counties; 4: Napa County; 7: Monterey and San Benito Counties; 8: San Luis Obispo, Santa Barbara, and Ventura Counties; 11: San Joaquin County north of State Highway 4, and Sacramento County south of U.S. 50 and east of Interstate 5; 12: San Joaquin County south of State Highway 4, Stanislaus and Merced Counties; 13: Madera, Fresno, Alpine, Mono, Inyo Counties, and Kings and Tulare Counties north of Nevada Avenue (Avenue 192).

Example 2: Grape Sourcing

If you find that vineyard development is not cost-effective, an alternative is to buy bulk juice or fruit and use it instead. However, the fundamental decision involved in making grape source decisions is similar to making vineyard development decisions financially. Most wineries are aware of bulk juice sources and grapes for specific varietals. Of course, there is a qualitative difference between buying bulk juice and contracting the grapes on the vine, but financially one must categorize costs and revenues specific to the grape purchase. The revenues from the grapes or bulk juice are not as manageable or as blatant. To a certain extent, your choice of bulk juice, grape, vineyard dictates the price of your bottle. This interdependency exists for most producers; it does not hold for all because of long-term contracts that change very little over time and also niche wines, whose price and market are somewhat stable.

The differences between these revenues and costs are, by definition, your expected cash flows. If you buy grapes before you see them (e.g., a winery contract with a new vineyard development where the first substantial crop is 3 years away), you are ostensibly engaging in a commodity futures contract, and are betting on the contract price versus the spot price of grapes in 3 years. There are costs and revenues associated with all these choices and you must find the choice that maximizes your profit, given the cost and market constraints your firm faces. To summarize, a basic model is:

$$\text{Grape or Juice Revenue} - (\text{Production Costs} + \text{Acquisition}) = \text{Grape ECF}$$

The pricing structure from winery to wholesale or sales force is normally freight-on-board (FOB), or cost-plus pricing. This price is determined to deliver a certain retail price that lands you (hopefully at a profit) in your targeted price point. Obviously, buying the lowest priced fruit may not necessarily provide the largest spread between revenue and cost, as the type of grapes you bottle affects the perceived quality of your wine. The true mathematical connection between the grape price and the bottle price may be a basic linear relationship, such as if grapes cost $2,000 per ton, the wine made from them should be $20 per bottle at retail. That relationship is ad hoc to point of fright. The grape cost is part of many input costs in the process that determine the price with production costs than does an ad hoc formula.

Once these expected cash flows are assembled, their present value can be assessed based on the timing of the revenue and costs. If you incur cost immediately, which is the most likely scenario in vineyard development, the costs are discounted differently than the revenues in that they begin in a different period. The message here is to set up your costs according to their timing. The viability of the present value calculation depends on how you set up these expected cash flows. Notice the use of the word "expected" continuously in this section. Unexpected costs can come at any time.

Two reasons exist for specific vineyard revenue uncertainty, especially if the vineyard is not established. First, the yield is unknown; it is unlikely a wine producer will contract to purchase your grapes unless regional or appellation characteristics dictate prices directly for certain varietals. If you have planted Russian River Pinot Noir in Sonoma County, it is likely that wineries will want to get your grapes under contract due to the relative lack of supply, as compared to other varietals. If you plant Cabernet Sauvignon in Lodi, you may have difficulty initially contracting the sale of your grapes. Second, until you get your first crop, no one knows the grape's viability or how well your claims of quality will hold up. Purchasing grapes under contract quickly is a key issue in budgeting. Varietal choice, location, vineyard history, winemaker scrutiny, portfolio fit, and other details lead to changes in revenue. Grape price generally cycle, as new plantings come online as overproduction of certain varietals takes place during industry booms and then depress prices as supply rises faster than demand. Growers must think about the revenue derived from the grapes once they are salable. If an existing vineyard is purchased, the second problem is not as large an issue.

New plantings also face timing problems. Unless contracted well in advance, the revenue to be derived from new plantings can be highly variable. It may depend on new market conditions, climatic changes, consumer perceptions shifting away from the vineyard's varietal, etc. Because of the lengthy time between a planting and subsequent viable crop, vineyard owners must take the present value of money into account, and also must remember that without a contract price, most revenue measurements are simply conjecture.

Real Estate Markets and Winery Investments

A major component of most winery investments is the underlying real estate, the land, buildings, and vineyards. Wine-related real estate has historically proven to be an attractive, stable, long-term investment, but it typically does not generate the rates of return sought for most winery operations. Also, the long period of time required for development of a winery and/or vineyard demand the devotion of the underlying land for many years with no return. With land in the highest quality areas commanding in excess of $100,000 per acre, the opportunity cost of that land, dedicated to a vineyard/winery development for several years, can become a significant additional cost. Conversely, however, the real estate can typically be financed (or leveraged, or geared) at attractive long-term rates, amplifying the return to equity.

The question of whether a winery should actually own the production vineyards has long been debated, with many examples of successful wineries who own no, or few, vineyards, choosing to outsource their grape supplies. While vineyard land may be leased, or contracted, to allow the winemaker adequate control of the wine-growing process, many industry participants feel vineyards must be owned to allow for proper long-term decision making. In recent years we have seen the emergence of alternatives to traditional ownership, with entities such as REITs buying vineyards, land, or even wineries, and then leasing those assets back to the wine company. If accomplished properly, these can be attractive alternatives for all parties.

In the wine business, Location, and Geographical Indexes (GI) may be the most critical factor impacting real estate values. Location in a defined Appellation (AOC) in France, or American Viticultural Area (AVA) in the US, may be an investor/winemaker's primary consideration in selecting a site or in considering the purchase of an existing winery or vineyard. The rights, or entitlement, to develop vineyards in such regions are tightly regulated. Hence, entitled land in the upper tiers of the recognized GIs will typically command high unit prices; however, such property will also tend to hold its value more consistently over the long term. Rates of return, then, tend to reflect the risk/reward factors of such investments.

Land and vineyards in the highest tiers of quality tend to be tightly held, with few opportunities for such lands to be acquired. The recent trend of consolidation in the wine business has amplified this trend, as the larger wine companies continue to acquire smaller entities and their properties, most often as a result of another dominant trend in the wine business, generational transition. As older family operations begin to address transition of the family business as the older generation looks toward retirement, the younger generations often lack the desire, financial fortitude, and/or skills to carry on the family business, and the business, and the properties, find their way into the hands of new owners.

External forces also play a key role in the value of most wine-related real estate, as land that is suitable for premium wine production is also likely to be attractive for other uses, most often residential estates. The value, or price, of land, then, is created by a complex matrix of location, climate, soils, water, and other legal and economic factors. In terms of pure production potential, the usual agricultural factors come into play, with climate dominating the matrix in most cases. Vignerons will seek very specific climate characteristics to plant specific varieties. Finding the proper climatic conditions, the next critical element is soil, as well as the various physical characteristics of the land, slope, drainage, aspect, etc. Quality and quantity appear to be inversely related in grape production, with the highest quality grapes coming from areas of the lowest yields and the highest yields commonly producing the lowest quality of grapes.

Water is perhaps the most interesting component of the equation, as vignerons seem always able to find water for sites meeting their climatic and soil criteria. Premium grapes are successfully grown without irrigation in many wine regions, albeit often at lower yields. Conversely, of course, in the commercial production regions, irrigation is a requisite component of the higher yields demanded. In many areas of the world, water may be the most critical factor limiting successful grape production, as our friends in Australia are painfully aware, as the extended 2006–2008 drought has driven their crop production lower and lower. California faces significant challenges in this regard, as the demand for water in this state significantly outpaces the available supply.

Given all of the above, we see that real estate is a necessary component of the winery business, one that can be a significant increment of total investment, but also one that may be highly leveraged to reduce initial capital requirements. However, even when leveraged, returns on real estate prove to be lower than those desired of the total winery business, and real estate carries its own elements of risk and reward.

Winery Economics and Finance

Principles similar to those above apply whether you are a vineyard manager, cellar master, or the sole proprietor/winemaker/tractor driver/janitor for your own label. Suppose you were going to buy new barrels to rid yourself of older barrels or to expand winery capacity. Would you not use the same principles as those used to calculate the viability of larger winery projects? Of course. Thinking of the winery as small pieces amalgamated into one, like any other business, provides insight into how each part of your wine business is related to the others. This is known as segment analysis, and should be employed any time you have joint processes taking place (which is classic in a winery), when projects have dual uses or feed into multiple products.

Segment Analysis

It is essential that the capital budgeting team for a winery subdivide the business into different parts and, if necessary, arbitrarily assign costs to different pieces of the business if not explicitly separated. There are two reasons why this should be done, under the auspices of generally accepted accounting principles (GAAP). First, it provides a more realistic view of your project's impact on the winery. Overstating or understating the cost distribution means a potential mistake in the net present value calculation for the project. This is true even if the project affects the entire winery, like the construction and use of a tasting facility. (A tasting facility encapsulates many ways one project can affect many different aspects of the winery without disturbing vineyard operations and having little effect on the cellar operations outside of inventory flows.) Most wineries have three major processes: *vineyard operations*, *winery operations*, and *marketing/public relations*, which include the tasting facility, sales, and distribution. Any winery's capital budget should begin with this trichotomy. From this, each operation should be split into different parts, depending on the company's organization. The organizational chart should follow this three-way split, but if there are 10 vineyards, there should be 10 segments identified under a header of "Vineyard Operations." If there are two wineries, there are two winery segments. There will be multiple vineyards and brands involved in the least as a diversification strategy, especially if there are multiple varietals. However, it is likely all the grapes flow through the same winery, as a quality and cost control measure, if not for nostalgia and marketing reasons. There is likely only one tasting facility as well.

Next is the issue of pricing both inside of the winery and outside to the consumer. While this is a marketing function, price determination does affect forecasting. Many wineries have a *transfer price*, or the price from winery to sales and marketing (including a tasting facility), which is bottle cost plus a markup. The FOB price is what the wine sells for at wholesale when the administration and marketing markup are added to the transfer price. Many wineries price in their tasting room at FOB times 2; this pricing scheme makes limited though practical sense for most wineries, so long as the price is competitive in their price point, as it is easy and everyone understands the pricing once the costs are announced.

However, is this retail price—FOB times 2—the optimal price? For most wineries it is not a price that maximizes profit. An analogy is throwing darts. Making arbitrary price choices are like aiming for the bull's eye: you might hit it by chance, but there are techniques that hit the bull's eye more often than not. To hit the bull's eye of maximum profit with price is tricky and takes some deeper analysis, but there are wineries that could do this analysis very easily by using historic revenue, cost, and quantity data for each brand or product that is priced separately. Statistically, the winery could get closer to economic optimization in price rather than just a guess, as $0.10 more or less on price could be the difference between no more profit and thousands of dollars of residual. This underscores the need to split the firm up and look at the winery or project as a series of smaller segments.

The best way to split capital budgets at wineries is into brands and then varietals for these reasons. The cost of the winery and sales/tasting facility can then be easily assigned to each brand by percentage sales or grape cost, depending on any large differences between brands. Any project that the firm seeks can now have its revenues and costs easily allocated to the appropriate parts of the firm and its impact analyzed accordingly. Table 13.2 shows some of the basic costs in a winery, assuming that the capital exists initially and the winery is a going concern or project or to be acquired. All other decisions are a subset of that larger acquisition.

For wineries, the issue is not budgeting as much as finding the source of funding. The next section is on funding options and finance, and linking the ideas above with the financial markets themselves.

Table 13.2. Winery Cost Categories

Raw grape juice or fruit	Winemaker
Nonadministrative labor	Cellarmaster
Maintenance	Waste disposal
Bottles	Breakage
Labels	Inventory costs
Corks	Brix testing
Electricity, water, sewage, and other utilities	Transportation
Cooling systems	Oak chips (optional)
Barrels	Boxes

Equity Versus Debt Financing, Mergers and Acquisitions

One of the key questions in finance over the last 50 years is determining the optimal source and mix of funding for projects (Table 13.3). Should a firm only use debt or loans, or should it go public, as just a few wineries and wine-related firms have done over the years? Should the firm use accumulated profits, or retained earnings? Does it matter? We begin with this final question because it motivates the reason why taking bank loans has been the majority choice in this industry.

The hours brooding over the choice of financing a winery project may be a waste of time. There is a very famous theorem in finance called the Modigliani-Miller Theorem that states the choice of financing mix is immaterial, if the cost of using each is the same. If the costs of each type of financing are equal, the theorem is intuitive.

However, it is unlikely that the choices will have the same exact cost, if certain financing choices are available at all. *Debt financing* seems to be constantly available to all firms, especially in the wine industry, where the market for ownership expansion or use of retained earnings is not as fruitful. However, we see later that a major concern of the industry may be finding enough financing through debt instruments to pay for large projects.

The ability to use *equity financing* is small in this industry outside of *mergers and acquisitions* (M&A). Table 13.4 provides a list of property acquisitions in California.

Table 13.3. Equity Versus Debt Financing

Equity financing	Money acquired from the small business owners themselves or from other investors.
Debt financing	Money that is borrowed to run the business. Many wineries use debt financing.

Table 13.4. Selected California Winery and Vineyard Property Acquisitions, 2004–2007

Property	County	AVA	Sale Date	Approx. Sale Price
Rabbit Ridge	Sonoma	Russian River	Sep-07	$4,000,000
EOS	SLO		Aug-07	$35,000,000
Firestone Winery	Santa Barbara	Santa Ynez	Aug-07	Undisclosed
Herrerias	Sonoma	Sonoma Coast	Mar-07	$2,000,000
Mastantuono	SLO	Templeton	Feb-07	$4,200,000
Lake Sonoma	Sonoma	Dry Creek	Dec-06	$7,000,000
Roche	Sonoma	Carneros	Dec-06	$7,000,000
Roshambo	Sonoma	Russian River	Nov-06	$9,000,000
Perry Creek	El Dorado	Somerset	Sep-06	$3,000,000
Ventana	Monterey	Greenfield	Sep-06	$14,000,000
Everett Ridge	Sonoma	Dry Creek	Aug-06	$6,000,000
Banderia	Sonoma	Cloverdale	Jun-06	$1,500,000
Green Valley	San Luis Obispo	Paso Robles	Jun-06	$1,500,000
Stimson Lane	Mendocino	Hopland	Mar-06	$6,000,000
Chateau Souverain	Sonoma	Alexander Valley	Mar-06	$30,000,000
Jekel	Monterey	Arroyo Secco	Jan-06	$7,500,000
Stevenot	Calaveras	Murphys	Jan-06	$7,500,000
Andrew Murray	Santa Barbara	Los Olivos	Jan-06	$10,500,000
Arrowood	Sonoma	Sonoma Valley	Jan-06	$10,500,000
Corbett Canyon	SLO	Edna Valley	Dec-05	$8,000,000
H.M.R.	SLO	Paso Robles	Sep-05	$2,000,000
Pietra Santa	San Benito	Hollister	Aug-05	$16,000,000
Mazzocco	Sonoma	Dry Creek	Jun-05	$5,000,000
Gan Eden	Sonoma	Green Valley	Jun-05	$4,000,000
Sycamore Creek	Santa Clara	Morgan Hill	Jun-05	$2,000,000
Viansa	Sonoma	Carneros	Jun-05	$11,500,000
deLorimier	Sonoma	Alex Valley	May-05	$6,500,000
Hartman Lane	Sonoma	Russian River	May-05	$4,000,000
Arrowood	Sonoma	Sonoma Valley	Mar-05	$11,000,000
Byron Winery	Santa Barbara	Santa Maria	Mar-05	$20,000,000
Firestone Winery	San Luis Obispo	City of Paso R.	Jan-05	$1,000,000
Old Byron	Santa Barbara	Santa Maria	Dec-04	$2,500,000
Goedeck	El Dorado	Fairplay	Aug-04	$1,000,000
Everett Ridge	Sonoma	Dry Creek	Jul-04	$2,500,000
Parducci	Mendocino	Hopland	May-04	$6,500,000
Bridlewood	Santa Barbara	Santa Ynez	Apr-04	$14,000,000
Gary Farrell	Sonoma	Russian River	Apr-04	$16,000,000
Stover Oaks	Tuolumne	Jamestown	Jan-04	$1,000,000
Stover Oaks	El Dorado	Placerville	Jan-04	$1,000,000

Firms such as Constellation have had great growth due their conglomerate strategy, but smaller wineries that are publicly traded have suffered greatly in the equity markets. In reviewing the list of publicly traded wine companies that have since been acquired by the larger firms, there are clear questions about the benefits of smaller wineries using an IPO strategy for growth. Following is a partial list of some of the more recent acquisitions of smaller publicly traded wineries by the major players: Mondavi, Beam Wine Estates, Chalone, Ravenswood, Southcorp, Vincor, AV Imports, Stag's Leap, etc. It is expected that merger and acquisition activity will only increase in the future, as the large global wineries continue to capitalize on economies of scale.

While a lot is made of share prices, and what they really mean, the key is not the stock's price today, but if it is volatile upward. A flat-moving stock makes for little in capital gains and may, for this reason, bar the ability of the firm financing future projects in the same way. The return on equity of wine firms is the key issue here. Table 13.5 shows some current publicly traded wine companies and their ticker symbols. The list has grown much shorter over the last decade.

The cost of using *retained earnings* can be very high. The key cost in using accumulated profits is what you could earn on the money versus the cost of debt financing. If you use retained earnings that could have earned 7% in a mutual fund and debt costs 5%, why risk the earnings when the spread between these two numbers is positive. If debt costs more than low-risk investments, then the use of retained earnings is less at issue, unless there are forecasted cash needs that retained earnings historically funds at your firm. Generally speaking, the risk of losing your retained earnings, or losing the availability of those profits for cash use, makes retained earnings the last re-

Table 13.5. Winery Stocks and Ticker Symbols

Company	Ticker
Brown-Forman	BFb
Concho Y Toro	VCO
Constellation	STZ
Diageo	DEO
Fosters	FBRWF.PK
Louis Vuitton Moet Hennessy	LVMH
UST	UST
Williamette Valley Vineyards	WVVI

sort. That leaves us the least costly and most used option: debt.

If rising levels of debt act as a signal to financial markets to increase the cost of debt financing to businesses, and interest rates begin to rise again over this decade, one saving grace is that many winery loans, except those to import and distribute, involve land. The high land prices in Sonoma, Napa, and San Santa Barbara counties are starting to affect (or infect depending on your perspective) all parts of California. General land prices provide a good picture to financiers that a vineyard or winery project will appreciate in value over time, if seen as a real estate deal in the least. Land has a finite supply, and vinable land even less, leading to high prices.

Demand augmentation, especially in certain counties and appellations, is another condition to sustained high prices. Most price bubbles, if you believe real estate is on that path, are burst by supply issues: equity market bubbles burst because stocks become available rapidly based on uncertainty about stock prices in the future. For these reasons, banks and financiers should feel confident in achieving a positive return on investment for winery project based on the land itself.

Conclusion: How Wine Financing Is Different

To summarize, wine finance is no different than other finance applications, except in two key ways. First, the nature of pricing in the industry, based somewhat on the subjective quality perception of the consumer, third-party experts, etc., forces the CFO of a winery to assume revenue streams in the future that can change quickly. Also, the lack of perfect overlap with cost and revenue, especially in a vineyard project, makes projections extremely problematic. Capital budgeting in this industry is very difficult, but it is generally like any other industry. Segmenting your project and identifying all costs is the key to a successful capital budget in the wine industry. In other industries, price forecasts are also imprecise, but the factors that go into pricing in this industry can make budgeting a great challenge.

The type of financing available may be a function of dollars need more than project viability. It seems that banks are less likely to finance large vineyard projects due to shrinking margins, and because wine is an agricultural by-product, national equity markets do not see much return to the projects. Unfortunately, historical data corroborate Wall Street's perception. However, merger and acquisition activity, as well as joint ventures, may be the best way for the industry to circumvent the whims of venture capitalists and keep financing within the industry. To a certain extent, that strengthens the equity and debt markets' perceptions of this industry's viability.

LEARNING OBJECTIVES:
- Define product costing terms
- Identify cost centers in the wine industry
- Explain how costs can be calculated in the wine industry using various examples
- Describe tax issues in the US wine industry

CHAPTER 14

WINE ACCOUNTING AND TAX

Jeff Sully
CPA, CMA, Dillwood, Burkel & Sully LLP

Terry Lease
CPA & Wine Accounting Professor, Sonoma State University

Jon P. Dal Poggetto
CPA & Managing Partner, Dal Poggetto & Co LLP

In a general sense, businesses in the wine industry use accounting information much like those in any other industry: 1) to issue financial statements to investors and banks or other creditors; 2) to file tax returns with federal, state, and local tax authorities; and 3) to assist management in planning and making decisions related to areas such as pricing and resource allocation. Rather than attempt to summarize basic accounting and tax principles in a few pages, this chapter focuses on issues that are unique or of particular importance in the wine industry. The purpose of this chapter is not to allow an owner or manager to do the winery's accounting; rather, it is to give the owner or manager an idea of how accounting information is developed and may be used. Its purpose is also to give the accountant a better idea of how to apply accounting principles and techniques to businesses in the wine industry. The point is not to turn winery owners or managers into accountants, but to help them communicate better with the competent accountants they engage or employ.

One of the most important but difficult issues a business in the wine industry faces is how to account for the costs of the various stages of wine production. Product costing affects the valuation of inventory and determination of cost of goods sold (thus, net operating income) on financial statements and tax returns, and influences any management decision that considers costs or profits. Accordingly, the chapter begins with that issue and devotes the majority of its space to it—providing a comprehensive costing example. It begins with an introduction to product costing, then describes the various cost centers for a wine business. From there it describes how to use cost centers, and then provides specific examples for farming and winery costs. The chapter concludes with an overview of the critical tax issues in the US wine industry.

Product Costing: Introduction

To understand inventory costing, the wine production process should be broken down into a series of steps within a continuous flow of processes. The winery farms, harvests, and crushes grapes at a desired degree of ripeness, ferments the juice or must, and clarifies the new wine. The wine ages until the flavors and balance achieve their greatest potential. Finally, the wine is bottled. At that point, the production process stops. The grapes arrive at the crusher and cases leave the loading dock.

Wine costing is the process of accumulating, allocating, and assigning both direct costs and overhead costs to wine inventory in a reasonable and consistent manner. Assigning a cost to this inventory, however, requires a slightly different view. The production process is consistent and definable. Within that process distinct stages of production activity can be defined. Each stage becomes a focal point for assigning costs.

For example, the *crush activity* has a collection of costs that are unique to crushing and fermenting grapes. *Bulk aging* is another area of unique costs, where labor activity may be minimal, yet utilization of building or barrel capacity may be significant. *Bottling* has costs associated with getting the finished wine into bottles. Finally, there are *overhead costs* that support all three areas. What makes each of these areas distinct from each other is that they have different activities from the same resource requirements.

During the harvest, there will be higher demands for cellar labor, utilities for tank refrigeration, specialized processing equipment such as crushers and presses, etc. The aging process requires barrels and space to store them, in addition to tank space for blending and chemistry adjustments. Labor will not be as high, although there will be a need for refrigeration during the warmer months. Bottling requires the use of equipment that has sat idle for 10 or 11 months out of the year. Now it will run 12 or 14 hours a day for a month or two, and during this time labor will be concentrated.

Each of the activities above deals with the same product. In some stages (such as crush), the product changes in substance; in other stages (such as bottling) its changes are more a matter of form. Yet each stage is a different activity and places different demands on the same resources, or *cost drivers* to support that activity. Stating it another way, even though an hour of labor is an hour of labor, the *cost objective* of that labor is different for crush, aging, and bottling. The cost objective is the accumulated measurement of cost drivers required to perform specific and related activities. Aging wines requires barrels; but once the wines are sent to the bottling tank, the barrels are empty. Identifying and separating the activities that occur in the production process is the first step in being able to segregate costs that are unique to that activity, which in turn allows for a more accurate determination of the costs that most directly relate to a specific wine.

Product Costing: Defining Cost Centers

Once the different activities have been identified and segregated, they become the basis for defining *cost centers. A cost center is an accounting construct designed to accumulate the cost drivers during the time frame over which costs are collected for reporting*. A given cost center should cover only one type of production activity. Cost centers are at the foundation of costing procedures.

"Thinking like a winemaker" will also help develop cost centers concepts. For example, farming costs are incurred to grow grapes. The costs associated with growing those grapes, plus harvest and transportation costs to the winery (along with costs for purchased grapes), become the initial costs for the resulting bulk wine. At harvest, the grapes come into the winery to be crushed, fermented, and made into wine. Crushing and fermenting add more costs to the new vintage, and cellar activities add further costs to the bulk wine in inventory. All of these costs become part of the bulk inventory, as measured by dollars per gallon. When these wines are bottled, the cost of bulk wine to be bottled comes out of bulk inventory and flows into bottled wine inventory, and bottling costs are added at this time.

Each of these stages represents four major cost centers: farming cost center, crush cost center, production cost center, and bottling cost center. Two more cost centers would be used for costs that support multiple stages of production: facilities cost center and payroll cost center. In addition, the business would have one or more cost centers for noninventory costs, that is: for costs that are associated with selling the wine, or with running the organization but are not associated with producing the wine.

In a well-structured chart of accounts, the accounts would be grouped and numbered by cost center to facilitate proper recording and efficient use of the cost information. Some costs, such as payroll costs and facilities-related costs (e.g., depreciation or rent, utilities, property taxes, insurance), would be recorded initially in the payroll or facilities cost center and then allocated to the cost centers for the different stages. This topic is covered in the next section.

PRODUCT COSTING: USING COST CENTERS

Each cost center contains the traditional grouping of cost classifications: *direct costs*, such as direct labor and materials, and *indirect costs*, such as supervisory labor and allocated overhead. The *direct costs* are those costs that are associated only with a specific wine. These costs include grapes, identifiable labor, and measured supplies.

Indirect costs are those costs that are associated with several different wines, because the same cost drivers may be shared by the different wines in the production process. The use of the resource by any specific wine is not measured directly, so the indirect costs must be allocated. The indirect costs may generate many different allocation methods. Some work fine, while others will generate inconsistent results. Different allocation methods are illustrated as part of the following examples of costing the different stages of production.

COSTING IN THE VINEYARD

Farming Costs

This cost center accumulates data pertaining to the cost of growing grapes for the current year. There will be direct labor associated with traditional farming operations, such as disking, pruning, thinning, harvest, etc. There will be direct materials purchases, such as fertilizers and other chemicals. Finally, there will be indirect costs that need to be allocated, such as supervisory labor, property taxes, depreciation, and insurance, to name a few. The question arises as to how to allocate these costs. Table 14.1 forms the basis for illustrating the most basic allocation method.

If all of the blocks are evenly spaced with 535 vines per acre, the cost of growing grapes could be viewed as being $4,365 per acre ($109,125/25). On the other hand, if each acre provides approximately the same yield, the cost could be viewed as being $1,180 per ton ($109,125/92.5) to grow grapes.

Table 14.1 takes the entire costs for this vineyard and divides it by the number of tons or acres to come up with a simple allocation rate. The allocation base is either tons or acres, but all grapes are treated equally. Changing the allocation base to vineyard blocks would refine this process because different varieties have different yields. As-

Table 14.1. Example of Basic Cost Allocation Method

Total acres	25
Harvested tons	92.5
Total farming costs	$109,125

sume that each block represents a different variety, with the acres and yield (in tons) for each variety presented in Table 14.2.

It still costs $4,365 per acre to farm. The new information in Table 14.2, however, indicates that the Cabernet block CS cost $28,373 to farm ($4,365×6.5 acres), for a cost of $1,065 per ton ($28,373/6.5 tons). The Zinfandel block ZN cost the same $4,365 per acre to farm, but the total cost for Zinfandel block ZN is $80,752 ($4,365×18.5 acres). Having harvested 65.85 tons of Zinfandel, the cost per ton is $1,226 per ton ($80,752/18.5 tons) versus $1,065 for the Cabernet. The new allocation takes into account that different varieties will yield different tonnages. However, it still assumes that the varieties expend resources at the same rate for each block.

But what happens if the vineyard decides to replace the CS block of 535 vines per acre with 1,100 vines per acre using recently developed technology and techniques? Does continuing to allocate cost as before provide a reasonable result? The required amount of some resources will be significantly different for blocks that are radically different from each other, as indicated in Table 14.3.

The yields have gone up from the denser planting, but so have total costs. Using the allocation method in Example 2 above, it costs $4,662 per acre to farm, with $30,303 allocated to the Cabernet ($4,662×6.5 acres) for $814 per ton ($30,303/37.25 tons). The Zinfandel is allocated $86,247 of costs ($4,662×18.5 acres) for $1,310 per ton ($86,247/65.85 tons). Does it make sense that the cost of

Table 14.2. Example of Cost Allocation by Block

Categories	Block CS	Block ZN	Totals
Acres	6.5	18.5	25
Block yield (tons)	26.65	65.85	92.5
Total farming costs			$109,125

Table 14.3. Example of Cost Allocation With Different Block Techniques

Categories	Block CS	Block ZN	Totals
Vines per acre	1,100	535	
Total acres	6.5	18.5	25
Block yield (tons)	37.25	65.85	103.1
Total farming costs			$116,550

the Cabernet would have decreased by almost 25% while the cost of the Zinfandel has increased by over 6%? This allocation creates a very large discrepancy between the cost of Cabernet and the cost of Zinfandel. However, the Cabernet's increased vine densities should require more labor, more times through with the tractor, more spraying, etc., while nothing changed with respect to the Zinfandel. Also, a larger investment is required (with more depreciation) for the Cabernet.

Assume an examination of the various accounts reveals that there are variable costs associated with activities such as general labor, pruning, harvest, etc. Practical experience suggests that the newer planting methods require more labor simply because there are more vines to deal with. A simple way to measure the additional resources required is to assume that the labor hours spent in an individual block is a measurement of all variable resources being applied to that block.

In addition to the costs described above, there are harvest costs that are more closely related to the tons of grapes harvested than to labor hours. Finally, there are fixed costs, such as property taxes and insurance, which do not relate to the vine spacing of the different blocks or the amount of grapes harvested. In order to arrive at an equitable allocation of the costs for this vineyard, as defined by the two blocks, three allocation bases are used to arrive at the allocation of costs, based on the information in Table 14.4.

Table 14.4. Example of Cost Allocation Based on Block Farming Costs

Categories	Block CS	Block ZN	Totals
Total acres	6.5	18.5	25
Vines per acre	1,100	535	
Yield (tons) per block	37.25	65.85	103.1
Labor hours per block	1,602	2,783	4,385
Total farming costs			$116,550
Labor-related costs			$57,757
Harvest costs			$23,611
General overhead			$35,182

The labor hours provides the basis for the allocation percentages for the variable costs in the vineyard. The Cabernet block uses 36.5% of labor hours (1,602/4,385) and is allocated that percentage of the labor-related costs. The remaining 63.5% will go to the Zinfandel block, even though the Zinfandel makes up 74% of the acreage. Consequently, the Cabernet would receive $21,081 ($57,757 × 36.5%), or $565.93 per ton. The Zinfandel would receive the remainder, $36,676, or $556.96 per ton.

The $23,611 of harvest costs is allocated based on tons. Using that basis, each ton harvested costs $229 ($23,611/103.1 tons). Finally, the $35,182 of other general overhead needs to be allocated. Because the general overhead applies to the entire vineyard in total, irrespective of the number of vines planted or grapes harvested, the remaining overhead is applied at the rate of $1,407 per acre, or $245.53 per ton for the Cabernet and $395.38 per ton for the Zinfandel. The final cost per ton calculation is listed in Table 14.5.

Different allocation methods or bases could be used, if management felt that a different approach would be more appropriate. The important thing to remember is that the allocation method must be understandable, cost-effective, and truly reflect the production in the field.

Table 14.5. Example of Per Ton Calculation

	Cabernet Sauvignon	Zinfandel
Variable labor costs	$21,081	$36,676
Harvest costs	$8,530	$15,081
Fixed overhead costs	$9,146	$26,036
Total costs	$38,757	$77,793
Total tons	37.25	65.85
Cost per ton	$1,040	$1,181

Figure 14.1 provides a big picture overview of a vineyard accounting worksheet. This is useful in understanding how all of the various components can be calculated.

COSTING IN THE WINERY

Overhead costs are those costs associated with keeping the winery operation viable. Overhead is the term for costs that are required to support production, yet are not associated with a specific wine. These are costs that the winery will incur no matter how much wine is produced. For example, a winery may have a building that cost $2,000,000 to build and can store several thousand barrels. That building will cost approximately $50,000 per year in depreciation, regardless of the volume of inventory stored in it. In any given year, one third to one half of the barrel inventory may be empty for several months during the year. Those barrels could have a significant cost associated with each year.

A winery may have varying levels of activity, yet they still relate to the same basic cost objectives. While the assets (building and barrels) have varying utilization rates during the year (e.g., number of months the barrels are full this year compared with last year), accountants usually make an assumption that will simplify costing, so as to avoid getting bogged down in the process while providing reliable and meaningful information. One of the simplifying conventions that accountants adopt is to use the longest

BlackAcre Vineyards				
FARMING COST ANALYSIS				
Variable farming costs			$	61,189
Harvest costs				23,611
Fixed farming costs				31,750
	Farming costs		$	116,550
Variable farming costs by variety				
Labor hours, Cabernet	1,602	36.5%	$	22,355
Labor hours, Zinfandel	2,783	63.5%		38,834
	4,385	100.0%	$	61,189
Harvest costs by variety				
Tons, Cabernet	37.25	36.1%	$	8,531
Tons, Zinfandel	65.85	63.9%		15,080
	103.10	100.0%	$	23,611
Fixed farming costs by variety				
Total acres, Cabernet	6.5	26.0%	$	8,255
Total acres, Zinfandel	18.5	74.0%		23,495
	25.0	100.0%	$	31,750

Per variety analysis			
	Cabernet	Zinfandel	Total
Variable costs	$ 22,355	$ 38,834	$ 61,189
Harvest costs	8,531	15,080	23,611
Fixed costs	8,255	23,495	31,750
Total	$ 39,140	$ 77,410	$ 116,550

Common size analysis (per ton)			
	Cabernet	Zinfandel	Average
Variable costs	$ 600	$ 590	$ 593
Harvest costs	229	229	229
Fixed costs	222	357	308
Total	$ 1,051	$ 1,176	$ 1,130

Fact pattern 4

Figure 14.1. Vineyard accounting worksheet.

period of time to report costs that provides the desired level of information. The selected time frame is then used for all cost allocations. For the examples that follow, this chapter assumes a 1-year time frame.

Note that utilizing a yearly time frame "averages out" the nonspecific production cost differences that show up from month to month. Costing on a monthly basis would result in swings in per gallon costs from vintage to vintage, depending on differences in the cost objective for each bulk wine in the cellar. While perhaps technically accurate, the information presented may not be very meaningful. Because the production cycle is generally seen in terms of years, the costs per unit should reflect that same time frame. To summarize, the time frame chosen should be one that provides the needed level of detail without becoming overly burdensome.

One of the important types of overhead costs is the category of *facilities costs*. Facilities costs are those costs incurred to keep the building open, without producing a single drop of wine, such as utilities, insurance, property taxes, depreciation, and security. These costs are recorded in a facilities cost center.

These costs need to be allocated to the various activities within the business. Determining an allocation method begins with delineating the use of the facilities in terms of cost objectives. For example, assume that an entire winery is located in one building. Further assume that 55% of the floor space covers production prior to barrel storage. This will include the crush deck, the fermentation tanks, the lab, dry goods storage, etc. Another 25% of the winery building is devoted to barrel storage, and 5% to bottling. This will not include cased goods storage. Remember, the production process (and cost accumulation) stops when the wine is bottled. Nonproduction activities, such as case storage, tasting room, sales, and administrative offices, make up the remainder of the floor space.

In other words, in this example, 85% of the building supports production activities. Of all costs associated with keeping the building open and in good operating condition, 85% of those costs relate to production. Thus, using floor space as an allocation basis, 85% of all of the facilities costs are allocated among the various bulk wines in inventory, as well as wines bottled this year. The other 15% of the costs would be expensed as a period expense (Figure 14.2)

These percentages, once established, will not change unless the production flow is rearranged or the building is expanded. The percentages shown above contain some generalizations, but the purpose of the allocation process is to generate reasonably accurate but cost-efficient estimates of the costs of production. It begins with the assumption that 85% of the facilities costs go into production and that 15% are expensed as period costs (e.g., with 10% for retail activities and 5% for administrative activities). Assuming that facilities costs total $87,980, production would be allocated $74,783, retail activities would be allocated $8,798, and administrative activities would be allocated $4,399, as shown in Figure 14.3.

The $74,783 applied to bulk wines is the estimated amount of facilities costs that the winery incurred over the course of the year to support crush, production, and bottling.

Payroll Costs

Before the example moves into the cellar, the payroll cost center deserves a brief mention. It is very common to have payroll entered directly into the respective payroll accounts

FACILITIES ALLOCATION SUMMARY			
	Sq. Ft. %		
Production costs		**Period costs**	
Crush	10%	Tasting room	10%
Cellar	45%	Offices	5%
Barrel storage	25%		
Bottling	5%		
	85%		15%

Figure 14.2. Facilities allocation summary.

FACILITIES ALLOCATION			
Total facilities costs	$	87,980	
Less: period costs		(13,197)	
Allocated to production	$	74,783	
Allocated to production (sq. ft.)			
Crush	10%	$	8,798
Cellar	45%		39,591
Barrel Storage	25%		21,995
Bottling	5%		4,399
	85%	$	74,783

Figure 14.3. Facilities allocation.

residing in cellar, crush, marketing, etc. While there is nothing wrong with this method, it results in the loss of an important management tool. Winery owners are shocked to see what their total payroll costs are when presented as a whole. It's easy to miss when seeing payroll only in parts. The payroll cost center collects all direct payroll costs from the various activities, *plus* the payroll overhead accounts, such as taxes, health benefits, workers' compensation, and then allocates those costs to the other cost centers. A typical allocation basis would be a percentage based on direct labor costs.

Crush Costs

Many wineries fail to use a crush cost center and include the cost of crush activity with aging or production cost centers. As a result, white wines tend to be undercosted, while red wines will be overcosted. The reason is that a wine that remains in the cellar for more than 1 year (such as a red wine) will have cellar costs allocated twice. If the cellar costs include the crush costs, the wine will receive an allocation of the crush costs in the second year even though it did not benefit from those crush activities.

The problem is compounded because wines that remain in the cellar less than 1 year will receive less than a full allocation of crush costs, as some of the crush costs went to wines that have already been in the cellar for a year. The way to solve this problem is to have a crush cost center.

For purposes of the ongoing example, assume that crush costs total $66,655. (Remember that the crush costs would include an allocation of the facilities costs.) A simple way to allocate the crush costs would be on a per-gallon basis. For example, if Cabernet made up 25% of the new wine by volume (gallons), then it would receive $16,664 of crush costs. (For the sake of conserving space, the example will now track the costs allocated only to Cabernet Sauvignon.)

A variation on this allocation method could be used if the wines used labor disproportionately to their volume. In the variation, labor costs would be separated from other crush costs and allocated based on the labor hours employed for each type of wine. The remaining costs would still be allocated based on the total number of gallons. The important thing is to try to make the allocation assumptions accurately reflect what is going on in the production process while keeping the allocation process relatively simple to administer and understand.

Production Costs

This cost center covers the production process once the wines have been declared off fermentation. It involves cold stabilization, fining, filtering, barrel aging, blending—everything up until bottling. These costs apply to all wines in inventory, and in principle the allocation is straightforward, based on gallons in inventory. Simply take each wine's number of gallons and divide by total gallons to come up with an allocation percentage. A complicating factor in practice, however, is the time frame covered. Different wines will come off fermentation and into the production cost center and then out of the production cost center and into bottling at different times.

The allocation of production costs to inventory should take into account the amount of time a given lot spends in the cellar. For example, 5,000 gallons held 6 months will require fewer resources and, therefore, should absorb less cost in total than 4,200 gallons held a full year. To accurately allocate cellar costs, a weighted average system is needed. The allocation calculation also needs to consider that, by the end of the year, some wines may have been bottled and not appear in inventory at all! Table 14.6 shows a weighted average calculation and resulting allocation of production costs. Note that wines from the 2000 vintage that spent less than 12 months in the cellar were bottled and would not show up in the bulk wine inventory at the end of the year. The actual year-end gallons in inventory are 44,742, comprising all 2002 wines and those 2001 wines that were in the cellar for 12 months.

Administrative Costs

In some cases, where indirect costs have not already been accounted for in the production departments, it may be necessary to allocate a portion of the costs reported as

Table 14.6. Bulk Wine Cost Summary

Variety	Year	Gallons	Months in Cellar	Extended	Wtd. Avg. (%)	Yearly Cellar Costs
Cabernet Sauvignon	02	6,146	2	12,292	3.2%	4,649
Cabernet Sauvignon	01	5,701	12	68,412	17.7%	25,877
Cabernet Sauvignon	00	5,845	6	35,070	9.1%	13,265
Zinfandel	02	11,326	2	22,652	5.9%	8,568
Zinfandel	01	9,455	12	113,460	29.4%	42,916
Zinfandel	00	10,560	4	42,240	10.9%	15,977
Chardonnay	02	4,365	2	8,730	2.3%	3,302
Chardonnay	01	4,999	12	59,988	15.5%	22,691
Chardonnay	00	4,502	1	4,502	1.2%	1,1703
Sauvignon Blanc	02	2,750	2	5,500	1.4%	2,080
Sauvignon Blanc	01	3,289	4	13,156	3.4%	4,976
Total		68,938		386,002	100.0%	146,006

administrative costs. For example, materials need to be ordered, compliance reports need to be filed, and repairs and maintenance need to be scheduled. To the extent these costs are included in administrative costs, part of the administrative costs should be allocated to inventory through the production cost centers. This could be a percentage of total actual administrative costs, or the sum of separately identified accounts within the administration cost center.

As a general rule, wineries try to mimic the amount of administrative resources devoted to production management, expressed as a percentage allocation. The part of the administrative costs allocated to inventory would, naturally, reduce the administrative expenses recognized on the income statement. The part of the administrative costs allocated to inventory should be further allocated to the different production cost centers (vineyard, crush, production, bottling) based on management's estimate of how much time they spend on each area.

Bottling Costs

Bottling marks the end of the production process. At this point, all of the production costs accumulated so far in the bulk wine inventory for the wine to be bottled are transferred to bottled wine. This cost center would contain account balances that tend to follow bottling activity. In other words, one would not expect to see a lot of accounting activity until the actual bottling, save for some depreciation and overhead allocations.

One of the bottling cost center accounts would be for bottling materials. All bottling materials would be added to this account and spread among all of the wines bottled. For example, if Bottling Materials has an account balance of $137,001 and 6,387 cases are bottled for the year, the cost per case would be $21.45. That method works and is fairly common.

A more accurate method, however, would reflect differentiated packaging costs for special or more expensive wines requiring a more "luxurious" package. In the Prepaids section of the balance sheet, there would be separate accounts for glass, corks, labels, etc. The recorded cost per unit for each item ($0.42 for corks, $7.05 for glass, etc.) would be used to transfer costs from prepaid bottling supplies to the bottling cost center for the materials used in the bottling run. Differentiating packaging materials in this way allows those costs to be analyzed in a meaningful way (e.g., compared to budgeted costs).

This is a good place to discuss an idea that applies in all areas, but probably has the biggest impact in bottling: *overhead should be allocated to the production cost centers based on budgeted costs and in such a way that equal activity bears equal costs.* Otherwise, the allocation of costs could be badly disproportionate to the use of resources.

Bottling, for example, generally occurs a few times a year, and the heaviest bottling may be early in the year, with a smaller run just before harvest. If overhead costs were allocated on a monthly basis as the costs are incurred, the first bottling run would be light on overhead, and the later one would be excessively burdened with overhead costs.

From an annual bottling budget, the overhead costs may be used to develop a set of standard costs that would be applied to each bottling run on a per case basis. Any difference between the overhead applied based on the budget-derived standards and the actual overhead allocated to the bottling cost center should be minor and would show up as a variance in cost of goods sold.

The next step is to bring in the costs from the bulk wine inventory and add them to the packaging costs. During the course of bottling, some wine will be lost from what was last recorded as the gallons in inventory. Keep in mind that the costs accumulated to date will become part of the final bottle cost. There is no need to expense spillage or filtration losses as they will be absorbed by the bottled wine as part of the final cost. The following example looks at the bottling of the 2001 Cabernet Sauvignon. Assume an estimated 5,701 gallons, with $121,477 of costs accumulated from other cost centers, will be bottled. The bottling report shown in Figure 14.4 below brings forward all costs from previous activities and adds the final costs from this last stage of production.

In the worksheet in Figure 14.4, a percentage of the costs of bulk wine is designated as fixed costs, based on a previous study of cellar costs, but the percentages may vary from winery to winery. Breaking costs into fixed and variable components allows for contribution margin analysis and variable budgeting. Also note that cased goods storage is not included in these calculations. Cased goods storage is a period cost, and should be expensed as a distribution or marketing cost. It is important to remember that once the wine is bottled, all production costs stop. In the example worksheet, the product cost is $73.71. That is the cost that should be used to reduce inventory and charge cost of goods sold for each case of the 2001 Cabernet Sauvignon sold.

TAX ISSUES IN THE US WINE INDUSTRY

Book to tax differences in the wine industry are focused around several critical areas. Generally, uniform capitalization rules (Unicap) apply under IRC §263A, whereby all farming and production costs, as well as interest and storage costs, must be capitalized. However, there are some exceptions that allow wineries and growers to use the cash method to expense certain costs in the current year, as opposed to the accrual method where costs will not be recognized until the wine is actually sold.

Example 5
BOTTLING INFORMATION

Date Bottled	January 15 - 16, 2003					
Product Bottled:	2001 Cabernet		Code	CS750-01		
Cases Bottled		2,343	*2.3775 =	5,570		
					Cost Analysis	
Gallons removed from bulk		5,701			Fixed	Variable
Estimated Losses		(131)				
Cost of Gallons removed from bulk			Cost	121,477.00	66,812.35	54,664.65
Glass (cases)	2,348	Unit Cost	$7.350	Cost 17,257.80		17,257.80
Packaging		Unit Cost		Cost —		—
Corks	28,500	Unit Cost	$0.420	Cost 11,970.00		11,970.00
Capsules	28,500	Unit Cost	$0.260	Cost 7,410.00		7,410.00
Front label	30,500	Unit Cost	$0.180	Cost 5,490.00		5,490.00
Back label	28,250	Unit Cost	$0.080	Cost 2,260.00		2,260.00
			Direct Materials	44,387.80		
Other						
Direct Labor			Cost $	4,428.27		4,428.27
Supervisorial			Cost		—	
Facilities Allocation			Cost $	2,399.23	1,758.64	640.59
Other bottling overhead			Cost $	2,066.53		2,066.53
Total Bottling Costs			$	51,215.30	68,570.99	106,187.84
Total Costs			$	172,692.30		
Total Cases Bottled				2,343		
Materials			$	18.94	FOB	186.00
Labor				1.89		
Overhead				1.02	Cont Margin	140.68
Bottling Cost per case				21.86	Cont Margin %	75.6%
Total Bulk Wine Cost per Case				51.85	Gross Margin	112.29
					Gross Profit %	60.4%
Total Cost per Case			$	73.71		

Figure 14.4. Example of bottling information worksheet.

One unique tax advantage of farm accounting is the ability to deduct certain soil and water conservation expenses for vineyard development [IRC §175(a)]. Under the code, costs incurred for leveling, conditioning, grading, terracing, contouring, and soil conditioning need not be capitalized. Additionally, the construction of diversion channels, irrigation ditches, and earthen dams may also be expensed. Finally, the eradication of brush and planting of windbreaks also fall under this section.

In order to qualify, the conservation methods used must be consistent with a plan approved by the Natural Resources Conservation Services (NRCS) of the Department of Agriculture. There is a limitation on the amount of deduction that may be taken in any one year. The deduction is limited to 25% of gross income from all farming operations, with a carryover of any unused deductions. It is also important to note that the election must be made in the first year the soil and water conservation expenses are incurred, otherwise the deduction is lost.

Generally, Unicap applies to farmers with respect to establishing a vineyard, in that the preproductive period is more than 2 years [IRC §263A(d)(1)(A)(ii)]. If farmers are not required to use the accrual method of accounting, they may have the option to expense or capitalize certain preproductive expenses (including interest cost) when establishing a new vineyard [Reg. §1.162-12(a)]. Capital expenditures cannot be deducted. However, if the election is made to expense (which must be done in the first year) all depreciation must follow the alternative (straight-line) depreciation methods of IRC §168(g)(2).

Up until recently, wineries had to be on the accrual basis for accounting for tax purposes. A majority of wineries also grow some grapes. Any farming costs incurred would become part of inventory and not recognized until the wine was sold. This could be as long as 4 years later, depending on the wine. Individuals, or certain small businesses, may use the cash method for farm accounting. Reg. §1.263A-4-(a)(4)(ii)(B) provides that the farming of a commodity up to the time of processing qualifies for cash basis, even if the entity is a winery using accrual accounting. Thus, a winery may expense farming costs. The result is that the tax-based inventory would be at a lower basis than the GAAP-based inventory. The expenses would be recognized up to 2 years earlier. The down side is that the tax-based gross profit would be higher when the low-basis wines are sold. However, if the winery continues producing wine each year the profit would be affected by the current deduction. Only when the winery stops production does the tax bill come due.

A recent exception to Unicap was Revenue Procedure 2000-22. It was available to taxpayers whose average annual gross receipts from all sources during the last 3 years was less than $1,000,000. This revenue procedure allowed wineries to use the cash method and not account for inventories. However, there was also a conformity requirement, in that the financial books had to be reported on the same basis. The banking industry cried "foul" immediately. Because of the flaws in 2000-22, the IRS issued Revenue Procedure 2001-10; 2001-10 removed the conformity requirement and refined the inventory accounting issues in that only *purchased* raw materials and supplies should be capitalized.

There is also a simplified production method available whereby the additional §263A costs divided by the traditional §471 costs times the ending §471 inventory costs equal the amount of §263A costs to add to ending inventory. For example, if §263A costs were $150,000, §471 costs were $500,000, and §471 ending inventory costs were $450,000, the amount of §263A costs to add to inventory would be ($150,000/$500,000) times $450,000 equals $135,000. There is also a de minimus rule. If total indirect overhead costs are less than $200,000 for the year, then only direct materials and labor need be capitalized. Everything else would be expensed.

Tax inventories must apply interest capitalization under the avoided cost rules if 1) the production period is greater than 2 years, (2) the production period is greater than 1 year and the cost is more than $1,000,000. The production period is measured from the time the grapes are crushed until the wines are made available for sale. The avoided cost rules assume that interest expense from any borrowings must be applied to inventory first. For example, if inventory requiring more than 2 years' production were $1,500,000, and the net book value of equipment used to produce those wines were $2,800,000, and your average borrowing rate was 6.2%, the amount of interest to capitalize would be as listed in Table 14.7.

As a final step, the capitalized interest calculation cannot exceed the interest actually paid. The calculation above takes into account debt required for production assets. The interest rate is a weighted average borrowing rate for all debt.

In addition to book-tax inventory differences, where capitalized production costs and costs of goods sold are generally higher, there are significant differences in depreciation. While GAAP uses straight-line and is preferred (although MACRS is acceptable), tax depreciation follows the Modified Accelerated Cost Recovery System (MACRS). Tables for MACRS depreciable lives can be found in Revenue Procedure 87-56. Table 14.8 illustrates some comparative depreciable lives.

Many wineries use the *last-in, first-out (LIFO)* method of inventory accounting to reduce their current taxable income. The LIFO method charges operations with current production costs. In periods of increasing costs, inventories are stated at prior, lower costs and cost of sales includes higher costs.

Table 14.7. Example of Capitalized Interest

Wine	$1,500,000
Assets	$2,800,000
Total	$4,300,000
Interest rate	6.2%
Capitalized interest	$266,600

Table 14.8. Comparative Depreciable Lives

Asset	GAAP Years	MACRS GDS Years	MACRS ADS Years
Vines	20	10	20
Trellis & irrigation*	10	7	10
Tractor	10	7	10
Building	40	39	40
Crusher	10	7	10
Barrels	5	3	4
Refrigeration	10	7	10
Bottling line	10	7	10

*In a recent tax court case (see Trentadue) the IRS has determined that the above ground trellis and irrigation costs should continue to be depreciated over 7 years and the underground portion should be depreciated over 15 years.

The most common LIFO method used is the double extension method, in which inventories are extended at base year costs (the costs of cased goods and bulk wine at the point in time LIFO was first adopted) and the resulting base year cost of the current inventory is compared to the

Table 14.9. LIFO Calculation Example

Inventory	Quantity	Base Year Cost	Extended at Base Year Cost	Current Cost	Index
Bulk	50,000	$10	$500,000	$600,000	
Cased	15,000	$45	$675,000	$750,000	
Total			$1,175,000	$1,350,000	1.15
Beginning inventory at base year cost	$1,075,000				
Increase at base	$100,000				
Index for increment	1.15				
LIFO value of pool addition	$115,000				
Beginning LIFO value	$1,075,000				
Ending LIFO value	$1,190,000				
Ending FIFO value	$1,350,000				
LIFO reserve	$160,000				

current FIFO cost to compute an index for the current year LIFO pool. An example for a LIFO calculation is illustrated in Table 14.9, assuming LIFO was elected the previous year. (In the example, LIFO has resulted in a cumulative reduction of taxable income of $160,000.)

Taxpayers using LIFO for tax purposes must also use it for financial reporting purposes; however, supplemental FIFO information can be provided to users of the financial statements. Not all inventory items need be under LIFO. If a winery was producing its own wines, as well as importing bulk wine subject to foreign currency translation adjustments, it might choose to value only the domestic wines under LIFO.

Special thanks to Tyler Comstock as contributing editor on this chapter. Tyler is CFO and Executive Vice-President at Kendall-Jackson Vineyards & Winery.

LEARNING OBJECTIVES:
- Describe how historical wine industry labor issues impact wine business today
- Identify and describe the four major job categories for a winery
- Define human resource management (HRM)
- Identify and explain the five buckets of HRM
- Explain future wine industry HRM issues.

CHAPTER 15

MANAGING HUMAN RESOURCES IN THE WINE INDUSTRY

Liz Thach
Wine Business Professor, Sonoma State University

Lillian Bynum
Vice President of Human Resources, Delicato Family Vineyards

Many of the finest wines around the globe proudly list on the label that the grapes are hand-picked. Yet with looming labor shortages in some parts of the world as well as rising labor costs and increased concerns over vineyard worker treatment, human resources in the wine industry has become increasingly important.

Thousands of people are employed in the wine industry. Many are attracted to it because of a love for wine and all its beneficial properties, such as being a unique product of nature, art, and science; enhancing food and health; and inspiring collaboration and friendship. However, working in the wine industry is not always filled with romance. It is still a global business, like many other businesses, and, as such, is bound to management policies that are linked to legal and regulatory control. In many cases, these policies protect the welfare of workers and the environment, but in the successful companies, they are also part of the very culture and are based on the premise that satisfied employees not only are more productive, but produce a higher quality product for the consumer.

With this in mind, this chapter focuses on how to manage human resources within the wine industry. It begins with a historical perspective on wine industry labor and then moves forward in time to modern-day practices. Specifically, it 1) provides an overview of the common employee positions within a vineyard and winery, as a business grows in size and production; 2) defines human resource management (HRM) in the wine industry; 3) describes the five major categories of HR (staffing and recruiting; training and development; employee relations, compensation and benefits; and record keeping and legal issues) as well as the special issues regarding these HR categories in the wine industry; and 4) highlights future HRM issues.

Wine Labor Issues Through the Centuries

It is theorized that wine was discovered by accident sometime around 6000 BC by a woman. Supposedly she picked some wild grapes, put them in a large clay pot to be consumed later, and then forgot about them for a while. Subsequently, when she looked into the clay pot, she discovered the grapes had naturally fermented and turned into a unique liquid that tasted lovely and provided a rather euphoric effect.

Books on the history of wine all agree that this event took place somewhere near Mesopotamia, though authors argue about the exact location, with different experts suggesting modern-day Turkey, Iraq, and even Georgia as the actual birthplace of wine (Grist, 2003). Regardless of the location, once wine was discovered, people began to cultivate it, and immediately the issue of labor in making wine arose.

This issue was not easy to resolve, as tending the vineyards, bringing in the harvest, and overseeing fermentation and storage of the wine took an incredible amount of people and work. However, eventually a solution was found, and that was the use of *slaves* in making wine. Indeed, in ancient Egypt, where we find the earliest "wine labels" dating from 3200 BC, there are also many drawings of slaves working in the vineyards and bringing in the harvest. Interestingly, these drawings also depict the winemakers to be *priests* or someone of high knowledge and nobility (Grist, 2003).

The use of slaves to make wine on a large scale continued throughout much of early history and can be found in the records of early Syria, Canaan, Greece, and the Roman Empire. Indeed, as the Romans expanded into Gaul—modern-day France—in the early first century AD, in many cases they used slaves to plant the first of some of the famous old vineyards.

As wine made its way to the New World, the practice of using slaves to tend the vineyard continued in almost every country. In South Africa, offspring of the local tribes and European settlers were obligated to work the vines of the old vineyards. In Argentina, Chile, and California, the native Indians were required to work the vineyards, which were established around the old Spanish missions. Ironically, here again, the priest served in the role of winemaker and overseer of the production process.

A notable exception was Australia, where they never succeeded in convincing the native Aborigine tribes to work in vineyards. During some of the early years in Australia, they did employ the use of convict labor in the vineyards. Indeed, convicts originally planted the famous Dalwood vineyards along the Hunter River outside of Syndey in the late 1820s (Simon, 1967).

Historical Impact on Wine Labor Issues Today

The history of the labor situation is still felt in New World wine operations today. For example, because Australians could not rely on a cheap local labor source, they became very innovative in developing mechanized methods to harvest their vineyards. Today, they are one of the world's premier wine-producing countries, using more than 90% mechanization. This practice lowers labor costs, so the production cost of the wine is less.

In Chile, Argentina, and South Africa, much hand-harvesting is still done in the fields, but the labor costs are lower than those found in Canada and the US. Because of this, their final wine production cost is also lower. At the same time, the issue of Fairtrade wine is heating up in Europe, and both Chile and South Africa have addressed this issue by ensuring that some of their wineries are being certified as implementing Fairtrade practices for employees. Certified wineries help workers buy houses, obtain school buses to send their children to school, pay for health coverage, and provide training among other services.

The US and Canada rely primarily on migrant labor forces from Mexico, who work in the vineyards for part of the year and then return to their homeland. Strict regulations on pay keep labor costs higher in the US and Canada and, therefore, the total production cost of the wine is higher in these countries. Also in these countries there is increased concern about the welfare of the workers, and many wineries have voluntarily implemented socially responsible practices and completed self-assessments such as the California Code of Sustainable Wine Growing Practices. Examples of positive human resource practices include providing health care benefits, additional training, progressive performance management systems, and in some cases subsidizing farm worker housing.

The trend in many New World countries to use mechanized harvesting is not feasible in all vineyard locations. Steep hillside vineyards, or those in very prestigious appellations, still use hand labor, either because harvesting machines don't work on steep hills or because of the cachet of "hand-harvested" grapes. Many top winemakers in Europe and the US still insist on hand-harvested grapes for their wines, because of the perception that it results in higher quality fruit—even though some studies have proven that quality results are generally equal between mechanized and hand-harvested wines.

These same practices are reflected in Old World countries as well, such as France and Italy, where growers use mechanized harvesting for lower and medium-priced wines, but the high-end wine producers still use hand labor. A trend that is different in the Old World, however, is that they often hire "wine students" to work harvest and/or rely on "ecotourism participants." In both cases, they capitalize on people who are drawn to the romance of wine to provide inexpensive labor in return for being allowed to experience harvest in a famous wine region.

COMMON EMPLOYEE POSITIONS IN A VINEYARD/WINERY

Like many other industries, the number of employees in a wine business depends on the size of production in the venture. Although several large wine companies have operations in many countries around the world, such as Constellation, LVMH, Gallo, Diegeo, Fosters, and others, most wine businesses, at one point in time, began as a small family business. Indeed, in looking at wine countries around the world, many small and medium-sized family wine businesses still thrive. Though they may not have the global distribution might of the large companies, many do well serving a small niche within a state, region, or country. They may specialize in a certain grape varietal or style. Some may refer to themselves as a "boutique" winery and only sell direct to customers who visit their private tasting rooms or to local restaurants. Regardless of their size, certain work functions must be performed, and these dictate the number and skill level of employees needed.

Table 15.1 provides an overview of the four major job categories within an integrated wine business. These are vineyard, wine production, marketing/sales, and management/administrative. Within each category, various job titles and positions are listed. The existence of the position depends on the size of the business. In some cases, one person may perform the function of several positions. In other businesses, the position may be outsourced to a labor contracting or consulting firm. As a company grows in size and complexity, it usually adds more positions to handle the increased production and regulatory requirements. For example, in a recent Western Management Group (2008) survey for the US wine industry, they list more than 200 independent job titles. However, most of those titles are subsets of the four categories described in Table 15.1. The following paragraphs highlight some of the key positions within a wine business.

Employee Functions in the Vineyard

A common phrase in the wine industry is that "the best wine is made in the vineyard." Indeed, an old joke claims

Table 15.1. High-Level Job Categories in Vineyard/Winery

Vineyard
VP of vineyard operations
Client relations
Vineyard managers & supervisors
Vineyard workers

Wine Production
Winemaker
Asst. winemaker
Laboratory technicians
Enologists
Cellar masters & supervisors
Cellar workers

Marketing/Sales
VP of marketing & sales
Marketing & sales management & supervisors
Sales reps
Public relations
Hospitality staff (tasting room or cellar door)
Direct sales

Management & Administration
General manager/CEO
CFP/controller
Accounting
Human resources
IT
Administrative staff

that the best winemakers "will set up a tent and camp in the vineyard," because the quality of the grape is the most important component of a fine wine. Obviously, the making and blending of wine requires much skill, but the health and operation of the vineyard is critical. Thus, the job of the *vineyard manager* (and *supervisors*) is one of the most important. This individual oversees the annual cycle of the vine and the performance of the *vineyard workers* who prune in the winter, sucker in the spring, thin in the summer, and harvest during the fall. During harvest, vineyards usually hire additional part-time vineyard workers to help bring in the grapes.

Throughout the year, additional work includes potential applications of fertilizer, pesticides, and water—depending on the country and legal issues regarding viticultural practices. For example, in some vineyards in South America, flood irrigation is still practiced, whereas in the US, drip irrigation is prevalent. Furthermore, some vineyards utilize organic or biodynamic farming methods, which don't allow for the use of most pesticides, but call for more labor in terms of weed control, cover crops, managing beneficial insects and other predators, as well as applications of natural field sprays, as is the case with biodynamic vineyards.

Finally, a new trend is the installation of scientific equipment and monitors in the vineyard, such as weather-monitoring stations, neutron probes, and GPS systems. These new technologies cause the work of the vineyard managers, supervisors, and workers to become more highly skilled in computer applications and data analysis. However, the traditional vineyard skills remain very important. These include not only the physical ability to complete the required work, but knowledge of viticulture, grape diseases, and other environmental conditions. A worker with good knowledge of the vine, who has a long history in a particular vineyard, can be a very important asset to a wine business.

Other roles in vineyard operations may include *client relations*. If a vineyard is privately owned and not part of a winery, the vineyard owner needs to sell grapes to a winery. This task calls for skills in sales and relationship management with winemakers and winery owners. In many cases vineyard owners establish long-term contracts with wineries, which require the vineyard to be farmed in a way that meets the winery's criteria. For example, the winery may expect a certain tonnage of grapes from the vineyard, which could cause the vineyard owner/manager to thin the crop more strictly than he/she would on their own. This requirement may demand additional work in the vineyard, and therefore increases labor costs. Customer relations, sales, negotiations, and contract management skills are required for this function.

If the winery owns their own vineyard, the vineyard manager will farm the vineyard according to the needs and specifications of the winery. In a large corporation, this function is usually assumed by the *VP of Vineyard Operations*, who oversees many vineyards, each with their own vineyard manager and crew. They may manage the vineyards in such a way that specific grapes are designated for specific wineries, as is often the case in large global companies.

In many cases, a winery will manage their own vineyards and also buy grapes from independent producers. This arrangement calls for a *manager or VP of Vineyard Relations* who negotiates long- and short-term grape contracts with a portfolio of independent grape growers. Finally, some wineries choose to outsource the complete vineyard operations side of the business to a vineyard contractor, who supplies all the labor and equipment, and oversees the complete vineyard farming process.

Employee Functions in Wine Production

Perhaps one of the most crucial positions for any winery-based business is the *winemaker*, who is often compared to a star player or quarterback. Indeed, the names of top winemakers are referred to in reverent terms and are considered celebrities in their own right. In a small family-run wine business, the owner is often the winemaker, general manager, and vineyard manager combined into one—with perhaps a son, daughter, or spouse helping with some of the other administrative and sales roles. Another option in a small family-run business is to hire a consulting winemaker of top billing to assist in making the wines. This practice occurs commonly in newly established wineries, in which the owner has sufficient funding to pay the high salaries demanded by the top consulting winemakers.

Larger wineries usually have a head winemaker, as well as several *assistant winemakers*. The global wine corporations may have a different winemaker for each winery in their portfolio, or have a cadre of winemakers who work across facilities, specializing in certain price points or varietals.

Winemakers possess a very specific skill set. Generally, they must have graduated with a bachelor's or master's degree from a top winemaking university, or have worked years as an assistant winemaker, learning the craft from an expert. Their background usually includes chemistry, math, and viticulture, plus practice in the art of blending. Some describe winemaking as an art; others as a science; some as a little of both. Regardless of the definition, the skills of a winemaker and quality of the grapes are both very important to the excellence of the final product.

Another important wine production function is the *enologist*. This is someone who runs the various lab tests on the wine to test for quality, consistency, and specific components. In smaller wineries, the winemaker may also fulfill this function or it may be outsourced. Larger corporations may have whole staffs of enologists and *lab assistants*.

Finally, the cellar crew in a winery is very critical. Again, depending on the size of the operation, there could be a *VP of Operations*, *Cellar Masters* (managers), *supervisors*, and a large crew of *cellar workers*. The work involves a variety of processes, ranging from handling the crush during harvest season to cleaning barrels and tanks, racking the wine, testing, topping, blending, fining, and eventually bottling. Other functions may include inventory control, warehousing, and shipping logistics. Again, in a small family-run winery, these functions might be completed by the winemaker and family/friends. Other companies choose to outsource such work to a custom crush operation. It is a labor-intensive operation, requiring physical strength and flexibility, as well as good math and logic skills. Managerial and supervisory roles require leadership and organization skills, plus being willing to pitch in and help do the work when needed. During crush, when the grapes are being processed after harvest, the size of a cellar crew swells with part-time workers to handle the additional work.

Employee Functions in Marketing and Sales

Marketing and selling wine is a very complex process, and is an issue that many wineries grapple with on a daily basis. Most countries, with the exception of the US, produce more wine than their population can consume. Therefore, exporting wine to other countries is necessary for survival. Fortunately, in most of these countries, the governments provide assistance by establishing policies which assist in exporting. Because of the importance of marketing and sales, every winery has someone who performs the function of *VP of Marketing/Sales*. In a small family-run winery, the owner may perform this function and also serve as winemaker and general manager. In larger wineries, it will be a full-time job, and in global corporations it may be broken into two separate positions, including a large staff of marketing plus sales *managers, supervisors*, and *representatives*. Some wineries may also choose to outsource some or all of this work to a marketing/sales consulting agency.

Sales and marketing skills in the wine business require more than traditional sales and marketing knowledge.

Professionals also must have a good knowledge of wine, understand the competition, but even more important, be well schooled in the legal regulations of shipping and sales in certain regions, states, and countries. This knowledge includes not only special label requirements, but also the distributor network through which wine is sold in different countries. For example, in the US where wine is still sold through the three-tier system in many states, the winery *sales professional* (Rep) must use relationship skills with distributors to get their wine placed in prominent retail locations. This is because the winery cannot be directly involved with sales to the customer in certain states. Many wineries hire a cadre of sales reps to work certain states and develop relationships with the distributors and retailers who serve those areas. This is also similar with international sales, in which a sales rep who speaks the language of the country and understands its culture is located in-country to promote positive relationships.

Traditional wine marketing activities, which include market research, brand development, pricing, and advertising, are usually performed by a different group of employees located within the winery headquarters. Finally, another area of marketing and sales for many wineries is direct sales to consumers. This is often performed through the use of a tasting room or cellar door sales, and requires employees who are skilled in wine hospitality and sales. A *Tasting Room or Hospitality Manager* is usually hired to oversee the operations of the tasting room. *Tasting room employees* are trained in how to let customers taste wine and encourage them to purchase both wine and wine-related merchandise, if available in the tasting room. This requires direct sales and customer service skills. In the past, these positions were viewed primarily as public relations positions, but in more recent years wineries have begun to see the benefits of training these employees in sales skills to improve revenues. For small wineries, this is often a primary method of wine sales.

Finally, other marketing and sales jobs within the wine industry, which are newer in practice, may include *wine club manager, Internet sales manager, public relations manager*, and *special events coordinator*. Again, these functions can all be performed by the same person in a small winery or be separate positions with direct reports in a larger operations. Finally, these can also be outsourced to an external consulting firm or agency.

Employee Functions in Wine Management and Administration

Oversight of a winery calls for the position of a *general manager or CEO*. This position requires traditional leadership skills, as well as good operations management, strategic thinking, marketing, and financial skills. The general manager oversees the smooth functioning of all the operations within a wine business, including collaboration and coordination between vineyard, wine operations, marketing, sales, and other administrative functions. In a small winery, the owner may assume this role. In larger companies, there may be a general manager for each winery in the portfolio, as well as a CEO for the whole company. In addition, many wine businesses have a *Board of Directors* to which the general manager and other executives report. The Board of Directors is often comprised of investors and/or shareholders in publicly held companies.

Reporting into the general manager or CEO is usually someone from the other three categories described in the previous paragraph, as well as three other important areas. The first is the *CFO* and/or *controller* function, which is responsible for finance and accounting. In larger companies, there will be a department of accountants responsible for tracking finances, paying invoices, and balancing the books. In smaller operations, this role may be handled by the spouse and a hired CPA.

The *human resources manager* is another function reporting to the general manager. This role includes overseeing the hiring, training, salary/benefits administration, employee relations, and legal record-keeping process for all employees. In small companies, it is often performed by the general managers and other supervisors, and/or may be outsourced. In larger corporations, there is usually a VP of HR, as well as directors, managers, supervisors, HR reps, and HR assistants. Also, there is often a *legal representative* on staff or a retained service agreement with a law firm to support the HR function regarding labor laws.

Another important administrative function is *IT (information technology) professional*. The role of this function is to oversee the effective operation of the company's computer and software infrastructure. As wine companies grow, it is becoming increasing critical to be networked globally to suppliers, distributors, and customers. The role of the IT function is gaining increasing importance, and in larger wine businesses this is reflected in new titles of CIO (chief information officer), as well as full-time directors, managers, supervisors, and reps to support the function. In small companies, this role is often outsourced.

Finally, every well-performing wine business has a cadre of *administrative support staff* who support all of the various functions mentioned above. They can be compared to the glue that holds everything together by effectively handling office communications, paperwork, computer input, and a variety of customer service functions.

Defining Human Resource Management in the Wine Industry

Human resource management (HRM) can be defined as *the management policies and practices in the wine industry that impact employee satisfaction and performance.*

Specifically, this covers the five categories of: 1) staffing and recruiting; 2) compensation and benefits; 3) training and development: 4) employee relations; and 5) record-keeping and legal issues. Figure 15.1 illustrates these major categories—or buckets—of HR, which are ideally linked to the company strategy.

Like other industries, the function of human resources in the wine industry is one that has only achieved prominence in the last several decades. Because so many wine businesses begin as family businesses, they often treat employee as "extended family." Indeed, when interviewed, one of the major reasons employees enjoy working at small and medium-sized wineries is because of the "family-like culture" (Thach, 2002). Because of this, most of the HR functions have traditionally been handled by the owner, or general manager. *Good HR practices are actually good management practices.* Companies that treat their employees with respect and dignity are naturally engaged in world-class HR practices.

Figure 15.1. The five categories of human resource management.

However, once a wine company has at least 100 employees on the payroll, it is highly recommended that a full-time human resource professional be hired to oversee the function (Davison, 2003). This is because increased size brings increased complexity regarding legal issues and performance management, but more importantly because there is a need for more intense focus on the strategic issues of human resource management. The costs of labor are high, and the impact of satisfied and productive employees on customer service and revenues has been proven time and again. Therefore, savvy wine businesses realize they need to professionalize by hiring people with HR knowledge and experience. Indeed, many of the large global wineries have recruited HR professionals from other industries to help them remain competitive in the following five major HR categories.

Staffing and Recruiting

Strategic staffing involves analyzing the strategy of the company and forecasting the type of skills and number of employees that will be needed currently and in the future to keep the company operating productively. For example, if a winery plans to double case production within 5 years, they will also need to determine how to handle the corresponding labor needs. Some of the questions they need to answer are: What types of skills will they need to accomplish this? How many people do they need to hire? Can they retrain and/or promote? Will it be more cost-effective to outsource? Or a more relevant question in today's consolidation mode is: Can I acquire another winery and use the skill set of the employees already working there? A reverse side to this issue is the potential need to downsize a company, and develop a plan to assist impacted employees in finding new jobs.

Once a strategic staffing plan has been put in place (and is updated on an annual basis), a wine business needs to focus on the day-to-day recruiting needs. This involves developing a strategy and procedure around how they will replace employees who may leave, as well as hiring for new positions. In the New World wine industry the most common recruiting method is *networking*, or word of mouth. Often by just getting the word out internally among a workforce, or via the supplier and distributor network, a wine business can solicit many resumes.

However, in cases where a specific skill set or years of experience are needed, the following types of recruiting methods are most commonly used: advertising in wine-related journals and newspapers; advertising via the Internet; and/or use of search firms. This is almost always the case with higher level management and executive positions.

Once a good selection of resumes has been received, the HR staff will evaluate them against the job requirements. In most cases, prior wine experience is a plus. Candidates will then be invited to interview with HR staff and the hiring manager(s). Next, testing and reference checking may occur, as well as analysis regarding any legal hiring requirements and "culture fit." More sophisticated companies use a standardized set of interview questions to evaluate a candidate on specific competencies, which may result in a numerical score. Once the candidate successfully passes the above processes, they are welcomed to the company through a series of orientation programs that may include training, tours of the facility, and matching with a mentor or buddy. In small companies, this is often an informal introduction to other employees and perhaps a welcome lunch.

Unique Wine Industry Issue
A major staffing and recruiting issue unique to the wine industry is the need to hire a large number of contingent workers during harvest and crush. This is the case in almost every New World wine country. Even those who

primarily use machines to harvest still need to hire some part-time workers to assist in the cellar with crush. Much time and effort goes into hiring employees to do this work each year. In some countries, wineries have established special housing for the contingent workers; in others they hire buses to pick up workers at their homes in near-by towns and then return them at the end of the work day. Another practice is to hold a Harvest Job Recruiting Fair, as well as to recruit at local universities for "part-time harvest interns."

Compensation and Benefits

Designing fair and competitive compensation and benefits systems is another major function of HR professionals. This often begins with a compensation and benefits strategy, ideally developed jointly by line management and HR. The strategy describes how the wine business will position itself in comparison with the competition. For example, a strategy may state, "we will pay 5% over market average." To establish a strategy, most wine businesses will participate in *salary surveys* to establish ranges for key job positions. In addition, HR staff will monitor any advertised salaries and benefits for both domestic and international job listings posted by their competitors.

Salary generally includes base salary, and any bonuses, overtime, stock options, savings plans, commissions, and other perks, such as company car, loans, and memberships, which impact salary. Benefits may include health, dental, vision, life, and disability insurance, as well as vacation, sick time, holiday pay, child care support, flextime, counseling support (EAP), gym, meals, employee discounts, and other options.

Designing a compensation and benefits system requires much effort, and is often regulated by law in different New World countries. For example, Australia, New Zealand, and

Canada have socialized medicine, so benefit plans look much different there than in the US. Also, currency, exchange rates, minimum pay, overtime, and cost of living differences between countries impact system design. All of these salary and benefit issues impact both employee satisfaction and the overall production costs.

Unique Salary and Benefits Practices in Wine Industry
One practice that is unique to the wine industry in this area is the use of creative perks and bonuses. For example, many wineries in New World countries give employees free or discounted wine, and often schedule celebrations and parties with wine and food around the cycles of the vine, such as pruning parties, budbreak, harvest, barrel tasting, and release parties. Related to this are special bonuses for working through harvest, returning early for pruning season, or selling a targeted amount of wine in a tasting room (Thach, 2002).

Training and Development

Training and developing employees in the wine industry is very important to worker satisfaction and productivity. The most common type of training is *technical or job related*. Examples include vineyard maintenance, cellar operations, wine knowledge, computer software training, as well as a variety of training on different job procedures. This type of training is offered either in a classroom format or one-on-one with a supervisor or peer demonstrating the proper technique.

One of the most important types of training in the wine industry is *safety training*. Winery workers are exposed to a variety of workplace hazards and must know programs like respiratory protection, confined space, hazard communication, and lock-out/block-out that are designed to keep them safe. Past practices, which assumed that all workers have sufficient common sense and will endure a significant level of pain and suffering just to be in the work force, have dramatically changed. Today, winery and vineyard employees need to be taught and learn the correct common sense ways to work safely.

In addition, in most countries, employers also have the ethical and legal responsibilities to ensure the correct job safety information is provided to every worker. For example, in California, safety training "tailgate sessions" often occur once a week. These involve the vineyard manager or supervisor teaching vineyard workers a "safety tip" for the week. In many cases the training is conducted in both Spanish and English, as many of the workers speak only Spanish.

Teaching safety to employees can be a big financial investment for wine businesses, but it provides a tremendous positive return. When employees work safely and without injuries, then they are happier, healthier, and more productive; medical and workers' compensation insurance costs are reduced; and employee replacement costs are reduced. By keeping everyone working safely, the efficiency of the company increases and profits can be made, which creates a win–win experience for both employees and owners.

Other types of wine business training include *customer service* and *sales, supervisory and management skills*, and updates on *industry and regulatory issues*. *Language* training is also important in some New World countries. For example, in both North and South America, winery employees are sometimes offered training in either English or Spanish language skills, so they may be come bilingual.

Wine Industry Training Issues
There are several training and development issues in the wine industry. One of these is *career development* and *succession planning*. Because there are so many small, family-owned businesses, employees who are not related to the

family sometimes complain that there are no upward career opportunities for them. Therefore, some employees will transfer to another winery in order to advance. Related to this is the issue of succession planning, because family members do not always want to take over a wine business. Therefore, determining who will succeed as head of a wine business is a common concern.

Another issue is the need for *advanced skill levels*. As the industry becomes more global and complex, additional skills are needed, such as computer knowledge, data analysis, global sales and marketing skills, and even basic literacy skills such as reading and writing for some field workers. Because of this, in some areas of the industry there are talent shortage issues. For example, in many wineries a cellar worker is an entry-level labor position. Applicants for such positions often lack the ability to advance to higher skill levels without a structured, well-designed training program.

Leadership and management skills are also an issue in this culturally diverse, ever-changing, competitive industry. Highly sought-after technical skills can be offset by a lack of leadership and management skills, which can put a company in a compromising position. Therefore, wine businesses continue to seek and develop employees who have both the technical and leadership skills to move their businesses forward in the global market (Thach & Shepard, 2001).

Employee Relations

Employee relations includes providing clear goals, feedback on performance, and two-way communication with a goal of creating a positive and productive work environment for everyone in the company. Ideally, this is a joint partnership between line managers and HR staff. Clarity on goals is provided via employee meetings and other communication tools, such as an employee handbook and notices. Feedback on performance is provided via informal discussion, performance appraisals meetings, discipline conversations, and recognition and rewards for a job well done. Two-way communication is encouraged as part of the culture, and may be formalized in "open door" policies, grievance processes, and/or standard meetings between employee(s) and manager. Wine businesses that implement these processes well and treat employees with respect and dignity find that employees are more satisfied, productive, and contribute innovative ideas to help the business succeed.

For example, in some wine businesses, practices have been established to encourage employees to continually look for ways to streamline processes and improve quality. They do this out of a sense of ownership and pride, because of the open culture and communication in the company. They will go above and beyond to ensure success, and truly are a company's competitive advantage.

Employee Relations Issues
One issue in this area, which is not unique to the wine industry, is *union activity*. Though some New World wine countries are required to use union employees, others are not, and prefer to keep owner management control of employee policies. The best way to do this is to treat employees better than union-required practices. Another issue is increased *turnover* in the wine industry, especially among certain jobs, such as tasting room employees, assistant winemakers, and some field worker positions. The cost of turnover, in terms of recruiting and retraining workers, can take a toll on company profits. Turnover is also a morale issue that can lead to decreased productivity and increased production costs also affecting a company's bottom line.

Record Keeping and Legal Issues

Governments in wine countries have different policies on how employee records are to be filed and maintained, as well as laws regarding employee relations. Many agree that the strictest legal policies are in the US, where auditors can review employee records at any time and verify that proper permits and identification are on file for each employee (e.g., social security card, I-9 permit, etc.). If errors are found, employees can be dismissed and the business fined.

There are also laws about discrimination and worker treatment that vary by country. One interesting example is in South Africa, where workers have lived for years in winery-provided housing on the vineyard property. Now the South African government had introduced a policy that requires property owners to give the house and the land on which it resides to the workers—if they had lived there for a certain number of years. This prompted many South African winery and vineyard owners to buy houses for workers in a nearby town, so they could retain ownership of their vineyard property.

This is different from practices in the US, where housing for migrant workers from Mexico is a growing social issue. Because many of the workers migrate to the US during harvest and pruning season and then return to Mexico, they do not have permanent housing. To save money, some will rent a small apartment or hotel room and then crowd 10 or more other workers in to sleep on the floor. Others may sleep in cars, under bridges, and in barns. Community concern over this has lead some groups, such as the Napa Vintners Association, to build migrant worker housing so the workers have clean, warm, and safe quarters in which to stay.

In Chile, the worker treatment situation is a bit different. For years, workers lived on the large "hacienda style" grounds of the winery and vineyards. Often they were paid in wine as well as currency. However, after a social movement within the country to limit alcohol consumption, many workers left the vineyards to move to the cities and acquire different jobs. This caused winery owners to hire buses to go to the local cities and towns to pick up workers and transport them to and from the vineyards to their home each day.

Worker treatment and discrimination issues continue to be a subject of debate in many countries, and are one of the human resource issues to which special attention should be paid. Positive efforts are under way in many wine-producing countries, such as the Fairtrade wine movement in South Africa and South America, as well as Sustainability and Social Responsibility assessments and certification programs in other countries. However, a consistent focus on these issues should be emphasized. In the long run, this will benefit wine businesses by not only providing more positive public relations press, but also in improved worker satisfaction and morale, which can lead to increased productivity and company profitability (Davison, 2003; Ulirch, 1997; Yeung & Berman, 1997).

Future HRM Issues

As the wine industry becomes more globally competitive, additional human resource issues will become more visible, in addition to those listed above. Following is a short list of some of these potential future HRM issues.

- **Implementation of Fairtrade Wine and Other Certification Programs for Workers:** As described above, it is expected that there will be more consumer pressure to verify that vineyard and winery operations are treating workers with respect and dignity—and implementing HR programs to insure their safety and well-being. Already in the UK, there has been a huge increase in Fairtrade wine sales. According to Fairtrade Fortnight (2007), the Fairtrade wine category has grown 77% in the past 2 years, and sales are estimated to double in the near future. The premium charged for these wines directly benefits workers and the local communities, as it is invested in social and economic projects. Furthermore, individual wineries and vineyards implement progressive HR programs such as health, school, and housing support. For example, the Los Robles Winery in Chile has purchased a school bus so local children can now attend school, provides supplementary health insurance, and has enabled 30 workers to buy a house.

- **Increasing Use of Mechanization and Impact on Workforce:** The rise of technology in the vineyard and the use of mechanization are expected to continue to grow in the future. This is similar to other industries, which have replaced workers with technology. The impact is a forced "reskilling" of the labor pool in these industries. Though there will always be some wineries doing hand-harvesting and pruning of vineyards, the rising labor costs coupled with the increased quality of mechanized methods are forcing this issue in many New World countries.
- **Rising Costs of Worker Benefits:** Related to the above issue is the rising cost of workers benefits—many of them government mandated. The main issue is around health care costs. Even wine businesses in countries with socialized healthcare systems often find they still have to purchase private healthcare as a benefit for some employees, especially those in the professional ranks. In California, new legislation is being pushed to force all employers with 50 or more employees to buy health care for each employee, even though this is not legally required now. The burden of this cost may force many small businesses to go under.
- **The Impact of Vineyards in Asia:** Currently both China and India are planting hundreds of acres of vine-

yards, and are using inexpensive labor to do so (Gastin, 2004). Though many argue that these Asian countries currently do not have the climate and appellation to produce world-class wines, each year the quality improves. In addition, wine consumption rates are rising in this part of the world, whereas they are declining in many other parts. China has an excellent past record of being able to produce other agricultural crops of good quality at much lower costs because of its inexpensive labor force. A good example of this is the apple crop, which China now produces more efficiently than any other country in the world. This has had a very negative impact on the apple industry in the US and other countries. It is possible that China and India could do the same with the wine grape industry.

- **Continued Industry Consolidation:** A final issue impacting HR in the wine industry is that of increasing merger and acquisition (M&A) activity all along the wine business chain (Gilinsky, McCline, & Eyler, 2000). Large wine corporations continue to purchase and add other wineries to their portfolio. In addition, vineyard management firms, distributors, and wine retailers are also consolidating through mergers and acquisitions. Each time this occurs, there is an impact on the human resource function, as some employees become redundant in their roles and must be let go or retrained to assume other needed positions. In addition, current HR systems, such as record keeping, pay structures, and benefits, often must be consolidated into one system—just as other financial and business systems in the companies must be merged. Many experts predict that M&A activity will continue in the wine industry for many years to come. Therefore, this will have an ongoing impact on human resource management.

LEARNING OBJECTIVES:
- Describe the historical context of US wine laws
- Identify tax and compliance issues
- Identify legal issues associated with trademarks and labels
- Describe farming and land use issues in the US
- Gain insights on global legal matters, such as trade issues and shipping internationally

CHAPTER 16

THE LEGALITIES OF WINE

Cyril Penn
Editor, Wine Business Monthly

Wendell Lee
Attorney, Wine Institute

Due in part to their popularity but more to their inherent intoxicating properties, alcoholic beverages are often regulated and taxed more heavily than other products in many countries around the world. Because of this, interesting legal production, distribution, and sales systems have been developed to control and monitor not only the sale of high alcoholic beverages such as gin, whiskey, and vodka, but also wine. These regulations identify provisions for many aspects of wine, including labeling regulations, level of alcohol, trade restrictions, winemaking practices, documentation, currency conversions, and many other legal issues that must be considered in the marketing, sales, and distribution of wine in a global environment.

With this in mind, this chapter provides an overview of some of the major issues regarding the legalities of wine. It begins with an historical look at US wine laws, and then moves forward in time to describe taxes and compliance issues, trademark and label matters, and farming and land use. The chapter then expands to a global view of some of the specific legal issues impacting New World countries. It concludes with useful information on trade and international shipping issues.

THE HISTORICAL CONTEXT OF US WINE LEGALITIES

There is a saying in the wine industry that it is often easier to ship wine between two different countries than it is to ship wine between two different states in the US. This is because, in the US, alcoholic beverages are regulated and taxed more heavily than other products. Until recently, wine, beer, and spirits were regulated, for example, by the same federal agency responsible for regulating firearms. The legalities of wine in America start with Prohibition and the eventual Repeal of Prohibition embodied in the 21st Amendment to the US Constitution. In the US, the 21st Amendment is the mother lode of all US state and federal regulations and laws. All roads come from it, and all roads lead to it.

The temperance movement in the US started in the 19th century as various states passed laws restricting alcohol. By the time the 18th Amendment to the US Constitution was passed in 1919, a majority of 33 states already had laws on the books prohibiting the sale of alcohol within their borders. Repeal came with the 21st Amendment in

1933. By then, Prohibition had devastated the US wine industry.

Had the 21st Amendment simply ended Prohibition, much of today's wine regulations would not exist. While Section 1 of the 21st Amendment effectively repealed Prohibition, Section 2 of the 21st Amendment gave plenary power to individual states to determine the manner of alcoholic beverage production and distribution within their own boundaries. Over the years, Section 2 of the 21st Amendment has been interpreted to mean individual states have absolute power to regulate and control alcoholic beverages in their boundaries and that the federal government must take a "hands off" approach. That assumption is why the regulation of wine, beer, and spirits is so disparate among various states. It is why Pennsylvania and Utah exert such control that they act as exclusive importers, wholesalers, and retailers of alcoholic beverages; why other states have created elaborate licensing and tax systems for alcoholic beverages; why statutes require alcoholic beverages to go through wholesalers; why elaborate "Tied House" restrictions separate producers from wholesalers and retailers; why some laws prevent wineries from choosing what distributor they work with; why Texas has "wet" and "dry" areas scattered across the state; why grocery stores sell wine in some states but not others; and on and on.

Alcoholic beverage regulation may have started out with the noble purpose of addressing temperance issues in the individual states, but these interests are fueled largely by politics; regulation has evolved into a form of political manipulation, if you will, of the *assumption*. State statutes are justified, at times, in terms of public safety but are also too often also motivated by profit.

In the topsy-turvy world of US alcoholic beverage law, the Supremacy Clause, which holds that federal law reigns supreme over state statutes that may conflict, at times appears to be standing on its head. While some of these state alcoholic beverage regulations would be seen as violating antitrust laws in any other industry, these laws exist today due in large part to the 21st Amendment. Franchise security laws, unique to the alcoholic beverage industry, provide statutory protections to wholesalers of alcoholic beverages and places suppliers at a contractual disadvantage. These laws perpetuate a three-tier distribution system (Figure 16.1) that strongly protects the wholesale tier, maintaining a wholesaler's economic interests as a matter of law by making it difficult for a supplier to terminate its relationship. Indeed, if one found a way to effectively and profitably eliminate the middle distribution level in another business it would be considered a stroke of genius: in the wine sector it would be considered criminal.

Wineries need distribution and distributors. The three-tier system looks to be here to stay but the wholesale tier has seen rapid consolidation. There were 100 wineries shortly

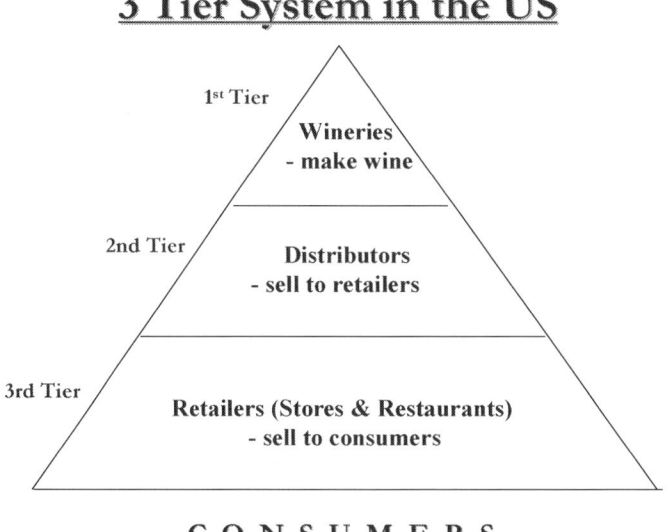

Figure 16.1. The three-tier wine distribution system in the US.

after Prohibition and more than 50 distributors, for example, in California. The top 10 wineries make about 90% of the wine, yet there are over 2,000 bonded wineries just in California (over 4,800 bonded wineries nationally) while two large distributors account for most of the business. Nationally, the 10 largest distributors account for about 90% of all wine sold. Wineries must compete for distributors to help them gain exposure and increase sales. The largest wineries have an advantage here. Smaller wineries can have trouble hooking up with distributors and are often forgotten in large distributor portfolios. Sometimes a distributor must decide what winery's numbers they're going to make and who's going to grow and who isn't. Wineries can depend on their distributors to take orders but not to sell the wine. They must do the selling themselves. At the same time, consumer demand for hard-to-find wines and the higher margins they provide make direct to consumer shipping increasingly attractive to the wineries.

The *assumption* has led to a mish-mash situation where about half of the states now allow people to order wine and have it shipped to their doors but the rest of the states do not. In some states, shipping directly to a consumer remains a felony. Wineries have been making some progress in the legislative arena as well as in the courts. In 2005, the U.S. Supreme Court ruled that Michigan statutes prohibiting out-of-state wineries from shipping wine directly to in-state consumers, but permitting in-state wineries to do so if licensed, discriminated against interstate commerce (**Granholm v. Heald**, 544 U.S. 460, 125 S.Ct. 1885, decided May 16, 2005). The Granholm decision did not overturn State's rights under the 21st Amendment, however. States post-Granholm still have the power to completely prohibit out-of-state wineries from shipping to consumers within the state as long as in-state wineries are also prohibited. Instead of discriminating against out-of-state wineries, some state legislatures have developed novel ways to avoid the results of Granholm by granting shipment privileges based on criteria which does not invoke interstate commerce directly, such as by granting rights to wineries that do not producer over a certain amount of wine per year. Granholm has spawned more, not less, litigation as the tug-of-war for wine availability continues.

Even before the Supreme Court announced its decision in the Granholm case last May, Costco had begun their challenge in Washington State, asserting that it was, among other things, unconstitutional and discriminatory to permit Washington wineries to sell directly to retailers, but forcing out-of-state wineries to sell through the three-tier system. In April of 2006, the trial court upheld most of Costco's claims that Washington State's liquor control system did little to promote temperance nor maintain an orderly market. While the case is on appeal, the Washington Legislature is seeking to make major changes to its alcoholic beverage control policies as a result of the trial court decision.

Taxes and Compliance

Alcoholic beverage "licensees" must be sensitive to federal and state regulations. Failure to comply with the Federal Bureau of Tax and Trade (previously the Federal Bureau of Alcohol, Tobacco and Firearms) and their respective state alcoholic beverage control agencies could lead, among other things, to the revocation of the required alcoholic beverage license. Also, the corporate, limited liability company and partnership companies some other industries use to achieve tax, liability protection, and other goals can also be problematic in the wine business, because these structures can violate state and federal alcoholic beverage regulations.

Some wineries hire outside consultants to assist with regulatory and compliance issues while some of the largest wineries have entire departments dedicated to this area. One such consultant has assembled the following list of agencies that a person establishing a new winery can expect to deal with. It is by no means exhaustive:

- Each state has a state control authority similar to the California Department of Alcoholic Beverage Control. A directory of these control authorities is maintained by the Federal Tax and Trade Bureau and can be found online at http://www.ttb.gov/wine/control_board.shtml
- The US Tax and Trade Bureau (TTB, previously the Bureau of Alcohol, Tobacco and Firearms) issues federal alcoholic beverage basic permits for the production and wholesaling of alcoholic beverages, including permits for alternating proprietors. The Bureau grants label approval, regulates production and trade practices, and collects excise taxes from producer and wholesaler permittees.
- The IRS assigns federal employer identification numbers.
- The county clerk registers fictitious business names for wineries using alternate brand names or for brand names specified by negociants.
- The U.S. Patent and Trademark Office registers trademarks.
- State taxing agency collects employer, income, and excise tax accounts.
- County or city planning departments provide information on zoning and use permits.
- Building departments issue building permits.
- State or county transportation authorities, depending on whether access is from state or county roads, administer parking and access requirements; may require possible improvements or encroachment permits.
- Forestry and Fire Services Department regulates fire protection requirements and hazardous materials use.

- State, county, or city water agencies regulate stream setbacks, underground tanks, wastewater monitoring, and issues required permits.
- Public/environmental health department regulates waste water disposal and issues a variety of permits relating to health concerns.
- Federal, state, and county agricultural agencies issue weighmaster licenses and grape purchase licenses, and oversee pesticide use.
- City business license tax office, if site is in city limits, issues business license.

Whether you hold a Tax and Trade Bureau-issued Bonded Winery basic permit or store wine in a licensed Bonded Wine Cellar (BWC), there are different Tax and Trade Bureau requirements governing the movements of wine in bond, tax paid removals, and wines shipped for export. When it comes to paying taxes, accurate record keeping is essential. Documentation and transfer records need to be held on site for 3 years or more and can be randomly chosen for review if a facility is selected for a random TTB audit. It is a challenge training staff on the changing compliance regulations, record keeping, and tax reporting requirements. TTB has a helpful Internet site and holds educational seminars that cover requirements for record keeping, reporting, and export documentation. If one's export documentation is inadequate, one may be held liable for the taxes and interest. Fraudulent reporting is subject to criminal penalties.

Many wineries hold their wine in storage after bottling in the bonded area of their winery or bonded wine cellar, and do not pay the required excise taxes until the wine actually is shipped. This helps to preserve cash flow, helps control inventories, and can prevent accidental mistakes. However, it is not unusual for wineries to pull bonded or unreleased wines for quality control checks, samples, or prerelease shipments. When this occurs, the wines must be transferred from in bond to tax paid status, and it is important for a winery to understand how to do this.

TRADEMARK AND LABEL ISSUES

One legal area any prospective wine industry professional will want to know about is trademark law. It is amazing how many trademark infringement suits wineries have been involved in. Infringement lawsuits arise when the use of a trademark, usually a wine brand name, conflicts with an already existing brand name. These situations can often be avoided with a trademark search conducted by a professional search firm or attorney. Searches can be extremely intensive and should be left to professionals, but there is a lot that can be learned about a potential trademark by searching three freely available online databases. While a search of these databases will not conclusively resolve whether a name is available, these searches can tell you a lot about whether a brand name is *un*available.

First, the US Patent and Trademark Office maintains an online database of all trademarks that have been registered or are pending registration. When searching this database, it is important to keep in mind first that the test for trademark infringement is whether its use would result in the likelihood of consumer confusion. Similar names in different product categories may not cause consumer confusion (e.g., Plymouth Arrow and Arrow shirts). The USPTO trademark database can be found at http://www.uspto.gov/main/trademarks.htm and the general home page is at http://www.uspto.gov/. Secondly, it is important to remember that registration of a trademark is not mandatory. There are countless trademarks that are currently in use as winery brand names that are not registered as trademark—a search of the USPTO web site alone will tell you only if a trademark is registered or attempting registration.

It is also important to remember that geographic names are generally considered generic and not capable of being a registered trademark. However, one can search the Tax and Trade Bureau's Public COLA (Certificate of Label Approval) database to determine whether a geographic brand name is currently in use. The Public COLA database actually contains data for all label approvals from 1990 to the present and can be found at https://www.ttbonline.gov/colasonline/publicSearchColasBasic.do. Unlike trademark registration, which is not mandatory, label registration is required. Still, there are limitations with a search of the TTB COLA registry because of its incompleteness and because there are many labels that are approved but yet never used in the marketplace. The fact that a brand name may be on a COLA may not necessarily preclude use.

Finally, a search of general online databases such as Google or Lycos may be able to uncover some potential conflicts with existing trademarks. Advanced search techniques on these databases can reveal information useful for a determination of whether a trademark is available for use.

In recent years we have seen many trademark lawsuits. The most well-publicized case involved Kendall-Jackson, who sued E.&J. Gallo claiming Gallo's Turning Leaf brand appeared too similar to its label (Gallo prevailed). There was Galleron vs. Galleron-Lane, Thunderbird (E&J Gallo) vs. Thunder Mountain Winery. In Michigan, Leelanau Wine Cellars sued the parent company of Chateau de Leelanau. Leelanau is the name of a sparsely populated county, a township within that county, a peninsula and a lake, all of which are just northwest of Traverse City in Michigan's northern Lower Peninsula.

The industry is place oriented and the debate over geographic brand names could fill its own chapter and even its own book. Conflicts arise partly because lawyers measure the legal strength of a mark by how readily it can be protected from knockoffs and near-miss imitators, and defended against charges of infringement or deceptiveness. Marketers, on the other hand, measure strength by the extent to which a mark enhances sales volume and margins—a goal that tempts many to seek a boost by making a prestigious geographic origin part of the brand name. There is sometimes some uncertainty about where the legal line lies.

The primary brand name on a wine label is a trademark. Slogans and package graphics are also trademarks, subject to the same legal criteria. Trademarks that say nothing about the nature of the product are strongest in the legal sense. To marketing and sales management, however, the ideal mark gives consumers additional information—typically, something about the product's characteristics or imagery.

Farming and Land Use Issues in the US

Water and air quality laws and land use regulations represent an entirely different set of legal issues for wineries. Winemaking is a manufacturing and packaged goods enterprise but is also an agricultural enterprise. Further, federal and state laws governing the daily operation of the wine industry often interrelate and overlap, making regulations hard to decipher.

There are laws relating to pesticide regulation, air pollution control, solid waste management and water quality, and noise control. In California, the EPA enforces compliance with California Air Pollution Control Laws, the California Clean Water Act, California Code of Regulations, and the Porter-Cologne Water Quality Control Act embodied in the California Water Code—which empowers the State Water Resources Control Board. In California, there are nine regional water quality control boards under the umbrella of the State Water Resources Control Board. They may impose waste discharge and storm water discharge requirements, and regulations specific to land disposal of wastes, groundwater discharges, and non-point source pollution.

The National Endangered Species Act of 1973 can restrict land use on agricultural land, evidenced by the endangered status recently given to the California tiger salamander, an ongoing issue that has delayed some vineyard and winery projects. In addition, wineries that grow their own grapes will need to comply with laws concerning farm labor contractors. Vineyard management company operators may need to register as farm labor contractors or have current contracts reviewed for compliance with the state enforcement.

A Global View on Wine Legalities

The US has the most complex web of legal requirements to navigate. Canada runs a close second. As in the US, Canada's federal government handles health and legal issues while most Canadian provinces have their own liquor control board governing the movement of alcoholic beverages. These entities are similar to "control" states in the US. Even hotels and restaurants in Canada must purchase wine through these liquor control authorities. As with distributors in the US, small and medium-sized wineries often have difficultly getting any attention from the massive governmental entities.

New World countries such as Australia and New Zealand are less restrictive, though, like the US, they have labeling requirements that include such things as health warnings about sulfites. There are distribution issues too, as it can be difficult, for example, to obtain an import license in Australia. Chile and Argentina are less restrictive in terms of their regulatory requirements and though Chile has taxed imports disproportionately when compared with the US, this trade barrier is being eliminated. There are also issues with respect to South Africa, with restrictions on licensing and labeling as well as steep import tariffs.

US producers do not ship much wine to these other New World wine-producing countries. These producers make good wine themselves and are not considered large target markets. Though building a winery can involve legal and regulatory hurdles in many countries, most New World countries are considered relatively easy to invest in. The most challenging land use regulations are probably found in California, particularly within Napa County.

It is important to note that virtually all of the US wine companies that have looked to expand their production into other countries have done so by entering through joint ventures. Having a joint venture partner to deal with local legal and regulatory as well as cultural hurdles is highly effective. It works particularly well when there is a reciprocal relationship. If a joint venture partner is shipping wine into the US, for instance, the US winery is likely to handle distribution, licensing, and trademark issues for the partner. This approach has worked well for a number of wineries but the Robert Mondavi Corporation is probably the best example. Joint venture partners include Opus One with the Baroness Philippine de Rothschild of Château Mouton Rothschild (France); Luce, Lucente and Danzante with Marchesi de' Frescobaldi (Italy); and Caliterra, Arboleda and Seña with the Eduardo Chadwick family of Viña Errázuriz (Chile).

TRADE ISSUES

International trade issues, while they are slow to change, are very important to the wine industry, which continues to globalize. Some of the most restrictive trade barriers on wine have been imposed by the European Union. This is a concern to New World wine producers because the United Kingdom is the largest market for imported wine. The US is continually working on trying to get more harmonization on common issues, such as how wine should be labeled. There are numerous ongoing trade issues. One of the largest and most complex involves "geographic brand names." The Europeans have long complained that the US should disallow the use of geographic brand names such as Chablis, Champaign, and Port, terms that are used generically in the US but that have more specific meaning to European producers.

One of the most recent international trade issues the wine industry is dealing with is the US Bioterrorism Act and its related regulations. The Public Health Security and Bioterrorism Preparedness and Response Act of 2002 was passed in that year by Congress to track America's food supply both in terms of its domestic distribution and imports into the US. The objective of the Act is to improve the ability of the US government to prevent and respond to bioterrorism. It appoints the Food and Drug Administration as the lead agency for developing regulations that will carry out the provisions of the Act. There are three main requirements of the Act. The Act requires members of the food industry (wine and other alcoholic beverages are considered "food" under the Act) to:

- register with the FDA;
- submit prior notices of imports with the FDA; and
- maintain and establish adequate records.

Violations of any provision of the Act can result in food imports (including wine) being held at the port of entry and can subject a company to both civil and criminal penalties. The Act also authorizes the FDA to administratively detain food that raises a threat of serious adverse health consequences or death.

Shipping Internationally

Wineries interested in conducting business internationally will need to be aware of international label requirements and will need to stay up to date on label changes and documentation for shipping internationally. They are going to need to pay excise taxes. Most countries impose excise taxes and this is typically based on alcohol content.

There are some good information sources available via the Internet and there are consultants who specialize in helping wineries with this. The US Department of Agriculture maintains a website which includes useful information about how to do business abroad. Many countries have resources that are available online, such as Australia's Wine and Brandy Corporation. Finally, WineVision has developed a very comprehensive website regarding how to export. It can be accessed at http://www.winevision.org/globalexporting/ or from their main portal, www.winevision.org.

LEARNING OBJECTIVES:
- Define sustainability
- Describe a successful implementation process by analyzing the California Sustainable Wine Growing Program
- Identify and describe five international sustainable wine-growing programs
- List key elements for successful sustainable programs in the future

CHAPTER 17

Environmental and Social Responsibility Issues

Jeff Dlott
President, SureHarvest

Karen Ross
President, California Association of Grape Growers

Allison Jordan
Executive Director, California Sustainable Winegrowing Alliance

Kari Birdseye
Former Communications Director, Wine Institute

Customers around the world are increasingly demanding products that are organic in nature and are produced in such a way that they are kind to the environment and people that help create them. Testaments to this growing movement can be seen on food product labels that verify they are grown organically; shampoos and cosmetics that state the products were produced without harmful testing of animals; and boycotts of coffee, chocolate, and clothing companies that have been accused of less than desirable human resource practices. This trend is also impacting the global wine industry, and already there have been several successful programs implemented in various locations around the world to assist wineries and vineyards in moving forward to embrace environmental and social responsibility issues.

This chapter describes this fascinating trend and resulting practices by providing an overview of sustainable winegrowing efforts. It is organized into the following four sections: 1) definitions of sustainability; 2) a description of California's Code of Sustainable Winegrowing Practices; 3) an overview of international sustainable wine-growing efforts; and 4) a forecast for sustainable wine-growing practices in the next 5 years.

DEFINING SUSTAINABILITY

Since the late 1980s there has been an exponential use of the terms "sustainability," "sustainable development," and "sustainable agriculture" in academia, government, non-profits, and the private sector. Most definitions of sustainable development and sustainable agriculture hold three key principles in common, approaches that are environmentally sound, socially equitable and economically feasible. The combination of these three principles has become known as the three "Es" for *environment, equity*, and *economics*. Figure 15.1 illustrates this principle. As the overarching principle of sustainability has become more widely adopted in the private sector, the three "Ps" has emerged as an increasingly popular way to refer to sustainability: people, profit, and the planet.

Much of the historical roots to the current widespread use of sustainability definitions that include environmental, social, and economic components can be traced to the 1987 report *Our Common Future* from the World Commission on

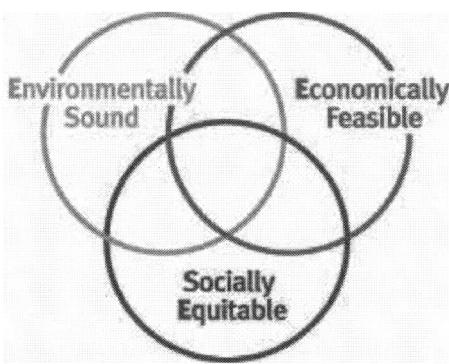

Figure 17.1. The three Es of sustainability.

Environment and Development. This influential report presented the following definition: "Sustainable development is development that meets the needs of the present without compromising the ability of future generations to meet their own needs." This definition was a guiding force in the formulation and adoption of Agenda 21 during the United Nations (UN) Conference on Environment and Development held in Rio de Janerio, Brazil, 1992. As defined by the UN, "Agenda 21 is a comprehensive plan of action to be taken globally, nationally and locally by organizations of the United Nations System, Governments, and Major Groups in every area in which human impacts on the environment."

Agenda 21 specifically addresses social and economic issues related to the environment including poverty, human health, housing, and consumption patterns; conservation and management of resource issues including impacts on land, water, and air resources, agriculture, forestry, fisheries, transportation, material use and waste disposal, etc.; the roles of the public, private, and nongovernmental sectors; and implementation issues including the roles of science, education, financial resources, decision-making, and legal systems. Ten years later at the UN World Summit on Sustainable Development held in Johannesburg, South Africa in 2002, the commitment to fully implement Agenda 21 was reaffirmed.

Given the diversity of individuals, organizations, and sectors working on sustainability, no one definition has emerged to adequately address this diversity of interests. In fact, the number of sustainability definitions has grown steadily to the point where there are many online resources that provide multiple definitions of sustainable development (e.g., http://www.unescap.org/drpad/vc/orientation/awareness/sustainable_development/sd_definition.htm), sustainability (e.g., http://www.ecy.wa.gov/sustainability/more_defns.htm), and sustainable agriculture (e.g., http://www.nal.usda.gov/afsic/AFSIC_pubs/srb9902.htm). Regardless of the definition one chooses or creates, turning principles into practices is far more challenging that deciding upon the "right" words.

CALIFORNIA SUSTAINABLE WINEGROWING PROGRAM

One excellent example of how environmental and social responsibility practices are being implemented in the wine industry can be found in the award-winning California Sustainable Winegrowing Program (SWP). This innovative program gives growers and vintners educational tools to increase adoption of sustainable practices and to measure and demonstrate ongoing improvement.

The SWP was launched in the fall of 2002, when Wine Institute and the California Association of Winegrape Growers (CAWG) began holding workshops throughout the state to introduce the California wine community to the *Code of Sustainable Winegrowing Practices*, a 490-page workbook self-assessment workbook with information on how to conserve natural resources, protect the environment, and enhance relationships with employees, neighbors, and local communities.

The workbook includes 14 chapters of practical guidelines, with a self-assessment tool for California vintners and growers to evaluate their vineyard and winery operations and voluntarily contribute data to measure adoption of sustainable practices. It was designed to be updated periodically to reflect current industry advancements, and a second edition of the workbook was released in fall 2006. Table 17.1 illustrates the list of the 14 chapters included in the Code.

The Joint Committee, comprised of more than 50 members of the Wine Institute and the CAWG, worked on the document for 2 years. Environmentalists, regulators, university educators, and social equity groups provided expertise to the project as well. In 2003, Wine Institute and CAWG established the California Sustainable Winegrowing Alliance (CSWA), a 501(c)(3) nonprofit organization to assist with program implementation.

Statewide Sustainability Reports document results, identify strengths and opportunities for improvement, and set goals to increase use of sustainable practices. The first Sustainability Report was issued in fall 2004, followed by a progress report in 2006, with the second full report to be released in 2009. New workshops targeting the most challenging areas are under way, and follow-up reports will track ongoing progress.

Table 17.1. The 14 Chapter Topics in the Code of Sustainability

Viticulture	Wine Quality
Soil management	Material Handling
Vineyard Water Management	Solid Waste Reduction Management
Pest Management	Purchasing
Ecosystems Management	Human Resources
Energy Efficiency	Neighbors & Community
Winery Water Conservation	Air Quality

The California wine community has provided a majority of the financial support for the program through direct and indirect contributions since its inception. Grants from government, foundations, and companies both underscore the importance of this program and enable CSWA to build on the program's success. For instance, the California Department of Food and Agriculture (CDFA) awarded a grant to the program at an early stage for widespread implementation of the Code's sustainable practices. Subsequent grants from American Farmland Trust, National Fish & Wildlife Foundation, U.S. Department of Food and Agriculture's (USDA) Natural Resources Conservation Service, USDA's Risk Management Agency, CDFA, and Pacific Gas & Electric Company have been used for projects focused on Integrated Pest Management, ecosystem management, air and water quality, risk management, climate change, and energy efficiency. CSWA, Wine Institute, and CAWG work closely with regional winegrowing groups, several of which have sustainability programs of their own, to hold self-assessment and targeted educational workshops throughout the state to help the industry adopt the Code and improve practices.

As of fall 2007, more than 1,300 vintners and growers—who produce over 60% of the wine and farm over half of the vineyard acreage—have participated in over 100 self-assessment workshops. Over 5,000 have participated in another 100 educational workshops targeting specific practices such as energy efficiency and air and water quality.

Another major improvement to the program has been the introduction of a newly launched website (www.sustainablewinegrowing.org), which includes a secure and confidential Web-based self-assessment and benchmark reporting system. In addition to offering participants the option to complete and submit their self-assessments online, the new system offers vintners and growers the ability to generate their own customized sustainability reports, to link to other Web-based resources, and develop and save action plans for improving practices.

The primary audience for this workbook is California winegrowers and vintners. The workbook and website contents are also useful to a wider audience including employees, suppliers, wine grape and wine buyers, neighbors and local community members, members of the environmental and social equity communities, policy makers, regulators, and the media.

A key desired outcome for the SWP project is the widespread development and execution of sustainability strategies in the California wine-growing community. Business strategy is often defined in terms of an operation's mission (the business purpose and fundamental reason for existence), vision (future desire, long-term goals), and values (core ideals, beliefs, and actions). It is important for all businesses committed to corporate social responsibility, from the small family-operated vineyard and winery to the multinational corporation, to clearly define and implement a sustainability strategy.

The Sustainability Mission

The mission for the development and implementation of the workbook is to provide winegrowers and vintners with a tool to voluntarily:

- assess the sustainability of current practices;
- identify areas of excellence and areas where improvements can be made; and
- develop action plans to increase an operation's sustainability.

The overall, long-term mission the SWP includes:

- Establishing voluntary high standards of sustainable practices to be followed and maintained by the entire wine community.

- Enhancing winegrower-to-winegrower and vintner-to-vintner education on the importance of sustainable practices and how self-governing will enhance the economic viability and future of the wine community.
- Demonstrating how working closely with neighbors, communities, and other stakeholders to maintain an open dialogue can address concerns, enhance mutual respect, and accelerate results.

Sustainability Vision

The vision of the SWP project is long-term sustainability of the California wine community. To place the concept of sustainability into the context of winegrowing, the project defines sustainable wine growing as growing and winemaking practices that are sensitive to the environment (environmentally sound), responsive to the needs and interests of society at large (socially equitable), and are economically feasible to implement and maintain (economically feasible). The combination of these three principles is often referred to as the three "Es" of sustainability (see Figure 17.1).

These three overarching principles provide a general direction to pursue sustainability. However, these important principles are not easily translated into the everyday operations of wine growing and winemaking. To bridge this gap between general principles and daily decision making, the workbook's 14 self-assessment chapters translate the sustainability principles into specific wine-growing and winemaking practices.

Sustainability Values

This project is guided by the following set of sustainability values:

- Produce the best quality wine and/or grapes possible.
- Provide leadership in protecting the environment and conserving natural resources.

- Maintain the long-term viability of agricultural lands.
- Support the economic and social well-being of farm and winery employees.
- Respect and communicate with neighbors and community members; respond to their concerns in a considerate manner.
- Enhance local communities through job creation, supporting local business and actively working on important community issues.
- Honor the California wine community's entrepreneurial spirit.
- Support research and education as well as monitor and evaluate existing practices to expedite continual improvements.

To date, the SWP has been very well accepted by wine businesses, policymakers and regulators, environmental and community groups, and consumers. The California wine industry has recognized the need to adopt progressive environmental and social responsibility practices in order to remain competitive in a global environment. The SWP has earned the California wine community a reputation as a leader in sustainability, and has been recognized by several awards including the Governor's Environmental and Economic Leadership Award, the CA Council for Environmental and Economic Balance's Governor Edmund G. "Pat" Brown Award, and CA Environmental Protection Agency's IPM Innovator Award. The program has also been used as an example by other US wine regions and agricultural sectors.

Overview of International Sustainable Winegrowing Efforts

The US is not the only country to embrace sustainable winegrowing practices. Indeed, there have been several other commendable efforts in both the New and Old

World. In 2006, FIVS, an international trade association for all sectors of the alcohol beverage industry, announced a set of principles for the global wine sector that satisfy the triple bottom line of economic, environmental, and social sustainability, among others. California, and programs in New Zealand, South Africa, Australia, and parts of Europe, described below, incorporate the FIVS Global Wine Sector Environmental Sustainability Principles.

Sustainable Winegrowing New Zealand Program (SWNZ)

After conducting a review of international sustainable viticulture schemes, the New Zealand industry developed a program using the Wadenswill (Swiss) scheme as a model. A working group of growers and industry representatives developed a pilot Integrated Winegrowing Program scheme first implemented in 1995–1996 on a trial basis in five vineyards. With a Sustainable Management Fund grant of $150,000 and additional support from Winegrowers of New Zealand, membership in the Sustainable Winegrowing New Zealand (SWNZ) program grew to 120 vineyards in 1997–1998. Similar to the California program, the SWNZ provides a framework for improving all aspects of performance in terms of environmental, social, and economic sustainability in both the vineyard and the winery. A winery program was added in 2002, with current accredited membership levels total more than 360 vineyards and nearly 40 wineries.

The positive points self-audit scorecard has 77 questions covering practices associated with all of the major production issues. A ranking on each practice evaluates whether it has negative impacts (either unsustainable or –10 points); is sustainable (ranked at 0 points); or is an area for desired improvement over the current practice (ranked +10 or +20 points). Scorecards are collected from all members at the end of each season and an analysis of the regional trends in vineyard management practices is reported back at regional member meetings (Gurnsey, Manktelow, Manson, Walker, & Clothier, 2004).

The SWNZ program has a national coordinator and is guided by a steering committee whose goal is to see the program adopted as a minimum production standard across the whole New Zealand wine industry. Plans are under way to develop a database and management and reporting tools to help identify key production issues that will enhance the long-term sustainability of the industry. Additional details about the program can be found at http://www.nzwine.com/swnz/index.html.

South Africa's Integrated Production of Wine (IPW) System

Guidelines for South Africa's Integrated Production of Wine (IPW) System were first published in 1993 for growers and, after several changes, such as the inclusion of wine making and packaging, were finalized in 1997. At the start of the 2005–2006 harvesting season, 400 cellars, producing wholesalers, and bottlers had signed up for the IPW system. This represents 90% of all wine grape production (Tromp, 2006).

South Africa's IPW program for grapes includes: IPW training; farm and vineyard management; soil and terrain; cultivars' rootstocks; vineyard layout; cultivation practices; nutrition; irrigation; pruning and trellising; crop and canopy management; growth regulators; Integrated Pest Management; handling of chemicals; and record keeping. Practices are rated as either good (5 points), average (2–3 points), or poor (0 points). To qualify as IPW grapes, the grower must achieve 75 points out of a possible 150 (or 50%). Other criteria include: only registered chemicals may be used within the specified safety periods; residue analysis must show no prohibited substances; and at least one representative of the operation must attend an IPW course.

South Africa's IPW program for wine includes: IPW training; grape quality; harvesting and transportation; equipment; SO2 levels; substances added to wine; fermentation; cooling; waste water management; disinfectants and cleaning agents; management of solid waste; noise and air pollution; and packaging materials. To qualify as an IPW wine, the grapes must be IPW produced; total points must be above 50% for the wine guidelines; no prohibited residues should be found upon analysis; and cellar records must be maintained (Broome, 2003).

An IPW Conformance Certificate is issued upon an annual third-party audit. Information about South Africa's IPW program is available at: http://www.ipw.co.za/.

Australian Wine Industry Stewardship (AWIS) Program

Although Australia has had sustainable viticulture research and other environmental projects underway for more than a decade, the Australian Wine Industry Stewardship (AWIS) program, commenced in 2005, provides a national framework for these efforts. The 2007 vintage marks the first industry-wide roll out of the AWIS grower survey, issued by participating wine companies to their growers and contractors.

AWIS participants, totaling approximately 40 wineries in 2006–2007, can participate in Australia's 56 regional natural resource management arrangements that are overseen by a multisector board that is responsible for identifying priority environmental issues, providing training and information as well as funding access, and are working towards Australia's eight National Environmental Outcomes. The AWIS survey, a key component of the program, addresses natural resource issues that are also regional board priorities: biodiversity, sustainable production systems, water quality, soil health, salinity, water use efficiency and climate change. Wineries can review responses to identify training needs and tracking improvements.

A national environmental committee for the wine industry is now working to increase the scope of AWIS to include winery practices and to investigate options for membership that could include two levels of participation: self-assessment and third-party verification. For more information, visit http://www.wfa.org.au/PDF/2007_AWIS_National_report.pdf.

European Programs on Sustainability

Modern-day advancements in sustainable viticulture can be traced to the European Integrated Fruit Production Systems initiated in 1977. Integrated Production (IP) standards were adopted in 1993. According to Broome (2003), wine grape-specific regional guidelines were developed by E.F. Boller in Switzerland (2nd edition, 1999, http://www.iobc.ch/IOBCGrapes.pdf).

The 10 sections of the European IP standards are: definition and objectives of IP; commitment of the grower; conserving the vineyard environment; site, rootstock, cultivars, planting system; alleyways and weed-free strips; irrigation; canopy management; integrated plant protection; efficient and safe application methods.

The European IP system is the basis of sustainable viticulture programs in two of the New World wine-producing countries described above: 1) the Sustainable Winegrowing New Zealand program; and 2) South Africa's Integrated Production of Wine program.

Another impressive European effort occurred in France in 2001. Here the Comité Interprofessionel du Vin de Champagne (CIVC) announced the intention of Champagne producers to avoid further harm to their environment by reducing the use of chemical fertilizers, pesticides, and fungicides by as much as 50% with adoption of "viticulture raisonée" (WineNews, 2002/2003). In announcing a commitment to more responsible vineyard practices, the region's 15,000 farmers acknowledged what winegrowers in other countries have come to recognize and embrace: Consumers care *how* the product is made.

United Nations Environment Program

As more agricultural programs participate in the United Nations Environment Program's Global Reporting Initiative, information should become more readily available and more standardization of sustainable viticultural practices could be expected. While the programs referenced here are not inclusive of all sustainable viticulture programs, this sample reinforces the importance of assuring that wines in the global marketplace are produced to meet consumers' expectations for environmentally, economically, and socially responsible practices.

SUSTAINABLE WINEGROWING PRACTICES: THE NEXT 5 YEARS

The global wine community faces a historic "tipping point" where it can become a leading model of sustainability in practice or the current international efforts could unravel and turn into public relations campaigns lacking on-the-ground actions. Other major economic sectors have been investing in sophisticated sustainability marketing campaigns—such as the energy, automotive, and electronics sectors—where the messages are on target but transparency, particularly science-based accountability, has not been the focus. The global wine community has the opportunity to establish and maintain a higher standard of conduct that, if fully executed, could thrust the wine sector into an international leadership position demonstrating that sustainability pursuits can lead to measurable improvements in environmental, social, and economic outcomes.

Key elements for the success of sustainable winegrowing practices in the next 5 years include:

- Leadership by winegrowers and vintners to ensure the widespread adoption of sustainable practices by small-, medium-, and large-scale producers.
- Transparent systems that allow for credible, yet efficient, collection and reporting of information on the true status of sustainable practices on a regional, national, and international scale.
- Improved science, technology, and management systems that improve the effectiveness and efficiency of sustainable practices.
- Increased national and international market-based competition for wines produced and distributed with sustainable practices.
- Increased performance-based local, state, and national regulatory systems that create real incentives to improve environmental and social performance and reduce public and private regulatory costs by utilizing improved technology and management systems.

In conclusion, environmental awareness and sustainability have never been more present in society than today. The issues of green business, protecting the environment, and monitoring social justice issues are in the news daily. The global wine industry is on the forefront of confronting and resolving issues surrounding these topics on many levels, and in many cases is a trailblazer and role model for other agricultural industries.

Note added in proof: *As this book goes to press two other notable sustainability efforts have been announced: 1) The International Wine Industry Greenhouse Gas Accounting*

Protocol developed in partnership with four nations and 2) Italy's Castello Banfi Winery is named the first winery in the world to be internationally recognized for exceptional environmental, ethical, and social responsibility (ISO 14001 and SA 8000).

LEARNING OBJECTIVES:
- Identify the eight major consumer trends impacting wine
- Describe the 11 future forecasts for the wine industry

CHAPTER 18

LOOKING TOWARDS THE FUTURE

Paul Dolan
CEO, Mendocino Wine Co.

Tim Matz
VP of International, Kendall Jackson

Liz Thach
Professor, Sonoma State University

Rich Cartiere (deceased June 2008)
Editor, Wine Market Report

The global wine business today has evolved over the last two decades in a similar fashion as the high-technology industry. Both are global industries that eventually matured to the point of facing competition from overseas producers, emerging country upstarts and numerous domestic rivals. Additionally, each industry is based on products that require specific instructions to use and enjoy, requiring at times a level of additional education that can be nearly overwhelming and daunting.

So it is easy to see that, like computers and other high-tech inventions that have revolutionized our lives, the seemingly simple act of drinking wine is bringing about an evolution in how we live. It ushers in "la dolce vita," or the good life of food and drink. This includes an appreciation for hand craftsmanship and products that reflect the regionality of where they are produced, as well as opportunities of conviviality and camaraderie.

This chapter explores some of these revolutionary and evolutionary changes in wine by presenting a vision for the future of global wine. It begins by examining existing competitive and strategic conditions within the industry that are moving us from a supplier-driven to a customer-driven perspective. Then the chapter describes several different consumer trends that are impacting the future of global wine. Finally, it presents a fairly positive forecast of how the global wine industry may look in the future.

From Supplier to Consumer Driven

For the most part, the global business of wine has most often been viewed from the supplier perspective rather than from the perspective of the consumer. It is most often reported upon, interpreted, and projected based upon how the producer, either grower or winery, sees the future.

But the reality today is that it is the consumer steering the global wine industry—suppliers are no longer in the driver's seat. High levels of production of both wine grapes and wine itself, and hypercompetition from the seemingly ever booming number of brands from around the world, have given the consumer an unprecedented level of power of choice at the retail end. The result is that the consumer is, and will for the foreseeable future be, the primary driver of evolution in the global wine industry.

That said, we should examine key consumer trends to understand the future of the wine business, both domestically and globally. We should remember that wine consumers also are buyers of many products. They live in a rapidly changing retail world that affects how they make decisions, and how they view both products and the supplier sources of those goods.

It is extremely difficult to find similarities between, and make generalized statements about, consumers in different countries. Indeed, most wine trade exporters look to find the differences so they can focus on adjusting their domestic approaches in that manner. But herein we will seek to identify some of the major emerging traits that are found in virtually all consumers in modern, industrialized nations.

Trend 1: Rising Interest in New Varietals From Around the World

In the past consumers have primarily stayed with the tried and true varietals they know such as Chardonnay, Merlot, and Cabernet Sauvignon. In the future, these mainstay varietals will continue to be popular, but there is also a trend of consumers venturing into new territories of grapes. We see this with the increase in global sales of Riesling, Sauvignon Blanc, and Pinot Noir. There is also a more adventurous consumer that enjoys trying signature varietals from key wine countries, such as Tempranillo from Spain, Gruner Vetliner from Austria, Malbec from Argentina, and Zinfandel from California. As newly emerging wine countries in Eastern Europe and Asia come online, they will also bring new varietals for consumers to experience. This trend is expected to increase over the next decade.

Trend 2: Increased Environmental Concerns

Consumers around the world are increasingly concerned with becoming more aware of environmental issues.

They are concerned about air, water, and soil pollution, global warming, wildlife, tree and plant protection, and many other ecological issues. Because of this they are worried about how products are made, and want to be assured that the environment was not harmed from any farming, harvesting, or production practices (Dolan & Elkjer, 2003). They are concerned about recycling of waste products, use of pesticides, and packaging that is biodegradable. Consumers have actually boycotted products they perceived as not being environmentally friendly.

In terms of wine, consumers are already asking for a broader selection of organic and biodynamic wines. In Europe some consumers look for wines that carry the Fairtrade label to verify that winery employees have been treated well in the creation of the wine. In addition, other consumers are asking about the carbon emissions of wine, which is driving more local wine purchases because of the extra fuel to import wine. There is also some concern about cork, because wildlife organizations have reported that the use of cork closures is helping to maintain the cork forests in Spain, Portugal, and Africa. If cork were to be replaced by alternative closures, there is the possibility the forests could be cut down for other crops—which would displace thousands of animal species. Consumers are increasingly concerned with these types of reports and some will change their purchase behavior to help protect the environment and employees.

As described in Chapter 17, a few wine businesses are being very proactive in pursuing sustainable winegrowing practices. However, there is ample opportunity for others to expand on this theme, and for the whole industry to communicate their efforts better. In addition, more innovative designs in packaging, which are reusable and recyclable, are needed.

Trend 3: Focus on Health and Responsible Drinking

The trend of focusing on healthy eating habits is increasing rapidly in many nations, and is expected to continue to do so. More and more consumers are conscious of the types of food and beverages they consume, and want to be reassured that they are healthy, organic, and not made from steroids or artificially created products. Consuming less fatty foods, exercising on a regular basis, and getting tested for various diseases or illnesses are common patterns and behaviors in many people's lives today. This is especially true in the US where the huge Baby Boomer population is aging and becoming much more health conscious.

Fortunately, wine has a reputation for being a healthy drink in most parts of the world. Since the advent of the now famous "French Paradox" on *60 Minutes* touting the health benefits of drinking wine in moderation, many consumers have looked to wine as a health additive (Perdue, 1999). Even though it is basically illegal for vintners themselves to advertise the health benefits of wine in the US, many consumers are still aware of the benefits of reduced heart disease linked to wine consumption.

This linkage of health and wine is still growing around the world. For example, the Chinese government has communicated to its citizens that wine is healthy. Countless doctors in Japan, Northern Europe, and the US prescribe some wine for people over 50 as a preventative aid. Every year new medical research is published with corroborates this fact.

Now new health care diets recommend the addition of moderate amounts of wine, emphasizing its low-fat, low-carbohydrate, low-calorie benefits. However, the wine industry has done little to emphasize these benefits. There is also opportunity to create new types of wine with vitamin additives, herbs, or other health-related flavorings like the bottled juice and water industries have done.

On the opposite end of the health continuum is the concern of binge drinking with wine, which is especially heightened in the UK and Europe. Governments and citizens are demanding that the wine, beer, and spirits industries take more responsible actions with their advertising and discounting, and to assist in ending irresponsible drinking habits.

TREND 4: THIRST FOR INNOVATIVE PRODUCTS

The most successful wine companies in the past were those vintners who sought out consumers and found out what their preferences were. They spent time discovering what their penchants for wine styles and flavors were and, based upon that, sought out innovative ways to deliver value, quality, and simplicity to them in a premium product.

Kendall-Jackson in the USA and Yellow Tail and Rosemount from Australia are good examples of wineries that have produced wines targeted to specific consumer tastes. All have produced new product lines of wine that are more fruit forward, have names that are easy to pronounce, and are priced competitively. Consumers have responded positively by buying these brands and making them top sellers.

A good role model of this premise is Coca-Cola, in the soft drink industry. When Coke saturated the US and western Europe three to four decades ago, one of their next big growth strategies was to go into less developed markets. They have proven successful in India, China, the continent of Africa, Russia, and other countries. All of these markets were consuming beverages of some kind prior to their entry, yet they were able to penetrate, educate, and develop an entire new consuming public.

The wine industry will follow suit, as long as the suppliers and wineries produce products and brands that meet the consumer needs, palates, styles, for each market. It is important to note that Coca-Cola, while the branding is similar around the world (although in different languages), the formula or blends of the product vary depending on the consumer taste profile preferences in each market. Some markets have consumers with a slightly sweeter palate than others. We have already seen the trend in China over just the last few years and really have only scratched the surface. This market alone could exceed all other markets combined in Asia.

There is still ample opportunity to expand in innovation for wine, especially around consumer taste preferences. However, there are other areas of innovation as well, such as packaging, simpler labels, easier openings, more portability, and other taste additives. Some of this has occurred in Australia and Chile, but there is still more room for expansion.

TREND 5: DESIRE FOR BOTH VALUE AND PREMIUM WINES

Consumers have a "big appetite" for discounts and deals as a result of their distrust of corporate claims and government regulations. But at the same time, consumers are searching for "affordable approximations," luxury items that they believe provide the highest quality at the most affordable price. For example, according to Leinberger (2003), 54% of Americans say they want to stay in a luxury hotel, and 47% want to eat in an expensive restaurant.

This provides an opportunity for wineries to prove they deliver value and capitalize on their brands as being such approximates of "the good, the very best in life." This doesn't mean that wineries should only sell cheap, inexpensive wine, but produce and market a high-quality wine, at a perceived good value, or price.

At the same time that consumers are searching for value wines, they are also buying up in wine price point. This is now occurring in most parts of the world—even Germany and the UK, which are known for their very low wine price points. Part of this is because of rising materials costs such as glass and changing grape supply situations, but the other side is the trend for consumers to choose to spend more money on wine. In the past they may have only purchased in the $7 to $10 range, but now are moving up to the $10 to $14 range. These dual trends of seeking value wine and also "buying up" in wine price are expected to continue in the future.

TREND 6: TRUST IN "WORD OF MOUTH"

A reference group refers to a group of friends or influential people that consumers respect. If the reference group recommends something, then other will follow. Due to the lack of trust in society today, the number of consumers who report that "word of mouth" is an important source of information and ideas is also rising. This is, and has been, the single most important influencer of wine choice by a consumer. Because of this wine producers and marketers would do well to concentrate on "influentials," key decision makers in the marketplace. Also, product placement of wine in relevant media advertising, movies, radio, and social networking websites can capitalize on this trend of reference groups.

Data supporting this trend can be found in a recent global study led by the University of South Australia and funded by the Grape and Wine Research Development Corporation (GWRDC) of Australia. The study was jointly conducted by researchers from eight different countries on wine

LOOKING TOWARDS THE FUTURE

consumer purchasing behavior. They found that the two most important factors consumers use to make decisions on which wine to purchase in retail stores are: 1) having tasted the wine previously and 2) having some recommend it (Figure 18.1).

TREND 7: INCREASED ONLINE PURCHASING

We now live in a virtual society—having moved from the rural to industrial; from postindustrial to digital. This started with radio, then TV. It has now moved to the Internet, cell phones, and PDAs. Consumers are more comfortable with researching and purchasing products via technology. This includes wine, and though the growth of wine sales through the Internet has been slower than other industries,

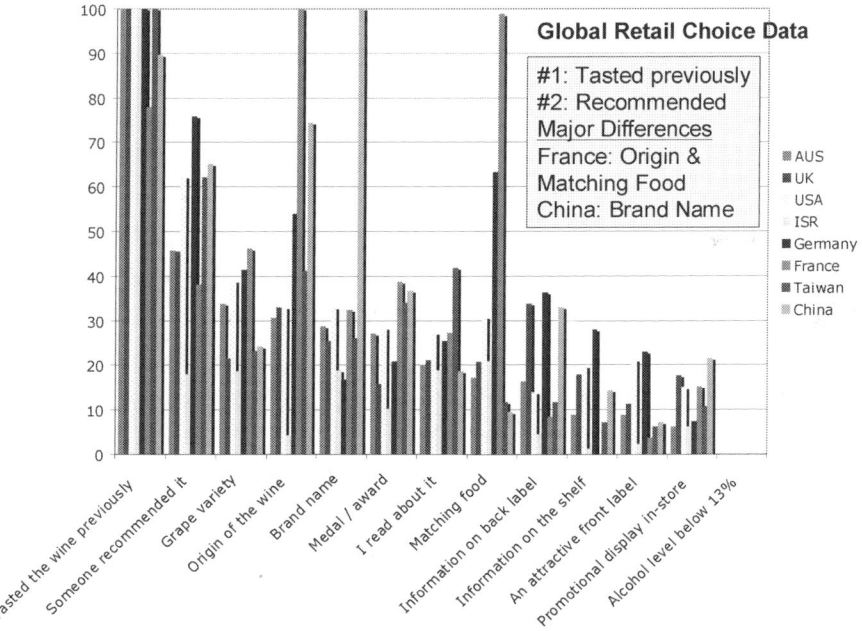

Figure 18.1. How consumers select wine in retail stores. Source: International Consumer Choice for Wine Study by Lockshin, L, Goodman, S., et al (2007)

it is still growing. It is expected that it will continue to increase in the future, especially with the advent of Wine 2.0 (see Chapter 9).

As the trend of purchasing products via the Internet continues to expand around the world, so will opportunities to purchase wines virtually. There are prospects for savvy wine businesses to establish partnerships with wine retailers in multiple countries who can ensure that wine is shipped directly to foreign addresses. Furthermore, current legal and tax restrictions on trade across all borders will continue to dissipate as consumers persist in pressuring global legislators to open trade borders. The rise of international wine clubs is a testament to consumers' desire to purchase wines from around the world and have the convenience of having them shipped to their door step.

Trend 8: Desire for Experiences Rather Than Possessions

Linked to environmental concerns and a growing disgust with the accumulation of material objects over the last few decades, this trend shows that consumers are seeking unique experiences rather than just possessions. This is seen in the increase in adventure travel around the globe, as well as substituting Christmas or other holiday gifts with a trip or money contributed to a charity in someone's name. Consumers want to experience something that they will remember, rather than just add another object to a full closet. This trend is more pronounced in the US and Europe than in Asia.

In terms of wine, this trend is witnessed in the increased number of wine tasting groups as well as a rise in wine tourism. It is also linked to "buying up" in wine price point in order to share a special bottle with friends as a remembered experience, rather than just to store in a cellar.

Future Forecasts for the Global Wine Industry

By examining these eight trends, it is possible to make some predictions on the changes that will occur in the global wine industry in the future. Some of these are just beginning to occur due to economic and competitive pressures that are forcing the more traditional wine businesses to change. Others are driven by innovative players in the industry who are listening to consumers and making changes based on their feedback. Still other companies are focused on cultivating new wine consumers, which creates additional opportunities and new directions. Taken together, we have used these activities, trends, and forces to forecast some of the following changes in the global wine industry.

Continued Growth in Wine Consumption

While wine consumption has fallen in Old World countries, such as France and Italy, it continues to rise in New World countries such as the US, Canada, India, and China. Indeed, the US is predicted to be the largest wine-consuming nation by the year 2010 (Long, 2005). It is predicted that wine consumption rates will continue to rise around the world, especially with a new focus on identifying new wine consumer segments. Research continues to show that wine is one of the fastest growing consumer beverage categories within grocery stores in the US (Byck, 2004).

Introduction of More Innovative Wine Products

As more wine businesses focus on understanding and meeting new consumer needs, intriguing new wine varietals will be introduced. In addition, it is possible that some wineries may begin to include ingredients such as vitamins, herbs, and other types of flavors targeted at matching consumer preferences. We predict that some wineries may focus on the health benefits of wines and promote the low-carbohydrate count found in wines. Others may pursue younger consumer segments with tastes for sweeter, fruit-flavored wines, with bright and humorous labels and names. Already the forerunners of such innovative products can be seen in the market with their fruit-forward wines and brightly colored labels with animal motifs. Successful examples include Yellow Tail (with a kangaroo), Rex Goliath (with a 47-pound rooster), Smoking Loon, and Leaping Horse.

At the same time, there will always be a place for the luxury wines produced by the traditional châteaux of both the Old and New World. The old, venerated labels should be able to hold their specific market niche, as long as they continue to focus on high-quality wines with a mystic of scarceness and allocation.

Continued Creativity in Wine Packaging

The traditional glass wine bottle with a cork closure will continue to exist, but we predict that many new packaging options will continue to be introduced into the global market. Innovative packaging is already available in most countries, such as Australia, Chile, and France, and is now becoming prevalent in the Old World wines. Consumers are asking for portable wine packaging, and more wineries have responded by introducing PET bottles for wine (plastic bottles, similar to the bottled water industry). In addition, more boxes, bags, and closures that do not require a cork screw will be introduced. Also smaller size packaging, such as wine in one- or two-glass servings will become more available to all markets.

Emphasis on Environmentally Friendly and Socially Responsible Wines

Related to the area above will be more wineries that emphasize and communicate their environmentally friendly practices. The consumer expects wineries and suppliers to take responsibility for their respective roles in contributing to these causes. Hence, recyclable packaging will become more important. Already more wineries are including information on their labels regarding sustainable and organic farming practices, carbon emissions, as well as socially responsible practices with employees and other community members. Because the wine business is an agriculture-based business, the sustainability of the land, the protection from pesticides and chemicals, and the promotion of organic soils and grapes are all positive and prevalent behaviors expected in the future. Also, if not handled in a proactive fashion, the wine industry may be forced to implement more responsible drinking programs and advertising.

Continued Consolidation, But More Newcomers

With all the recent mergers and acquisitions one wonders if consolidation will ever stop. We predict that it will continue for a while, not only on the winery side, but also on the distributor and retailer levels. Due to competitive cost pressures, the large global players will need to combine to achieve economies of scale. It is also possible that other beverage industries, such as soft drinks and American beer companies, may enter the wine arena.

At the same time, we see more newcomers entering the market. One of the most interesting and sometimes perplexing trends in the wine industry is how small wineries/companies and small brands can continue to penetrate through the entire system and end up in the consumer's hands growing exponentially without heavy advertising or even distribution clout. Examples include Yellow Tail, that grew from zero to over 12.4 million cases in less than 7 years, and Kumala in the UK market, which grew from zero to 2.5 million cases in just a few years. It is also important to note that brands emerge that have been in existence yet stagnant. For example, La Crema, which recently regained consumer interest with a focus on Pinot Noir, grew from less than 100,000 cases to over 500,000 cases in a few short years—and at super premium prices.

Due to the fragmentation of brands in the wine industry, the consuming behavior of experimentation, discovery, and trial, and the gatekeepers, there will always be small brands that penetrate through the distribution and retail/restaurant tiers to the consumer. This, while challenging to many companies, is actually one of the very refreshing nuances of the wine industry. The small player has a chance to make it. In the soft drink or beer world, while it has and does occasionally happen, more often than not the small new entry gets squashed before it is given a chance.

The other phenomenon in the wine industry is the ability for a very small brand to niche themselves and sell just a few cases at a very high profit, to ensure their sustainability and existence. While wine quality must be there, and getting it into the right hands of wine enthusiasts and public relations contacts, there are many wines that have done this and have been successful. Hence, it is another strange yet rewarding nuance of the wine industry that a small winery can quietly have a very small niche of consumers willing to pay a very high price for a quality wine, which allows that winery to stay and prosper in their existence. This will not change, as it has been done in the Old World of wine and the New World of wine, for the entire history of the industry.

Stricter Wine Label Requirements

With consumer concerns over food safety and government concerns with bioterrorism, it is predicted that the wine industry will soon be forced to provide more information on labels. In the EU there is already serious discussion about requiring wineries to include all ingredients on labels, including calorie, carbohydrate, minerals, enzymes, and other elements. In addition, it is expected that information about environmental issues, such as carbon emissions in making products, may be required in the future, as well as social responsibility information about employee treatment.

It is possible that stricter warning language about the risks of binge drinking and other irresponsible drinking behavior will be expected. Already a few countries, such as New Zealand and the UK, include the recommended unit amount for safe daily consumption on wine labels. It is possible that this may become a requirement for all wineries. Finally, with the concern with quality standards, it is possible that wine could follow the direction of computers and other technologies and eventually need ISO or other quality manufacturing process certification. If this were to occur, it would have to be driven by the large retailers such as Tesco and Wal-Mart. Related to this is the need to have up-to-date barcode or RFD labeling so that a bottle of wine can easily be traced by to a specific lot and each ingredient supplier. This has already been implemented in many wine countries, but label tracking around this issue is expected to become more sophisticated.

The Emergence of a True Global Wine Brand

This is and has been a popular buzzword for a decade now and given all the recent mergers and acquisitions, it is inevitable that a true global wine brand will emerge. This would be a brand that is developed and sold around the world, but would source its grapes from a variety of countries.

Already there are some precursors in existence such as Lindeman's, which is now sourcing Chardonnay and Shiraz from different countries, but selling them under the Lindeman's label rather than create a second brand. Another example is Mateus, which is creating rose wines from red grapes in different countries and selling them in different markets. They have recently sourced tempranillo grapes from Spain to introduce a new rose in the UK market. Both of these companies are in the beginning phase of creating a true global wine brand, but they have much farther to go before they come up with a model that is similar to Pepsi or Coca-Cola, whose ingredients are sourced globally but customized to local tastes.

Continued Increased Cost and Efficiency Focus

As the global wine industry continues to become more competitive, we predict that wineries will need to focus more on reducing costs and increasing efficiency in vineyard, wine production, and marketing efforts. More sophisticated software that tracks and analyzes production, finance, and marketing program results will become more common. Even small and midsize wineries will begin to focus on these issues, as the rise in consolidation and new entry wineries continue to put financial pressures on the industry. The wine industry will need to adopt the more sophisticated business and technology practices of their brethren in other beverage industries.

More Direct to Consumer and Internet Sales

We predict that consumers will continue to demand to purchase wine directly via the Internet and other direct sales methods. This is already occurring at low levels via the Internet, but is expected to increase in the future. As global wine consumption rises, consumers will demand the right to purchase wines from a variety of appellations around the world. Many wine consumers are "discovery-oriented" and enjoy trying a wine from a different country, or a unique style or grape varietal.

Related to this is the rise of more direct to consumer cellar door or wine clubs. The number of wineries that have added clubs so they can ship wine directly to consumers on a monthly or quarterly basis has risen dramatically in the last 5 years. Some are even able to ship wine internationally to customer's homes. In addition, some of the global wineries, such as Constellation, have created international wine clubs, where consumers receive a selection of wines from their wineries around the world. The rise in retailer wine clubs, especially using an e-commerce base, is also predicted to increase. These may not only remain with wine shop retailers, but could expand to grocery stores and restaurants that could create their own consumer wine clubs.

Rise in Asian Wineries and Consumption

As described in other parts of this book, we forecast that wine consumption in Asia will increase, especially in China and India. China is the fastest growing wine-consuming nation now. Likewise, we predict the rise of more vineyards and some very good wineries in this part of the world. India is planting vineyards to develop a viticulture industry that will not only supply the local market, but the Indian restaurants around the world. Driving forces of this forecast include low labor costs, continued French and Australian winemaking consultation in these regions, an emphasis on wine as a prestige and health-related product, as well as governmental support of the Asian wine industry. The rise of this new Asian market force will create both positive and negative impacts on the global wine industry. On the positive side, wine consumption will increase; on the negative side, we see this driving more fiercely competitive cost issues for the industry.

Increased Wine Tourism

A final positive prediction for the wine industry is the continued popularity and rise of worldwide wine tourism. As the experiential, innovation, and environmental trends among consumers expand, the idea of visiting winemaking regions; participating in harvest; and even blending a wine in a "wine boot camp" will appeal to a certain segment of consumers. As described in Chapter 10, wine regions around the world have initiated programs to invite visitors to spend more time tasting and experiencing the good life of wine. There is a range of appealing vacation formats for wine tourism from wine culinary and wine ecotourism to wine paired with golf and spa vacations. We predict this will increase and be available to consumers in all the major wine-producing regions of the world.

CONTRIBUTORS

Jon Affonso is Winemaker and Owner of Rail Bridge Cellars in Sacramento, California. Prior to starting his own winery, Jon worked in the industry for 10 years in research and winemaking positions with Dry Creek Winery, Sutter Home, Renwood, and Château Angélus in St. Emilion, France. Jon holds a master's degree in enology from CSU, Fresno, a wine MBA from Sonoma State University, and a Bachelor of Science in Geology from California State University, Sacramento.

Jean Arnold is President of Hanzell Vineyards, a small family-owned winery that crafts Burgundian-style wines in the foothills of the Mayacamas Mountains in Sonoma Valley. Prior to joining Hanzell Vineyards, Jean was CEO of Jackson Family Farms and its nine independent wineries. She has over 25 years of experience in business leadership, executive management, marketing, and sales with notable wineries such as Chateau St. Jean, Chateau Montelena, Jordan Vineyards & Winery, Chalk Hill Estate, and Williams Selyem. In 1999 she founded her consulting firm, the Jean Arnold Group, to offer luxury positioning and management to private wine industry clients including Laurel Glen Vineyards, Rudd Vineyards & Winery, and the management firm of Motto, Kryla, & Fisher. At this time, The Jean Arnold Group includes Chalk Hill Estate Vineyards & Winery, independent private clients, and Jean's partnership in a new wine venture, Ottimino.

Thomas Atkin is Associate Professor of Operations Management at Sonoma State University, where he teaches Wine Business Operations, as well as general business operations. In addition, he is active in wine business research and has published several articles in this area. He joined Sonoma State in 2001 after receiving a Ph.D. in Operations and Sourcing Management from Michigan State University. His job experience encompasses 13 years as general manager of a manufacturing plant and 12 years in restaurant management.

David Beckstoffer is President and CEO of Beckstoffer Vineyards, an independent grape grower that owns approximately 3000 acres in the North Coast of California, including over 200 acres in the prestigious Rutherford Bench appellation. David has worked in this family business for the past 6 years, taking over the reins of daily farming operations from his father. Prior to entering the vineyard management business, David worked for 9 years at Bechtel in their Project Finance and Development group. He holds a B.S. and M.S. in Civil Engineering from Stanford, and an M.B.A. from the Wharton School of Business at the University of Pennsylvania.

Kari Birdseye is the former Director of Communications at the Wine Institute, where she oversaw the development of the Sustainable Winegrowing Program, in partnership with the California Association of Winegrape Growers. Kari also spent 11 years working for the Cable News Network (CNN) where she held several positions before becoming an Emmy award-winning Executive Producer. She has a B.A. in Journalism from San Francisco State University. Kari was also an author for the *Code of Sustainable Winegrowing Practices.*

Linda Bisson is a Professor of Viticulture & Enology at UC-Davis where she teaches classes in wine production. She is the holder of the Maynard A. Amerine Endowed Chair in Viticulture and Enology. Dr. Bisson is also a member of the advisory boards of the American Viticulture and Enology Research Network and has just accepted the position of Science Editor for the *American Journal of Enology and Viticulture.* She is lead principal investigator on the multidisciplinary, multiprincipal investigator program in stuck fermentations funded by the American Vineyard Foundation. She received her Ph.D. in Microbiology from the University of California at Berkeley.

Lillian Bynum is the Vice President of Human Resources for Delicato Family Vineyards in Manteca, California. Delicato is currently the 10th largest wine company in the US, with over 10,000 acres of vineyards and five major brands. Lillian oversees all human resource activities, including training and development, compensation, and safety. She is a certified Professional of Human Resources (PHR), member of the Society for Human Resources Management, and Past President of the San Joaquin Human Resources Association. Lillian is also a member of the wine industry's Western Management Group Compensation Advisory Board and certified trainer of Franklin Covey's "7 Habits of Highly Effective People."

Jack Carlsen (B.Econs, Ph.D., W. Aust.) is Co-Director of the Curtin Sustainable Tourism Centre at Curtin University, Western Australia. He conducts tourism research at Curtin, which includes tourism marketing, development, planning, and management. Jack has an excellent research track record, and has produced more than 100 scholarly publications in books and book chapters, academic journals, conference proceedings, and research reports to industry and government. He coauthored (with Donald Getz and Alison Morrison) the first book on *The Family Business in Tourism and Hospitality*, published in 2004 and has recently coedited (with Stephen Charters) *Global Wine Tourism: Research Management and Marketing* (2006) in addition to more than 20 scholarly publications on wine tourism-related topics.

Richard Cartiere was publisher of *Wine Market Report*, which reports on the "inside story on the wine business for more than 5,000 wine industry executives worldwide." In addition, he was also editor of *Global Wine News e-Monitor*. Rich had many years of experience in wine business writing, was a frequent presenter at wine conferences, and was one of the founding members of WineVision. His journalism career spanned 25 years, including stints as the West Coast editor of *Wine Enthusiastic* magazine, as content editor on *Smart Wine*, and as assistant business editor at the *New York Times* Regional Newspaper Group's wine country newspaper in Santa Rosa, CA. Sadly, Rich passed away in June 2008. He will be greatly missed by all in the wine community.

Tony Correia is an Accredited Rural Appraiser, and specializes in the appraisal of large, complex, agricultural properties, and difficult appraisal assignments. He is the president, and owner, of Correia-Xavier, Inc., and is an instructor of appraisal courses and seminars throughout the nation and Mexico, and is also a frequent public speaker on agricultural, appraisal, taxation, and estate planning issues, and the vineyard and wine industries. He has been a guest speaker for many organizations, including the American Bar Association and the Appraisal Institute. Tony is a founding member of the World Association of Valuation Organizations, and a graduate of CSU-Fresno, with a Major and postgraduate work in English, and a second major in Russian.

Jon P. Dal Poggetto, CPA, is Managing Partner of the Santa Rosa accounting firm of Dal Poggetto Company LLP. He has 29 years of public accounting experience in the wine industry. He served as National Director of Winery Services for the international accounting firm of Touche Ross & Co. before starting his own firm in 1992. He is the author of *A Practical Guide to Winery Cost Accounting* and served as chairman of the 1992 Wine Industry Conference for the California CPA Foundation. He was also a founder of the Deloitte & Touche Wine Industry Financial Survey.

Jeff Dlott is Owner and President of SureHarvest, a company that offers software and services for sustainable agriculture. The previous name of the company was RealToolbox. Jeff received his Ph.D. in 1993 in entomology at UC Berkeley where he combined ecological field research with social science program design and evaluation tools to understand and extend sustainable agriculture systems. Jeff continued his natural and social science sustainable agriculture research and taught insect ecology and agricultural ecology at UC Berkeley from 1994 to 1995. In 1996, Jeff founded Collaborative Research and Designs for Agriculture, a sustainable agriculture research and educational 501(c)3 nonprofit focused on the design, implementation, and evaluation of large-scale projects to improve the environmental performance of agricultural ecosystems. Jeff is also one of the authors of the Code of Sustainable Winegrowing Practices.

Paul Dolan is president of the Mendocino Winery Company, which is known for such award-winning wines as Parducci, Big Yellow, Roselle, Tusk'n Red, Sketchbook, Zingaro, and Paul Dolan. Prior to this job, Paul was the president and CEO of Fetzer Vineyards. A native Californian descended from a family of winemakers, he has successfully integrated his personal interest in sustainable business practices into day-to-day operations. Paul is very active in the California wine community, and a founding member of WineVision. He recently published the book *True to Our Roots: Fermenting a Business Revolution*, which describes how sustainable winegrowing practices have been applied at Fetzer. He holds a Master in Enology from CSU-Fresno.

Megghen Driscol is a California-based Public Relations & Corporate Communications consultant with more than 15 years of experience in the wine industry. Her prior positions include Director of Public Relations and Corporate Communications with Allied Domecq Wines, Director of Public Relations and Communications at Diageo Chateau & Estate Wines, as well as Vice President of Public Relations at Southcorp Wines for more than 6 years. Megghen has served on the Wine Market Council, the Australian Wine Bureau, the Wine Institute, and several local and international wine associations. She began her wine career at the Sterling Vineyards School of Service & Hospitality and is a graduate of the Australian College of Journalism.

Robert Eyler is an Associate Professor of Economics at Sonoma State University in California. He earned a Ph.D. from the University of California, Davis in 1998. He has published several scholarly articles on the California wine industry, and has acted as a consultant in winery litigation for free trade advocacy across states and in a trademark damage case. His academic fields of specialization are macroeconomic and monetary theory, applied econometrics, and economic history. He is also the director of the Center for Regional Economic Analysis at Sonoma State University, concentrating on providing information and doing research for the North Bay economy in Northern California.

Peter Gago is Chief Winemaker at Penfolds Wines Pty Ltd in South Australia. He joined Penfolds in 1989 after graduating from Roseworthy College with a Dux of the Bachelor of Applied Science (Oenology). He also holds a B.S. in Education from the University of Melbourne, and was a math and science teacher for over 8 years before becoming a winemaker. Peter has traveled extensively, and has written three books on wine: *Discovering Australian Wine—A Taster's Guide*; *Australian Wine—From the Vine to the Glass*; and *Australian Wines—Tastes and Styles*. Peter is currently head winemaker for the famous Penfolds Grange.

Denis Gastin is a wine writer who grew up in Australia's northeast Victorian wine regions and has had a lifelong interest in wine. He is a feature writer and Australian Correspondent for Japan's liquor industry newspaper, *The Shuhan News*, and Korea's leading monthly wine magazine, *Wine Review*. Over the past decade he has been a contributor to various other journals and wine reference books, including *The Oxford Companion to Wine*, *The World Atlas of Wine*, *Wine Report*, *Wine Companion*, *The Pocket Wine Guide*, *Australian & New Zealand Grapegrower & Winemaker*, among others. Currently, Denis resides in Australia and is working on a new book about the growth of the Asian wine industry. He holds a B.Com. degree from the University of Melbourne and was a postgraduate scholar at Concordia University, Montreal.

Donald Getz is a Professor in the Haskayne School of Business, University of Calgary, Canada. In addition to conducting research on wine consumption and tourism, and being a wine tourist himself, Don has developed an international reputation as a leading scholar and proponent of event management and event tourism. Related areas of expertise include destination and resort management, family business and entrepreneurship, rural tourism, consumer research, and special interest travel. He has authored a number of books including *Event Management & Event Tourism* (1997; second edition 2005), *Explore Wine Tourism* (2000), and *Event Studies* (2007). He is also a Visiting Professor at the University of Gothenburg in Sweden.

Mark Greenspan, Ph.D., is a viticulturist who currently owns and operates Advanced Viticulture, LLC, a viticultural consulting firm serving the premium winegrape industry. He has a master's degree in Horticulture/Viticulture and a doctorate in Agricultural Engineering, both from the University of California, Davis. He is one of very few private practitioners who have been elected as an honorary member of Gamma Sigma Delta, the Agricultural Honor Society. He is regarded as one of the world's leading experts in wine grape water management and has written scientific and trade journal articles on the subject. In addition to his command of grapevine irrigation practices, he has extensive experience in vineyard mineral nutrition, crop load management, vineyard uniformity, grape maturation, viticultural climate, and viticultural technologies. Mark is an active contributor to wine industry publications and writes a featured monthly column on viticulture for an internationally circulated periodical.

Bruce Herman joined Foster's Wine Estates (then Beringer Blass Wine Estates) in the position of Senior Vice President, Sales & Marketing, North America, in July of 2004, managing all sales and marketing activities for the US and Canada. Following the June 2005 acquisition of Southcorp Wines, Bruce became Senior Vice President, Sales, and leads the two US sales forces of Foster's Wine Estates, Beringer Estates Group, and Limestone Estates Group, as well as the FWE Canadian sales team.

Josh Hermsmeyer is a new father and, in his spare time, he moonlights as the Generalissimo of Capozzi Winery. Josh is blogging the birth of Capozzi, a new Russian River Valley Pinot Noir winery, at

PinotBlogger: the Capozzi Winery blog. Josh studied Economics and Winemaking at UC Davis, where he met and was swept off his feet by his lovely wife Candace.

Allison Jordan is executive director of the California Sustainable Winegrowing Alliance and communications programs manager at the Wine Institute. Previously, Allison served as vice president and acting executive director of Resource Renewal Institute, a San Francisco-based environmental organization. Allison is a transatlantic fellow with the German Marshall Fund's Marshall Memorial Fellowship Program. She holds a master's in Public Policy from University of California, Berkeley and a B.A. in Psychology from Allegheny College in Meadville, PA.

Tor Kenward spent 27 years as a senior executive with Beringer Vineyards working in winemaking and marketing, helping to build a wine company known to collectors worldwide for its outstanding Reserve and single vineyard Cabernet Sauvignons. He retired in 2001, and along with his wife Susan, started TOR Kenward Family Wines, a small wine company based on single vineyard Cabernet Sauvignons and Chardonnays.

Armen Khachaturian is Director of Wine Sales for Hanzell Vineyards, one of the oldest wineries in California specializing in pinot noir and chardonnay. In this position, Armen travels extensively serving high-end restaurant and retail accounts and working in partnership with Hanzell distributors. Prior to Hanzell, Armen worked as a Corporate Retail Specialist with The Henry Wine Group, a specialized wine distributor. Armen has over 12 years experience in sales, specializing in the wine and hospitality industry. He holds a B.S. in Wine Business Strategy from Sonoma State University, and was the Student Commencement Speaker of his graduating class.

Terry Lease is Professor of Accounting in the School of Business & Economics at Sonoma State University, where he teaches Wine Industry Accounting and Finance course and conducts professional development seminars in the Wine Business Program. He is a licensed CPA in California and Florida. His primary areas of interest in accounting are Taxation and Management Accounting.

Wendell Lee is General Counsel for The Wine Institute, headquartered in San Francisco, California, USA. The Wine Institute is the public policy advocacy association of California wineries. It brings together the resources of over 1,000 wineries and affiliated businesses to support legislative and regulatory advocacy, international market development, media relations, scientific research, and education programs that benefit the entire California wine industry. As counsel for the Wine Institute, Wendell focuses on the myriad of legal issues impacting wine around the world.

Larry Lockshin is Professor of Wine Marketing, Director of the Wine Marketing Research Group, and Director Post Graduate Programs in Marketing and Wine Marketing at the University of Australia. In this capacity he teaches "Managing the Wine Business for Profit; Optimising the Wholesale/Retail Relationship in Wine; and Retail Marketing Management." He, along with Tony Spawton, conducts seminars and executive programs in wine marketing in most of the wine regions of the world and consults with major Australian wine companies. Larry is also a very prolific writer and researcher in wine marketing, with over 60 academic articles and an equal number of trade publications on the topics of wine consumer behavior, wine business management, and wine distribution and retail. Larry holds a Ph.D. in Marketing from Ohio State University, an M.Sc. in Viticulture and Agricultural Economics from Cornell University, and a B.A. in Humanities from Ohio State. He served on the Strategy 2025 and the Marketing Decade Committees of the Australian Wine Industry and he currently holds a committee position on the Domestic Marketing Taskforce for the Australian Wine and Brandy Association.

Gary Long is an On-Premise Sales Representative with Young's Market—one of the largest wine distributors in the United States.

Gary joined Young's in 2004 as an Off-Premise Sales Rep in the Napa, California region servicing grocery stores and chain retail establishments. He was promoted to On-Premise sales in the Berkeley/Oakland region in 2006 and provides wine and spirits support to a wide variety of fine dining venues, bars, and casual restaurants. He is a professional wine-taster with a discriminating palate and an encyclopedic memory of global wine vintages. Gary holds a B.S. in Wine Business Strategies from Sonoma State University in California.

Linda Nowak is Professor of Marketing at Sonoma State University, and teaches in both the Wine Business Strategies and General Marketing programs. Her major classes are Wine Marketing, Principles of Marketing, and Marketing Management. Linda has published four articles in *The International Journal of Wine Marketing* entitled "Building Brand Equity: Consumer Reactions to Proactive Environmental Policies by the Winery"; "Effects of the Dietary Guideline Label Statement on Wine Purchase Intentions of Young Adults"; "Country of Origin Effects and Complimentary Marketing Channels: Is Mexican Wine More Enjoyable When Served with Mexican Food"; and "The Importance of Non-Financial Performance Measures in Wine Business Strategy."

Janeen Olsen holds the position of Professor of Marketing at Sonoma State University. Her international background has provided her the opportunity to conduct export seminars for business executives in many Latin American cities. She has published extensively in marketing and international business journals and presented papers at conferences in Europe, Asia, and Latin America, as well as in the US. She has developed international trips for the Wine Business Program at SSU and takes Wine Business students on a wine industry tour of Chile. She has served on the Global Task Force for Wine Vision. She is active in conducting marketing research for the Wine Business Program at SSU and has published and presented her research in industry publications and seminars. She is working towards a certificate in Vineyard Management.

CONTRIBUTORS

Cyril Penn is editor in chief of Wine Business Communications, headquartered in Sonoma, California. He joined the firm in September 1998 as editor of *Wine Business Insider*, and in January 2000 he was named editor in chief of *Wine Business Monthly*. Mr. Penn has over 15 years of wire service, magazine, and broadcast experience. Mr. Penn began his career as a journalist in New York at Reuters in 1987, where he covered the energy industry for more than 3 years. He moved on to become a freelance reporter specializing in energy, high technology, and biotechnology. Prior to joining Wine Business Communications, Mr. Penn was managing editor of the California Energy Markets newsletter in San Francisco. Mr. Penn holds a bachelor's degree in media studies from Fordham University in New York.

Elizabeth Rice works in the international sales division of Delicato Family Vineyards—headquartered in Napa, California, with some of the largest vineyard holdings in California. Her current position is Area Manager-Europe where she is responsible for Delicato wine sales in the European Union. Based in the United Kingdom, Elizabeth is very familiar with international wine sales issues and regulations. Prior to moving to Europe, she managed Canadian wine sales for Delicato and lived in Toronto. Elizabeth graduated from college in 2001 with a B.S. in Wine Business Strategies from Sonoma State University, California. She began working for Delicato as an intern while in college.

Karen Ross has been president of the California Association of Winegrape Growers since 1996. She is also the executive director for the Winegrape Growers of America, a national organization of state wine-grower organizations, and executive director of the California Wine Grape Growers Foundation, which sponsors scholarships for the children of vineyard employees. Karen is a co-editor for the Code of Sustainable Winegrowing Practices, and one of the founding members of WineVision, where she currently leads the Sustainability Taskforce. She graduated from the University of Nebraska-Lincoln, the Nebraska Agricultural Leadership Program, and the Graduate Institute of Cooperative Leadership, University of Missouri.

Mack Schwing is cofounder of WISE Academy LLC, an education company specializing in Wine Industry Sales Education for professionals who are selling directly to consumers. He was recently the Director of the Wine Business Program at Sonoma State University for over 4 years. Previously he retired from Deloitte Consulting after working there for almost 30 years. His last 5 years were spent as a Senior Partner and Global Director of Programs and Initiatives. During his career, he lived in Japan for 3 years where he was Chairman of the Wine Committee at the Tokyo American Club and was active in the Japanese wine import market. Mack holds an M.B.A. in production management and a B.S. in Applied Mathematics from Michigan State University.

Tony Spawton is Associate Professor of Wine Marketing and the International Director of the Wine Marketing Research Group at the University of South Australia. He was instrumental in developing much of the contemporary wine marketing curriculum while at Roseworthy College in the mid-1980s. Tony is a key contributor to programs, seminars, and workshops in wine marketing locally and in most of the wine-making regions of the world, as well as a teacher in the Masters In Wine Marketing offered by the University of South Australia worldwide via its on-line delivery platform. Tony teaches the Applied Wine Marketing and Global Wine Marketing courses. Tony was an economics and marketing expert at The International Organisation of Vine and Wine (OIV) from 1990, and was unanimously elected President of the Expert Group "Market Analysis and Networks" in 2000. He is a member of the Scientific and Technical Committee of the OIV, holding specialist subgroup positions in wine industry development, and consumer and professional education. He has published numerous papers and articles on wine marketing research and applications.

CONTRIBUTORS

Jeff Sully, CPA, CMA, is a founding partner in the accounting firm of Dillwood, Burkel & Sully LLP. He has spent more than 30 years in the wine industry. He has served in all facets of the business, having been a grower, winemaker, and retailer. In addition, he was Chief Financial Officer for a large Sonoma County winery. He has also been a consultant to the State University of New York, Binghamton, for wineries in New York State. He has lectured extensively in New York, Oregon, and California. He speaks regularly for the California CPA Education Foundation, having chaired several of their Wine Industry conferences.

Richard Thomas is a Professor Emeritus, Viticulture & Wine Education, Santa Rosa Junior College, Santa Rosa, California. He has taught and consulted in Northern California vineyards for more than 30 years. Currently he is a viticultural consultant and continues to teach and coordinate courses at Santa Rosa JC in viticulture and wine tasting. Rich has written numerous articles on viticulture and is author of a monthly column for *Vineyard & Winery Magazine*. He is a professional wine judge and coordinator for several major US wine competitions. Rich holds a master's in Viticulture and a B.S. in Vocational Agriculture from UC-Davis.

Roy Thornton is a Professor of Enology at California State University–Fresno where he teaches classes in viticulture and enology, and conducts research on wine genetics and yeasts. Before coming to California, he taught for 20 years in New Zealand at Massey University, and also worked for Gallo as a Senior Research Microbiologist. Roy has published numerous research articles on enology and viticulture. He holds both a B.S and Ph.D. in Applied Microbiology from Strathclyde University in Glasglow, Scotland. He organized the 1st International Wine Microbiology Symposium in 2006 and is working toward the second in 2009.

Paul Wagner is a Professor of Wine Marketing in Napa Valley College's Viticulture and Enology department, as well as the founder of Balzac Communications & Marketing. In his role at Balzac Paul has many wine clients, including Diageo Chateau & Estate Wines, the Canandaigua Wine Company, the Union des Grands Crus de Bordeaux, Vinitaly, The L.A. County Fair International Wine Competition, Pernod-Ricard USA, Trinchero Family Estates, The Court of Master Sommeliers, and a host of other wine and food specialists. Before starting his own firm, Paul was general manager of Barson/Armstrong, a communications agency. He is a frequent guest lecturer at Golden Gate University, Sonoma State University, The University of Trieste, The University of Beaune, and UC Berkeley extension in the fields of wine, wine marketing, and wine production.

BIBLIOGRAPHY

Academy of Wine Communications. (2003). *Wine industry public relations specialists working together.* Retrieved from http://www.academyofwine.org/

Agrain, P. (2003). Note de conjuncture mondiale. *Bulletin d'OIV, 76*(876–868), 424–453.

Anderson, K., & Norman, D. (2003). *Global wine production, consumption and trade, 1961–2001: A statistical compendium.* Adelaide: Centre for International Economic Studies.

Beverage Digest. (2007). *Top 10 CSD results for 2006.* Retrieved from http://www.beverage-digest.com/pdf/top-10_2007.pdf

Biodynamic Farming and Gardening Association. (2008). *What is biodynamic agriculture?* Retrieved from http://www.biodynamics.com/biodynamics.html

Bird, D. (2005). *Understanding wine technology: The science of wine explained.* San Francisco: The Wine Appreciation Guild.

Birdseye, K., Ross, K., & Delott, J. (Eds.). (2002). *The code of sustainable winegrowing practices.* San Francisco: The Wine Institute.

Broome, J. C. (2003). *Sustainable viticulture programs around the world.* Retrieved from http://www.sarep.ucdavis.edu/production/viticulture/asev2003.htm

Brown, G., & Getz, D. (2005). Linking wine preferences to the choice of wine tourism destinations. *Journal of Travel Research, 43*(3), 266–276.

Brown, G., Havitz, M., & Getz, D. (2007). Relationships between wine involvement and wine-related tourism. *Journal of Travel and Tourism Marketing, 21*(1), 31–46.

Bruwer, J. (2003). South African wine routes: Some perspectives on the wine tourism industry's structural dimensions and wine tourism product. *Tourism Management, 24*(4), 423–435.

Byck, P. (2004, January). *State of the industry: Setting the stage for 2004.* Presentation made at the Unified Wine & Grape Symposium, Sacramento, CA.

Cambourne, B., Macionis, N., Hall, M., & Sharples, L. (2000). The future of wine tourism. In M. Hall, L. Sharples, B. Cambourne, & N. Macionis (Eds.), *Wine tourism around the world*. Oxford: Butterworth-Heinemann.

Carlsen, J. (2004). A review of global wine tourism research. *Journal of Wine Research, 15*(1), 5–13.

Carlsen, J., & Ali-Knight, J. (2004). Managing wine tourism through demarketing: The case of Napa Valley, California. In *International Wine Tourism Research, 2004. Proceedings of the Margaret River, Australia, International Wine Tourism Conference*. CD produced by Vineyard Publishing Inc., Guildford, W.A.

Carlsen, J., & Charters, S. (Eds.). (2006). *Global wine tourism: Research, management and marketing*. Wallingford: CABI.

Carlsen, J., & Dowling, R. (2001). Regional wine tourism: A plan of development for Western Australia. *Tourism Recreation Research, 26*(2), 45–52.

Cartiere, R. (2003). Wine Internet Purchase Survey results. *Wine Market Report*.

Chaney, I. (2002). Promoting wine by country. *International Journal of Wine Marketing, 14*(1), 34–42.

Charters, S., & Ali-Knight, J. (2000). Wine tourism—a thirst for knowledge? *International Journal of Wine Marketing, 12*(3), 70–81.

Charters, S., & Ali-Knight, J. (2002). Who is the wine tourist? *Tourism Management, 23*(3), 311–319.

Constellation Wines U.S. (2006). Project Genome. Retrieved from http://www.cwinesus.com/pdf/PressReleaseGenomeUnified.pdf

Coppla, C. J. (2000). Direct marketing sales boom with proliferation of wine clubs. *Wine Business Monthly, 7*(6).

Cox, J. (1999). *From vines to wines*. North Adams, MA: Storey Books.

Davison, B. (2003). Reviewing corporate financials show how HR measures up. *Employee Relations Today*, 30(1), 7–17.

Dodd, T. (1995). Opportunities and pitfalls of tourism in a developing wine industry. *International Journal of Wine Marketing,*

7(1), 5–16.

Dodd, T., & Bigotte, V. (1997). Perceptual differences among visitor groups to wineries. *Journal of Travel Research, 35*(3), 46–51.

Dolan, P., & Elkjer, T. (2003). *True to our roots: Fermenting a business revolution.* Princeton: Bloomberg Press.

Duijker, H., & Johnson, H. (2000). *The wines of Chile.* New York: Spectrum.

Eccles, R. G., Lanes, K. L., & Wilson, T. C. (1999). Are you paying too much for that acquisition? *Harvard Business Review, 77*(4), 136.

Fairtrade Fortnight. (2007). http://www.fairtrade.org.uk/

Ferguson, S. (2002). Taking it direct. *Wine Business Monthly, 9*(1).

Foster, L. (2003). *A pilgrimage to the French Bordeaux wine country.* Retrieved from http://www.fostertravel.com/FRBORD.html

Gallagher, N. (2003, August 11). Squeezed by grape market. *The Press Democrat,* p. D1.

Gary, D. (2006). *RFID technology: A case study of implementation and the resulting effects on supply chain management.* Masters thesis, Sonoma State University.

Gastin, D. (2008). Asia. In T. Stevenson (Ed.), *Wine report.* London: Dorling Kindersley.

Getz, D. (2000). *Explore wine tourism: Management, development, destinations.* New York: Cognizant Communication Corp.

Getz, D., & Brown, G. (2006). Benchmarking wine tourism development: The case of the Okanagan Valley, British Columbia, Canada. *International Journal of Wine Marketing, 18*(2), 78–97.

Gilinsky, A., McCline, R. L., & Eyler, R. (2000, March/April). Best business practices in the northern California wine industry. *Industry Analysis,* 30–37.

Gimeno, J., & Woo, C. Y. (1999). Multimarket contact, economies of scope, and firm performance. *Academy of Management Journal, 42*(3), 239–264.

Goold, M., & Campbell, A. (1998). Desperately seeking synergy. *Harvard Business Review, 76*(5), 70–83.

Grist, J. (2003, July). *Ancient vintage: Wine in ancient Israel and*

Mediterranean lands. Presentation at the WineSpirit Meeting, Napa.

Gurnsey, S., Manktelow, D., Manson, P., Walker, J., & Clothier, B. (2004). *Sustainable Winegrowing New Zealand®: Technical developments and achievements.* Presented at the 6th International Conference on Integrated Fruit Production, Baselga di Pine, Trento, Italy, September 26–30. Retrieved from http://www.nzwine.com/assets/GURNSEYet_al_paper_final_version.pdf

Haleblian, J., & Finkelstein, S. (1999). The influence of organizational acquisition experience on acquisition performance: A behavioral learning perspective. *Administrative Science Quarterly, 44*(1), 29–57.

Hall, M., & Macionis, N. (1998). Wine tourism in Australia and New Zealand, In R. Butler, M. Hall, & J. Jenkins (Eds.), *Tourism and recreation in rural areas* (pp. 267–298). Sydney: Wiley.

Hall, M., Sharples, L., Cambourne, B., & Macionis, N. (2000). *Wine tourism around the world.* Oxford: Butterworth-Heinemann.

Handfield, R. B., & Nichols, E. L., Jr. (1999). *Introduction to supply chain management.* Upper Saddle River, NJ: Prentice Hall.

Harrison, J. S., Hitt, M. A., Hoskisson, R. E., & Ireland, R. D. (2001). Resource complimentarity in business combinations: Extending the logic to organizational alliances. *Journal of Management, 27*(6), 679–691.

Howell, J. (2001). *Chile experience travel guide.* Los Angeles: Turiscom Publishing.

Johnson, H. (1985). *The world atlas of wine* (3rd ed.). New York: Simon & Schuster.

Johnson, H. (1989). *Vintage: The story of wine.* New York: Simon & Schuster.

Johnson, H., & Robinson, J. (2001). *The world atlas of wine* (5th ed.). London: Octopus Publishing Group, Ltd.

Keller, K. L. (2003). *Strategic brand management.* New York: Prentice-Hall Publishers.

Khanna, T., & Palepu, K. (1999). The right way to restructure conglomerates in emerging markets. *Harvard Business Review, 77*(4), 125–135.

Korolishin, J. (2007). *The top 100 beverage companies.* Retrieved from http://www.bevindustry.com/Archives_Davinci?article=2120

Kotter, J. (1996). *Leading change.* Boston, MA: Harvard Business School Press.

Lang Research Inc. (2001). TAMS (Travel Activities and Motivation Survey). Retrieved from www.tourism.gov.on.ca/english/research/travel_activities/index.html

Laforet, S., & Saunders, J. (1999). Managing brand portfolios: Why leaders do what they do. *Journal of Advertising Research, 39*(1), 51–65.

Leinberger, P. (2003, July). *Consumer trends in the global wine industry.* Presentation made at the WineVision Conference, Napa, CA.

Lockshin, L., Macintosh, G., & Spawton, A. (1997). Using product, brand, and purchasing involvement for retail segmentation. *Journal of Retailing and Consumer Services, 4*(3), 171–183.

Lockshin, L., & Spawton, T. (2001). Using involvement and brand equity to develop a wine tourism strategy. *International Journal of Wine Marketing, 13*(1), 72–82.

Long, J. (2005, April 21). United States taking over as top wine consumer. Newhouse News Service.

Lukacs, P. (2000). *American vintage: The rise of American wine.* New York: Houghton-Mifflin.

MacNeil, K. (2001). *The wine bible.* New York: Workman Publishing.

Markides, C. (1997). To diversify or not to diversify. *Harvard Business Review, 75*(6), 93–101.

Marshall, J. (2001). Are mergers paying off? *Financial Executive, 17*(2), 26–33.

McCallum, K. (2007, July 17). Meeting retailer's needs. *The Press Democrat.*

Mitchell, R. (2006). Influences on post-visit wine purchase (and non-purchase) by New Zealand winery visitors. In J. Carlsen & S. Charters (Eds.), *Global wine tourism: Research, management and marketing* (pp. 95–109). Wallingford: CABI.

Mitchell, R., & Hall, M. (2006). A review of global wine tourism research. *Tourism Review International, 9*(4), 307–332.

Monczka, R., Trent, R., & Handfield, R. (1998). *Purchasing and supply chain management.* Cincinnati, OH: South-Western College Publishing.

Moulton, K., & Spawton, T. (1997). Can the wine industry survive regulation. In L. T. Wallace & W. R. Schroder (Eds.), *Government and the food industry: Economic and political effects of conflict and co-operation.* Boston: Kluwer Academic Publishers.

Noe, R. A., Hollenbeck, J. R., Gerhart, B., & Wright, P. M. (2004). *Fundamentals of human resource management.* Boston: McGraw-Hill.

Nowak, L. I., & Newton, S. K. (2006). Using the tasting room experience to create loyal customers. *International Journal of Wine Marketing, 18,* 157–165.

Nowak, L., Thach, L., & Olsen, J. E. (2006). Wowing the millennials: Creating brand equity in the wine industry. *Journal of Product and Brand Management, 15,* 316–323.

O'Mahony, B., Hall, J., Lockshin, L., Jago, L., & Brown, G. (2006). Understanding the impact of wine tourism on post-tour purchasing behavior. In J. Carlsen & S. Charters (Eds.), *Global wine tourism: Research, management and marketing* (pp. 123–137). Wallingford: CABI.

O'Neill, J. (2004, January 27). *The wine industry at the beginning of the 21st century: A fast ride in a new direction?* Presentation at Unified Wine & Grape Symposium, Sacramento, CA.

O'Neill, M., & Charters, S. (2000). Service quality at the cellar door: Implications for Western Australia's developing wine tourism industry. *Managing Service Quality, 10*(2), 12–122.

Office International de la Vigne et du Vin. (1995–2008). *The state of viniviticulture world report.* France: Author.

Palich, L. E., & Gomez-Mejia, L. R. (1999). A theory of global strategy and firm efficiencies: Considering the effects of cultural diversity. *Journal of Management, 25*(4), 587–607.

Peng, M. W. (2001). The resource-based view and international business. *Journal of Management, 27*(6), 803–830.

Perdue, L. (1999). *The wrath of grapes: The coming wine indus-*

try shakeout and how to take advantage of it. New York: Avon Books, Inc.

Perreault, W. D., & McCarthy, E. J. (2002). *Basic marketing.* New York: McGraw-Hill Publishers.

Poitras, L., & Getz, D. (2006). Sustainable wine tourism: The host community perspective. *Journal of Sustainable Tourism, 14*(5), 425–448.

Porter, M. (1985). *Competitive advantage: Creating and sustaining superior performanc e.* New York: Free Press.

Posert, H. & Franson, P. (2004). *Spinning the bottle: Case histories, tactics and stories of wine public relation.* Napa, CA: HPPR.

Professional Friends of Wine. (2004). *Regions and wineries.* Retrieved from www.winepros.com

Quester, P. G., & Smart, J. (1998). The influence of consumption situation and product involvement over consumers' use of product attribute. *Journal of Consumer Marketing, 15*(3), 220–238.

Rabobank International. (2002). *The major wine companies.* Internal company document.

Roberts, L., & Sparks, B. (2006). Enhancing the wine tourism experience: The customers' viewpoint. In J. Carlsen & S. Charters (Eds.), *Global wine tourism: Research, management and marketing* (pp. 47–55). Wallingford: CABI.

Robinson, J. (Ed.). (2006). *The Oxford companion to wine* (3rd ed.). New York: Oxford University Press.

Seth, A., Song, K. P., & Pettit, R. (2000). Managerialism or hubris? An empirical examination of motives for foreign acquisitions of U.S. firms. *Journal of International Business Studies, 31*(3), 387–404.

Shanken, M. (2006). *The U.S. wine market: Impact databank review and forecast 2006.* New York: M. Shanken Communications, Inc.

Sharples, L. (2002). Wine tourism in Chile: A brave new step for a brave new world. *International Journal of Wine Marketing, 14*(2), 43–54.

Simon, A. (1967). *The wines, vineyards, and vignerons of Australia.* London: Paul Hamlyn.

Spawton, A. L. (1990). Development in the global alcoholic drinks industry and its implications for the future marketing of wine. *International Journal of Wine Marketing, 2*(1). Reprinted in *European Journal of Marketing, 24*(4).

Spawton, A. L. (1997). *Globalisation and its implications to strategy development as the key to future success.* Paper presented at XX11eme World Congress of Vine and Wine, Argentina, December 1–5.

Spawton, A. (2003a). Supply chain management in the wine sector. *Bulletin d'OIV, 76*(876–868), 389–424.

Spawton, A. L. (2003b). *Will globalisation commoditise premium wine* (Working paper). Wine Marketing Group, University of South Australia.

Spawton, T., & Juniper, J. (2001). *Regional competencies and global trends in firm concentration within the wine industry.* Paper presented at XXVIth World Wine Congress, Adelaide, Australia.

Stanwick, S., & Fowlow, L. (2005). *Wine by design.* Seattle: Academy Press.

Stevenson, T. (2004). *Wine report.* London: Dorling Kindersley.

Taylor, R. (2006). Wine festivals and tourism: Developing a longitudinal approach to festival evaluation. In J. Carlsen & S. Charters (Eds.), *Global wine tourism: Research, management and marketing* (pp. 179–208). Wallingford: CABI.

Thach, L. (2002, July). Social sustainability in the wine community. *Wine Business Monthly.*

Thach, L. (2007). *Trends in wine tourism.* Retrieved from http://www.winebusiness.com/ReferenceLibrary/webarticle.cfm?dataId=50125

Thach, L., & Eaton, C. (2001, May). E-Commerce adoption in the wine industry. *Wine Business Monthly, 8*(5), 31–33.

Thach, L., & Shepard, J. (2001, November). The importance of supervisory & management training in the wine industry. *Wine Business Monthly.*

Thompson, A. A., & Strickland, A. J. (2003). *Strategic management: Concepts and cases* (13th ed., chap. 6, 9, 10). New York: McGraw-Hill/Irwin.

Tinney, M. C. (2007, September). Direct shipping update. *Wine Business Monthly, 14*(9), 68–70.

Tromp, A. (2003). *The South African system of integrated production of wines.* Retrieved from http://www.ipw.co.za/about_us_eng.pdf

Ulrich, D. (1997). Measuring human resources: An overview of practice and a prescription for results. *Human Resource Management, 36*(3), 303–320.

Voss, R. (2000, May issue). *The New South Africa.* Retrieved from http://www.winemag.com/ME2/dirmod.asp?sid=&nm=&type=Publishing&mod=Publications%3A%3AArticle&mid=8F3A702742 1841978F18BE895F87F791&tier=4&id=7E3A949BEF234ED194C 40E83DA643FDA

Wall, S. J. (2001). Making mergers work. *Financial Executive, 17*(2), 34–35.

Webb, M. (2005). *Adventurous wine architecture.* Mulgrave, Australia: Images Publishing.

Western Management Group. (2008). *Wine industry compensation survey.* Retrieved from http://www.wmgnet.com/dnn/SurveysUSA/WineIndustryCompensationSurvey/tabid/101/Default.aspx

Williams, P. (2001a). The evolving images of wine tourism destinations. *Tourism Recreation Research, 26*(2), 3–10.

Williams, P. (2001b). Positioning wine tourism destinations: An image analysis. *International Journal of Wine Marketing, 13*(3), 42–60.

Williams, P., & Dossa, K. (2003). Non-resident wine tourist markets: Implications for BC's emerging wine tourism industry. *Journal of Travel and Tourism Marketing, 14*(3/4), 1–34.

Williams, P., & Kelly, J. (2001). Cultural wine tourists: Product development considerations for British Columbia's resident wine tourism market. *International Journal of Wine Marketing, 13*(3), 59–77.

Wine Diva. (2002). *Australian wine: Geographic indication system.* Retrieved from http://www.winediva.com.au/regions/regions2.asp

Wine Market Council. (2006). www.winemarketcouncil.com

Yahoo Finance. (2003). *Factiva databases.* Retrieved from http://www.lib.uwo.ca/database/secure/jumpstart.shtml

Yeung, A. K., & Berman, B. (1997). Adding value through human resources: Reorienting human resource measurement to drive business performance. *Human Resource Management, 36*(3), 321–335.

INDEX

Aborigine, 362
Accounting, wine, 333–351
Acetic acid, 102
Acid, 78, 87, 97, 113, 137
Acquisitions, 55, 384
Adelaide, 64, 286
Adelaide Hills, 65
Africa, 291, 422
African-American, 185
Agenda 21, 403, 404
Aging, 96, 103
Aging, reductive, 110
Agricultural resort area, 251
Airplanes, 85
Alberta, 213
Alcohol level, 101–102
Alexander Valley, 65
Almaden, 10,
Alpine county 318
Alsace, 138
Alternatives, bottles, 138–139
American Express, 169
American Farmland Trust, 406
American Temperance Union, 181
American Viticulture Area, 8, 65, 322
American, barrels, 134
Amorin Cork, 44
Anthocyanins, 101
Antialcohol, 14
Antinori, 282
Anti-Saloon League, 181
Appelation d'origne contrôlée (AOC), 322
Appellation, 65
Arboleda, 397
Argentina, 65, 67, 76, 167, 170, 211, 397, 411, 419
Arizona, 200
Arrow shirts, 393
Art of winemaking, 105
Asia, 15, 277, 383, 419, 423, 426

Asia, labor, 383–384
Asian, 185
Asia-Pacific, 291
Australia, 4, 6–11, 50, 64–67, 76, 79–80, 103, 110, 112, 141, 163, 170, 211, 214, 281–282, 284, 323, 410, 412, 422–424, 428
Australian dollars, 180
Australian Wine Industry Stewardship Program (AWIS), 412
Australia's National Environmental Outcomes, 412
Australia's Riverland, 285
Australia's Wine and Brandy Corporation (AWBC), 7, 399
Austria, 419
AV Imports 329

Bacardi, 6
Bag in box, 138
Banfi, 416
Banrock Station, 285, 299
Barbera, 67
Baroness Philippine de Rothschild, 397
Barossa Valley, 65–66, 103, 113, 286–287, 300, 312
Barrel, life span, 135
Barrel, prices, 104
Barrels, 100, 103, 134–135
Barrels, oak, 135
Bartonvale, 286
Basic weekend wine drinkers, 164
Basket press, 135–136
Beckstoffer, Andy, 91
Beckstoffer Vineyards, 44, 91–94
Being consistent with brand image, 168–169
Benchgraft, dormant, 74
Benefits, HR, 376
Benefits, rising costs, 383

Beringer, 10, 15, 223, 281–282
Beverage Digest, 18
Beverage Dynamics, 269
Bhutan, 15
Binder, Rolf, 287
Biodynamic, 89–90, 119
Bioterrorism, 148–149
Bioterrorism, compliance, 148–149
Bioterrorism Preparedness and Response Act, 148
Bistro Wine Bar, 44
Blacket, Simon, 103
Bladder press, 135
Blending, 96, 105, 107
Blogger.com, 231
Bloggers, 230–231
Blogs, 220, 229–231, 233
Bloom, 77
Blossom Hill, 10
BMW, 166, 176
Board of Directors, 371
Boller, E. F., 413
Bon Appetit, 268
Bonded Wine Cellar (BWC), 392
Bonded Winery Basic Permit, 392
Bonny Doon, 50, 148, 170
Bootlegging, 182
Bordeaux, 49, 66, 68, 105, 111, 138, 274,
Bottle aging, 142
Bottle, alternatives, 138–139
Bottle costs, 109, 349
Bottle size, 109, 138
Bottles, 81, 138–139
Bottles per case, 81
Bottles per vine, 81
Bottling, 96, 108
Bottling line, 109, 136
Bottling line, mobile, 108–109, 136–137
Brand, 113
Brand, a tale of two, 173–174
Brand building, 168
Brand building, conversations, 229–230

Brand, global wine, 117–118
Brand, good names, 169
Brand image, 168–169
Branding, 155–176
Branding, definition, 157–158
Branding for export, 297
Branding, importance, 160–161
Branding, international, 171–173
Branding success, 161–162
Brand loyalty, 169–171
Brand manager, 217
Brand names, 169
Brand positioning, wine, 166-167
Brands, global, 10
Brand story, 172
Brand strategy, 20
Brazil, 215
Brett, 105
Brettanomyces, 105
British Columbia, 25, 65, 248
Brix, 78, 87, 97
Broker, definition, 185
Broker, finding one, 187–188
Broker, role of, 185–187
Brokers, 179
Brokers, working with, 188–191
Bronco Wine Company, 53
Brown-Forman, 18, 330
Build to sell, 56
Bulk wine, 128–129
Bulk wine cost, 348
Burgundy, 138, 174
Burjal Arab, 256
Bush vine, 76
Business decision, winemaking, 96-97
Business Excellence Model, 147-148
Business of enology/winemaking, 95–120
Business strategy, 20
Buyer's own brand (BOB), 288

Cabernet Franc, 67–68
Cabernet-Merlot, 301

INDEX

Cabernet Sauvignon, 7, 66–68, 105, 301, 309, 312, 419
Calgary, 250
California, 7–9, 24, 50, 53, 64, 68, 91, 105, 111, 117, 120, 174, 186, 200, 211, 214, 216, 223, 323, 327, 330, 410
California Air Pollution Control Laws, 395
California Association of Winegrape Growers (CAWG), 93, 404, 406
California Beverage Journal, 269, 287
California Clean Water Act, 395
California Code of Regulations, 395
California Code of Sustainable Wine Growing Practices, 363, 402, 404
California Council for Environmental and Economic Balance, 409
California Department of Alcoholic Beverage Control, 391
California Department of Food and Agriculture (CDFA), 406
California districts, 317
California Sustainable Winegrowing Alliance (CSWA), 405–406
California Sustainable Winegrowing Program, 404, 407, 409
California Water Act, 395
Caliterra, 397
Cambodia, 15
Canaan, 361
Canada, 4, 29, 65, 67, 181, 213–215, 427
Canopy, 75
Cantine Cooperative, 10
Cap, 101
Cape region, 65
Capital expenditure, 130
Capitalized interest, 355
Cardboard brick, 138

Career development, 377
Carmenere, 8, 67
Casella Wines, 10, 282
Cases, 81
Castel Freres, 6
Caviro, 6
Cellar door, 222–227
Cellar equipment, 136
Cellar Master, 368
Central Coast, 8, 64
Central Valley, 53
Certification, supplier, 146–147
Chablis, 398
Chadwick, Eduardo, 397
Chain accounts, 191–192
Chalk Hill, 173–174
Champagne, 49–50, 398
Channels, direct wine sales, 222–227
Channels, sales in US market, 199–200
Characteristics of a good brand name, 169
Chardonnay, 67, 80, 105, 160, 173–174, 194, 301, 309, 419, 431
Chardonnay-Semillion, 301
Charles Shaw, 10, 120, 265–266
Chateau de Leelanan, 394
Chateau Mouton Rothschild, 397
Checklist, sales plan, 202
Chemical profiling, 114
Chenin Blanc, 67–68, 309
Chianti, 105
Chief Financial Officer (CFO), 371
Chile, 4, 10–11, 64–65, 67, 74, 76, 88, 100, 110, 149, 167, 211, 214, 381, 423, 426
China, 4, 13, 15–16, 90, 134, 215, 384, 422–423, 427, 433
Chips, 135
Chips, oak, 104
Christmas, 426
Clare Valley, 113
Classics, 97

Clays, 107
Client relations, vineyard, 366
Climate, 64
Closure, 109, 141
Clubs, wine, 224–225
CNET News, 233
CO_2, 79
Coca-Cola (Coke), 18, 110, 166, 173, 176, 281, 422–423, 432
Cole Ranch, 8
Comité Interprofessionel du Vin de Champagne, 413
Commitment, to export, 178
Commodities, 135
Commodity strategy, 127
Companies, major wine, 6
Comparative depreciable lives, 356
Compensation & benefits, 376–377
Competencies, core, 126
Competition, 210–211
Competitive Focus Axis, 48–49
Competitive set, 189
Competitor analysis, 25, 27, 296
Competitors, 28, 166
Complementary strategic options, 54–55
Compliance, direct wine sales, 227–228
Concha y Toro, 6, 10, 52, 330
Congress, 398
Connecticut, 200
Consistency with brand image, 168–169
Consolidation, 137, 384
Constellation, 5, 10, 16, 18, 46, 52, 164–165, 281, 330, 364, 433
Consumer profiling, 120
Consumer Reports, 265
Consumer segmentation, 162–166
Consumer segmentation by consumption rate, 162–163
Consumer segmentation by motivation, 163
Consumer segmentation by shopping behavior, 164–165

Consumer, demand, 211
Consumers, 162–163
Consumers, core, 162
Consumers, enthusiasts, 164
Consumers, image seekers, 164
Consumers, marginal 162
Consumers, nonadopters, 163
Consumers, overwhelmed, 166
Consumers, satisfied shippers, 165
Consumers, savvy shoppers, 165
Consumers, traditionalists, 165
Consumer trends, 419–426
Consumption, 13
Consumption, wine, 90
Contract, grapes, 69–72
Contracts, longer term, 149–150
Control states, 183–184
Coonawara Bin 128 Shiraz, 113
Cooperage, 104
Co-pigmentation, 116
Core wine consumers, 162
Corking, 109
Corks, 109, 140–142, 146
Corks, synthetic, 141
Corporate strategy, 20
Corporatization, 5, 14
Cost allocation by block, 339–341
Cost-benefit analysis, 69–71
Cost centers, 336–338
Costing a vineyard, 338
Costing in a winery, 342
Cost of quality, 145–146
Cost Plus, 192
Costco, 390
Costs, 80, 92–93
Costs, administrative 348
Costs, barrels, 134–135
Costs, direct, 337
Costs, farming, 338
Costs, indirect, 337
Costs, payroll, 345
Costs, production, 347
Costs, tanks, 133
Costs, winemaking, 96
Costs, winemaking supplies, 137

INDEX

Costs, winemaking variable costs, 137–138
Costs, winery, 342–345
Countries, Old World wine, 12
Countries, top wine producing, 11
Critical success factor (CSF) analysis, 36, 38–40
Crush, 96–98
Crush costs, 346
Crush equipment, 98
Crush process, 99
Crystallization, 108
Cult, wines, 227
Cultured yeast, 100
Current issues, enology, 117–121
Custom crush, 98, 126, 130
Customer, experiences, 171
Customer focus, 118–119
Customers, 139
Customers, listen to, 171
Customers, vineyard, 93
Cycle of the vine, 77–80
Cycles of grape prices, 71

Dalwood Vineyards, 362
Database systems, 82
Decanter, 268
Decision matrix, supply chain, 128
Demand augmentation, 330
Denmark, 215
Department strategy, 20
Depletion goals, 202
Deutsch and Sons, 282
DHL, 227
Diageo, 10, 18, 46, 52, 330
Differentiated broad market, 48–50
Differentiated niche, 48–50
Dinners, wine, 207–208
Direct mail campaigns, 225–226
Direct sales, percentages, 221
Direct to consumer, 199–200, 208–209
Direct to consumer wine sales, 219–234

Direct to trade, 199–200
Direct wine sales, 219–234
Direct wine sales, advantages, 220–221
Direct wine sales, challenges, 220–221
Direct wine sales, channels, 222–227
Disease, 77
Disney, 166, 176
Distell, 6,
Distribution, 30, 57, 125, 143–144, 177–195
Distribution, channels, 191–195
Distribution choices, 179
Distribution, in the US, 181–182
Distribution, outside the US, 180–181
Distribution, process in US, 184
Distribution, restrictions, 212–213
Distribution, skills, 195
Distribution systems, 294
Distributor analysis, 25, 29–30
Distributor, consolidation, 221
Distributor, finding one, 187–188, 203–204
Distributor, questions to ask, 203–204
Distributor, role, 183–185
Distributors, 179
Distributor sales meetings, 205–206
Distributor sales rep, 191–195
Distributors, working with, 188-191, 204–205
District Community Economic Development Society, 250
Diver, 203
Divineware, 203
Division strategy, 20
Dollars per acre, 71
Dollars per hectare, 71
Dom Perignon, 173
Dormancy, vines, 79–80
Dormant benchgraft, 74

Dry Creek, 65–66, 129
Dust, oak, 135

Eastern Europe, 90, 172, 312, 419
Eastern European, barrels, 134
Eavior Societa Coop, 10
Economically viable, 119
Electronic data interchange (EDI), 149
egullet.com, 232
Egypt, 274, 361
Eight complementary strategic options, 54–55
Eighteenth (18th) Amendment, 181, 386
Emerging markets, 215
Emirates Airlines, 256
Emirates Palace Hotel, 256
Employee functions, 371
Employee positions, 363–372
Employee relations, 379
End of year reviews, 210
Engagement, marketing strategy, 230
England, 178, 274
Enologist, 368
Enology, 95–120
Enology, current and future issues, 117–121
Enthusiasts, 164
Environmental, 103
Environmental, conditions, 142–143
Environmental issues, 401-416
Environmentally friendly, 94
Environmental Protection Agency, 395
Enzymes, 137
EPC, 153
Equipment, cellar, 136
Equipment, winemaking, 137
Equity versus debt financing, 327
eRobertParker.com, 232
eSkye Solutions, 203
Esquire, 268

Estate Vineyards, 132
Euro, 180, 281
Europe, 89–90, 119, 138, 214, 291, 363, 422, 426
European Union (EU), 12, 212, 398, 430
European Economic Community, 274
European Integrated Fruit Production Systems, 413
European sustainability, 413
Eventscorp W.A., 256
Exchange rates, 211
Expected cash flows 315
Exporting, 210–217, 283
Exporting tactics, 289
Exporting, wine from US, 213-214
Export, markets for US wines, 214–215
Export Readiness Checklist, 292
Exports, 11
Export value chain, 308
External analysis, 21–22

Facebook.com, 232–233
Fairtrade, 382
Fairtrade Fortnight, 382
Farming costs, 338
Farm plan, 87, 93
Fast follower, 55
Fast moving consumer goods (FMCG), 278
Federal Alcohol Administration (FAA) Act, 182
Federal Bureau of Alcohol, Tobacco and Firearms, 390
Federal Bureau of Tax and Trade, 390–391
FedEx, 227
Fermentation, 96, 99–101, 115
Fermenters, rotary, 112
Fetzer Winery, 44, 214
FIFO, 357
Fighting varietals, 97
Filtering, 107, 112

INDEX

Filters, 137
Financial aspects of wine, 313–331
Financial decisions, 315
Financial foundations, 314
Financial objectives, 47
Finding a broker, 187–188
Finding a distributor, 187–188
Fining, 107
Finishing, 96, 105
Finland, 213
First Mover, 55
Five competitive wine strategies, 53
Five Ps of wine marketing, 158–159
Five Ss of wine appreciation, 108
FIVS, 410
FIVS Global Wine Sector Environmental Sustainability Principles, 410
Fixed costs, 133–135
Flame wars, 232
Flavor, 78, 97
Flavor profiling, 96, 116–117
Flood irrigation, 76
Freight on board (FOB) pricing, 199
Focus, 189
Foil, 109
Foils, 139–140
Food and Drug Administration (FDA), 398–399
Forecasts, wine, 427–434
Forestry and Fire Services Department, 391
Foster's/Foster's Brewery, 6, 10, 18, 44, 52, 138, 281, 330, 364
Foudres, 100
France, 6, 10–11, 53, 64, 66, 68, 76, 91, 274, 277, 281, 284, 305, 361, 413, 427–428
Franzia, 10, 53
Free run, 102
Free the Grapes, 227
Freight on board (FOB), 199

French barrels, 134
French Paradox, 421
French Rhone Syrah, 263
Fresno county 318
Fully independent tourists (FIT), 252
Fume Blanc, 7
Future forecasts for global wine industry, 427
Future HRM, 382
Future issues, enology, 117–121
Future trends, in wine, 417–434

Gago, Peter, 111–113
Galleron vs. Galleron-Lane, 394
Gallo (of Sonoma), 6, 9–10, 15, 18, 45, 52, 125, 167, 214
Gallons per ton, 81
Gaul, 361
General manager, 371
General market, 193
Generally accepted accounting principles (GAAP), 324, 355
Genetically modified grapes, 113
Geographical indexes, 322
Geographic Indications System, 8
Geographic information system (GIS), 87, 92
Georgia, 361
German, presses, 112
Germany, 6, 11, 141, 215, 424
Gewürtraminer, 67, 309
Glass stoppers, 141
Globalization, 15, 123–124
Global market, 28
Global marketing, 280
Global positioning system (GPS), 82, 84–87, 92
Global wine brand, 117–118
Global wine distribution, 178–179
Global wine market, future of, 311
Global wine sector trends, 275
GMO grapes, 113
Go viral, 232
Goals, long-term, 46–47

Google, 229–231, 233
Gourmet, 268
Government requirements, 212–213
Governor Edmund G. "Pat Brown Award", 409
Governor's Environmental and Economic Leadership Award, 409
Graham Decision, 389–390
Grand Cru Burgundy, 175
Grand cru, 175
Grands Chais de France, 6, 10
Grape and Wine Research Development Corporation (GWRDC), 424
Grape crush report 318
Grape grower, 30
Grape growing, 24
Grape price wheel, 71
Grapes, 66
Grapes per bottle, 81
Grape samples, 87
Grapes, blush, 99
Grapes, buying, 131
Grapes, buying from a grower 131-132
Grapes, red, 99
Grapes, white, 99
Grape varietal, 66–67
Great Experiment, 182
Greece, 361
Greenhouse gas, 415
Green vines, 74
Green, wine, 119
Grenache, 27, 312
Growing grapes, 132
Growth tubes, 75
Gruner Vetliner, 419

Haffner, 226
Hanzell Vineyards, 174–175
Hardy (BRL), 9–10, 15,
Hardy's Bar, 281–282, 285–286, 304

Harvest, 77–78, 87, 96–98
Harvest, by hand, 88–89
Harvester, 85, 92
Harvest Job Recruiting Fair, 376
Hazard analysis, 147
Hazard Analysis and Critical Control Point (HACCP) system, 147
Head pruned, 76
Hectare, 71
Hedge (box pruning), 80
Heineken, 18
Henkell, 6,
Hispanic, 185
Holland, 277
Hotel, Restaurant Café (HORECA), 279
Houston, 294
Human resource management, definition, 372
Human resource model, 373
Human resources, buckets, 373–375
Human resources, categories, 373–375
Human resources, in vineyard, 93
Human resources manager, 372
Human resources, wine, 359–384
Hunter River, 362

Iceland, 213
Idaho, 25, 27
Image-oriented wine drinkers, 164
Image Seekers, 164
Image, brand, 169
Image, brand, consistency, 168–169
Impact, 269
Importing, 210–217
India, 4, 15, 90, 215, 383–384, 422, 427, 433
Indiana, 8
Indonesia, 15
Information sharing, 149
Inglenook, 10

INDEX

Ingredient labeling, 148
In-house, 126
Innovative focus, 50
In-store tastings, 206–207
Integrated pest management, 406, 411–412
Integrated Production (IP) standards, 413
Integration, backwards, 55
Integration, forward, 55
Internal analysis, 21–22, 33, 36
International branding, 171–173
International markets, 214–215
International relationships, 172
International selling, 210–217
International selling tips, 215–217
International Wine Industry Greenhouse Gas Protocol, 415
Internet, 82, 144, 200, 214, 220, 226–227, 229, 375, 425
Internet discussions, 229
Internet marketing, 228–234
Internet sales, 226–227
Interstate shipping, 227
Inventory, 139, 145
Inventory control management, 145
Inyo county 318
IPM Innovation Award, 409
IPW Conformance Certificate, 412
Iraq, 361
Irrigation, 75–76, 83, 85, 131
IRS, 391
ISO, 90, 120, 431
ISO 14001, 90, 416
ISO 9000, 146-147
ISO certification, 120
IT professional, 372
Italy, 6, 10–11, 64, 91, 112, 277, 284, 416, 427

Jacob's Creek, 10, 291
Japan, 4, 215, 421
JIT, 139
Job positions, marketing & sales, 369–370
Job positions, vineyard, 364–365
Job positions, winery 367–368
Job positions, winery management & administration, 371–372
Johannesburg, 403
Johannisberg Riesling, 67
Johnson, Hugh, 16
Joint Committee, 405
JP Chenet, 10
Jug, 97
Juran, Joseph, 146

Kelowna, 250
Kendall Jackson, 14–15, 49, 394, 422
Kentucky, 8, 227
Key account, lunches, 206
Kings County, 318
Kingston Estate, 288
Know your customers, 170
Korea, 15
Korolishin, 18
Koshu, 8
Kotter, John, 45
Kumala, 304, 429
KWV, 11

Lab analysis, 137
Lab assistant, 368
La Baume, 281
Labeling, 96, 108
Labeling, ingredients, 148
Label requirements, 211–212
Label, second, 111
Labels, 109, 139–140
Labels, glue, 140
Labels, multiple, 110–111
Labor issues, 360–361
La Crema, 430
Lactic acid, 105
Ladies Home Journal, 268
Lake County, 92
Land preparation, 72
Land prices, 66
La Tache, 49

Latin America, 172
Leaping Horse, 428
Leelanan Wine Cellars, 394
Legal issues, 380–381
Legal representation, 372
Legalities, wine, 385–399
Lehman, Peter, 287
Library wine, 202
LIFO, 356, 357
Lindemans, 10, 50, 138, 431
Lochness-Ogopogo, 248
Logistics, 179, 211–212, 308
Long's Drugs, 192
Longer term contracts, 149–150
Long-term goals, 58
Long-term strategic goals, 46–47
Los Robles Winery, 383
Low-cost broad, 53
Low-cost broad market, 48
Low-cost niche, 48, 53–54
Loyalty, brands, 169-171
Luce, 397
Lucente, 397
Luxury brands, 168
Luxury wine, 167
LVMH, 18, 52, 330
Lycos, 394

Macroenvironment analysis, 25, 30–31
Madera County, 318
Magill Estate Shiraz, 113
Mail, direct, 225–226
Maipo, 65
Malbec, 8, 67–68, 419
Malic acid, 105
Malolactic fermentation, 102, 105
Management, supply chain, 123–154
Management, vineyard, 76–80
Managing human resources, 359
Manic-depressive wheel of grape prices, 71
Marchesi del Frescobaldi, 397
Marginal wine consumers, 162

Margins, 220
Marin County, 318
Market analysis, 25–26
Marketer, 217
Market, identification, 178–179
Marketing, 155–176
Marketing contract, 71–72
Marketing, definition, 177–178
Marketing plan, 188
Marketing strategy, engagement, 230
Marketing strategy, openness, 230
Markets, emerging, 215
Market target, 48
Market Watch, 269
Marlborough, 65, 78, 300, 312
Marlborough Sauvignon Blanc, 300
Martinborough, 304
Maryland, 200
Massachusetts, 200
Massachusetts Beverage Journal, 269
Master Sommelier, 264
Mateus, 431
Maturing vineyards, 80–81
Maule, 65
Mayacamas, 174
McLaren Vale, 65, 69
Mechanization, 79, 88–89, 92, 383
Mediterranean, the five regions, 64
Mendocino, 64, 92
Mendocino Wine Co., 50
Mendoza, 65
Merced County, 318
Merchandise, tasting rooms, 223
Mergers, 55, 384
Mergers and acquisitions, 327
Merlot, 67, 68, 105, 301, 419
Message boards, 230–232
MetaCork, 141
Mexico, 4, 362, 381
Michelin, 166
Michigan, 389, 394,
Microbiological stability, 108

INDEX

Microclimate, 83
Micro-oxygenation, 105, 112, 115–116
Microsoft, 233
Middle East, 291
Mildare Blass, 281
Millennials, 167, 228, 233
Mis-en-Bouteille, 276
Miss Sandalford, 255
Mission, 58
Mission statement, 43–45
Mobile bottling line, 136–137
Modified Accelerated Cost Recovery System (MACRS), 355
Modigaliani-Miller theorem, 327
Moldavia, 277
Mondavi, 9, 15, 100, 214, 224, 329
Mongolia, 90
Mono County, 318
Monopoly states, 183–184
Monterey, 50
Monterey County, 318
MovableType.com, 231
Muddled strategy, 53–54
Multiple labels, 110–111
Murray River, 286
Muscat, 67
Must, 99
Myanmar, 15
MySpace.com, 232–233

Napa, 64, 67, 92, 100, 111, 125, 330
Napa County, 318
Napa Valley, 49, 65–66, 91, 221
Napa Valley Grape Growers Association, 93
Napa vintners, 125
National Endangered Species Act, 396
National Fish and Wildlife Foundation, 406
National sales director, 203
Native Americans, 361
Natural Resources Conservation Service (NRCS), 353, 406
Neck tags, 207
Negotiation, 132
Net present value, 315
Netherlands, 215
Networking, 375
New vine logistics, 227
New World, 119–120, 141, 168, 179, 274, 276, 279, 361–363, 376, 380, 396, 409, 428, 430
New World grape varietals, 67
New World wine, 1–16
New world wine, countries, 4
New world wine, definition, 1–3
New World wine, drivers, 4–5
New world wine, style, 3
New York, 186, 227
New York City, 13
New York State, upper, 312
New Zealand, 11, 65, 67, 78, 90, 119, 141, 211, 410, 431
News Corp, 233
Newsletters, 225–226
Niagara Peninsula, 312
Niche market, 48
Nike, 118
Nonadopters, 163
North America, 291
North Coast, 64, 91
Northern Europe, 421
Northwest, 25
Norway, 213

Oak, additives, 135
Oak, aging, 104
Oak flavor, 105
Oak staves, 104
Oanzante, 397
Office International de la Vigne et du Vin (OIV), 4, 11, 16, 148
Off-premise, 184, 191–195
Off-premise, distribution, 194–195
Off the grid, 119
Off-trade, 199
Ohio, 8

Okanagan Valley, 65, 236, 248–251
Old World, 68, 274, 279, 363, 409, 427–428, 430
Old World, varieties, 7
Old World wine, style 3
Oliver, 250
OND, 192
One hundred (100)-point system, 264
Online video, 232
On-premise, 184, 191–195, 206
On-premise, distribution, 193–194
Ontario, 65
On-trade, 199
Open-minded, 217
Openness, marketing strategy, 230
Open states, 183–184
Opus One, 111, 397
Oregon, 25, 200
Organic, 89–90
Organic wine, 119
Orlando Wynham, 281, 291
Our Common Future, 402
Outsourcing, 55, 126, 149
Overhead, 342
Overture, 111
Overwhelmed, 166
Oxidation, 138
Oxygen, 105

Pacific Gas and Electric Co. (PG & E), 406
Packaging, definition, 159–160
Palliser Estates, 304
Pauillac, 49
Payment, 180
Payments, patience, 212
Payroll costs, 345
Peer reviews, 229
Penaflor, 6
Penfolds, 266
Penfolds Grange, 110, 111–113, 287
Pennsylvania, 387
Pennsylvania Beverage Journal, 269

Pepsi-Cola, 173, 281, 432
Performance review, 189
Pernod Ricard, 6, 9–10, 18, 281
Peroxide, 142
Personal data assistant (PDA), 86, 425
Personality, brand, 168
Perth, 237, 252, 255
Perth Convention Bureau, 256
Peru, 4
Peter Vella, 10
Petite Sirah/Durif, 8
Petite Syrah, 67
Petite Verdot, 68
pH, 70, 78
Phases of strategy, 22
Phillips, Dan, 287
Pinot Gris (grigio), 67, 80
Pinot Noir, 20, 54, 67, 174, 301, 309, 419
Pinotage, 8, 67
Placement, definition, 159
Planting vines, 73–74
Plymouth Arrow, 393
Point of sale (POS) material, 207
Polyethylene terephthalate (PET), 138, 428
Polyvinyl chloride (PVC), 138
Pomace, 102
Popular premium, 97
Porter, 48
Porter's Five Forces, 25, 32–35
Porter-Cologne Water Quality Control Act, 395
Portfolio strategy, 48, 51–52
Portugal, 274, 277, 284
Positioning, 166–167
Positioning strategy, 167
Postbottle maturation, 110
Pounds per vine, 81
Premium, 97
Premium wine drinkers, 163
Prendiville, Peter and Debra, 251
Preparation, 216
Presses, 135–136

INDEX

Price, definition, 158
Price per ton, 70
Prices, vineyards/grapes, 93
Prices, wine categories, 97
Pricing, 98, 129, 150, 296, 302
Pricing strategy, 167
Primativo, 12
Product costing, 334, 336–337
Product, definition, 158
Product line expansion, 55
Production costs, 347
Production, wine, 13, 129–130
Professional selling, 198–199
Profits, 164–165, 211
Prohibition, 182, 386, 387
Promotion, definition, 159
Promotion strategies, 306
Proteins, 107-108
Pruning, 80
Public Certificate of Label Approval, 394
Public Health Security and Bioterrorism Preparedness and Response Act, 148, 398
Public relations (PR), 206, 230, 259, 266, 271
Publicly traded wine companies, 330
Pumping over, 101
Punching down, 101

Quality, 80, 172
Quality control, 109, 145–148
Quality control, management, 145–148
Quality control, systems, 145–148
Quality focus, 49
Quality/value equation, 120

Rabobank, 6
Rack and return, 101
Racking, 101, 105,
Radio frequency identification, 86
Rain, 79
Rapel, 65

Ravenswood, 170, 329
Real estate markets, 321
Record keeping, 380–381
Record keeping, legal issues, 381
Refrigeration, 108
Regrafting, 80–81
Regulations, 7
Regulations, direct wine sales, 227–228
Regulations, grape, 67–68
Regulations, varietal amount, 106
Relationship, marketing via Internet, 228–234
Relationships, 216–217
Relationships, international, 172–173
Reliability, 216
Renmano, Berri, 9
Repeat of Prohibition, 386
Restaurants, 206
Restaurant Wine Magazine, 264
Retailers, 206
Reverse osmosis, 114–115
Rex Goliath, 428
RFD, 431
RFID, 86, 150–154
RFID, benefits, 151
RFID data flow system, diagram, 153
RFID tag, 150
RFID technology, 150–154
Riesling, 67, 301, 309, 419
Rio de Janeiro, Brazil, 403
Ritual-oriented wine drinkers, 163–164
Ritz Carlton, 264
Riunite, 10
Robert Mondavi Corp., 397
Robert Parker, 201, 264, 287
Robinson, Jancis, 2
Robots in vineyard, 84
Role of wine media, 262
Roman Empire, 361
Romans, 361
Rootstock, 73–74

Rosemount Estate, 10, 49, 422
Rossi, Carlo, 10
Rotary fermenters, 112
Russia, 13, 422
Russian River, 65, 320
Rutherford, 65
Rutherford Bench, 49, 91

SA 8000, 416
Sabmiller, 18
Sacramento County, 318
Safeway, 111, 169, 192
Salary surveys, 376
Sales channels, 199–200
Sales, direct, 219–234
Sales, direct to consumer, 219–234
Sales meetings, distributor, 205–206
Sales plan, 188, 200
Sales plan, checklist, 202
Sales plan, creating, 200–201
Sales plan, evaluation & revisions, 209–210
Sales plan, implementation, 203-209
Sales plan, revisions, 209-210
Sales price, 211
Sales, tracking, 202–203
Sales, wine, 197–217
San Benito County, 318
San Francisco Chronicle, 264
San Francisco Wine Competition, 225
San Joaquin County, 318
San Luis Obispo County, 318
Sandalford, 236, 248, 252–253
Sandalford Cellar Door and Emporium, 252
Sangiovese, 67, 105
Santa Barbara, 330
Santa Barbara County, 318
Santa Rita Winery, 110
Sante, 269
Santiago, 64
Satellite receivers, 84

Satisfied Sippers, 165
Sauvignon Blanc, 67–68, 80, 160, 309, 312, 419
Savvy shoppers, 165
Scandinavia, 277, 295, 305
Schloss Wachenheim, 6
Screw auger, 136
Screw cap, 141
Search firms, 375
Sebastiani, 9
Second label, 111
Section 2 of the 21st Amendment, 387
Segment analysis 324
Selling, professional, 198–199
Selling tips, international, 215–217
Selling wine, online, 233
Semillon, 67–68, 309
Sena, 397
Sensor technology, 82, 84, 86–87
Shanken, 9–10
Shared premise, 130
Shelf life, 138
Shelf space, 210–211
Shelf talkers, 207
Shipping, states allowed, 225
Shipping, wine, 144–145
Shiraz, 7, 67, 113, 263, 286–287, 300, 312, 431
Shiraz-Cabernet, 301
Shiraz-Grenache, 301
Shock, bottle, 142
Sicily, 282
Sierra Foothills, 24, 37, 40
Sight, of wine, 108
Signature varietals, 7–8
Silver Oak, 49
Sin tax, 14
Single varietal, 106
Sip, 108
60 Minutes, 421
SKU, 152
Slap and ship system, 153
Smell, 108
Smith, Peter A., 286

INDEX

Smoking Loon, 428
Social media, 230–231
Social media, components, 230–232
Social networking, 220, 228–234
Social networking sites, 232
Social networks, 231, 232
Social platform, 228
Social Responsibility Assessment, 382
Social responsibility, issues, 401–416
Social wine drinkers, 164
Soil, 64, 70
Soil sample report, 70
Sommeliers, 208
Sonoma 330
Sonoma County, 65, 67, 125, 129, 168, 174, 226, 318, 320,
South Africa, 4, 6, 8, 11, 64–65, 67, 69, 141, 275, 277, 281, 284, 304, 312, 361–362, 397, 381–382, 403, 410
South Africa Integrated Production of Wine System, 411, 413
South America, 312, 382
South Coast, 64
Southcorp, 9, 15, 329
Spain, 11, 64, 105, 274, 284, 419, 432
Spanish, 93
Special events, direct to consumer, 224
Special events, marketing, 224
Special events, selling, 224
Spinning cone, 114–115
Spit, 108
Spraying, 85
Spring in the Valley Committee, 255–256
Sri Lanka, 15
Stabilization, 96, 103, 105
Staffing and recruiting, 374
Stag's Leap, 329
Stanislaus County, 318

Star system, 264
States, shipping, 225
State Water Resources Control Board, 395, 396
Staves, oak, 104
Steiner, Rudolf, 89
Stellen Bosch Wines, 282
Stelvin, 141
Stocks, wine, 330
Stonegate, 225
Stonewall Vineyards, 287
Storage, 96, 109–110, 142–143
Strategic direction, winemaking, 97
Strategic group map, 28–29
Strategic objectives, 47
Strategic options, 54–55
Strategies, definitions, 53
Strategies, wine business, 47–50
Strategy, 17–62, 167, 178
Strategy, commodity, 127
Strategy development phase, 43–44
Strategy, development process, 21
Strategy document, 47, 56–60
Strategy implementation and evaluation, 22, 58
Strategy, levels, 20
Strategy, muddled, 53–54
Strategy reassessment, 22–23, 60–61
Strategy, wine business star, 22
Strategy, wine, definition, 19
Stryker, 225
Stuck fermentation, 100
Successful branding, 161–162
Succession planning, 377–378
Sucker, 77
Sulfur dioxide, 108
Supplier certification, 146–147
Supplier, winery, 25
Supplies, 127–130
Supplies, common winemaking, 137
Supply chain, 123–154

Supply chain, definition, 124
Supply chain, diagram, 125
Supply chain management, 123–154
Supply chain management, definition, 124–125
Supply chain management trends, 148-149
Support staff, 372
Supreme Court decision, 227–228
Supremacy clause, 387
Sustainability, 118–120
Sustainability assessment, 382
Sustainability, definition, 402–403
Sustainability mission, 407
Sustainability values, 408
Sustainability vision, 408
Sustainable Winegrowers New Zealand (SWNZ), 410, 413
Sustainable winegrowing practices, 414
Sutter Home, 10
Swallow, 108
Swan River, 252
Swan Valley Branding Campaign, 255
Swan Valley Food and Wine Trail, 256
Sweden, 29, 213
Swirl, 108
Switzerland, 413
SWOT Analysis, 36, 38, 40–43
Sydney, 362
Syrah, 309
Syria, 361

Taiwan, 15
Talca, 64
Tanks, 100, 103–104, 133–134
Tannat, 8, 24, 27–28, 32–34
Tannin management, 115–116
Tariffs, 211
Tartrate, 108
Tasting room/hospitality manager, 370

Tasting rooms, 144, 222–223, 233
Tastings, in-store, 206–207
Tavernello, 10
Taxes, 211
Tax issues, 351
Tax laws, 14
Tax, wine, 351–357
TCA, 102
TCA taint, 102
Tea bags, oak, 104
Technology in vineyard, 81–85
Technology, sensor, 82, 84, 86–87
Technology, winemaking, 114–117
Telemarketing, 226
Telemetry, 82
Telephone sales, 226
Temperature, 100, 142–143
Tempranillo, 27, 67, 105, 419
Terrior, 97, 118
Tesco, 111, 281, 431
Texas, 239
Thailand, 4, 15
Thompson, 48
Three-tier system, 125, 143, 183, 201, 388
Thunder Mountain Winery, 394
Thunderbird, 394
Ticker symbols, 330
Tierd House, 387
Tiffany, 158
Time management, 216
Tips, international selling, 215–216
Toast, oak, 104
Top wine producing countries, 11
Torres, 15
Total quality management (TQM), 149
Tourism Council Western Australia (TCWA), 256
Tracking sales, 202–203
Trade, 191
Trade issues, 397
Trade, section on website, 191
Trade shows, 205
Trade visits, 208

INDEX

Trader Joe's, 10, 266
Traditionalists, 165
Training and development, 377
Training, leadership, 377
Training, managerial, 377
Training, safety, 377
Training, technical, 377
Training, wait staff, 207
Traverse City, 394
Trellis, installation, 75–76
Trends, 419, 426
Trends, supply chain management, 148–149
Trinchero Family, 10
Triple bottom line, 94
Tulane county 318
Turkey, 361
Turning Leaf, 394
Turnover, 380
Turrentine, 71
TV, 425
Twenty-first (21st) Amendment, 182, 386–389
Twenty (20)-point UC Davis system, 264
Two-Buck Chuck, 266
Ty Comstock, 357

Ultra-filtration, 107
Ultra-premium, 97
United Kingdom (UK), 111, 214–216, 398, 422
Unicap, 353–354
Union activity, 380
Unique attribute, 49
United Arab Emirates (UAE), 256
United Nations Conference on Environment and Development, 403
United Nations Environment Program Global Reporting Initiative, 414
United Nations World Summit on Sustainable Development, 403
United States (USA), 4, 6, 10–11, 24, 29, 53, 67, 76, 88–90, 110–111, 120, 148, 164, 178, 181, 187, 191, 212, 214, 216, 277, 281–282, 287, 291, 362, 377, 381, 384, 386, 398, 422, 426–427
United States Bioterrorism Act, 398
United States Constitution, 386
United States Department of Agriculture, 399
United States Department of Food and Agriculture (USDA), 406
United States Patent and Trademark Office, 391–392
United States Supreme Court, 14
United States Tax and Trade Bureau (TTB), 391
Unwooded, 105
UPC, 211
UPS, 227
Uruguay, 4, 8, 24
USDA Risk Management, 406
US dollar, 180, 214
User-generated content, 228–229
UST, 330
Utah, 387

V. Sattui, 221
Value chain analysis, 36–38
Value chain, grape grower, 37
Value chain, wine industry, 36
Values, 45–46, 58
Varietal, 66
Varietal, single, 106
Varieties, unusual, 221
Vats, 100
Vendor managed inventory (VMI), 149
Ventura County, 318
Veraison, 77–78
Verdelho, 8
Veritas Winery, 287
Vertical shoot positioned, 75
Vermont, 200

Veuve-Cliquot, 49
Viansa, 221–223
Vidal Blanc, 67
Vietnam, 15, 215
Vina Errazuriz, 397
Vincor, 329
Vine pricing, 74–75
Vines, planting, 73–74
Vineyard accounting worksheet, 343
Vineyard acquisitions, 94
Vineyard, biodynamic, 89–90
Vineyard databases, 85–86
Vineyard designate, 68
Vineyard development 316
Vineyard, growth, 90
Vineyard, layout, 72–73
Vineyard maintenance, 77–80
Vineyard management, 76–80
Vineyard manager, 365
Vineyard operations 325
Vineyard, organic, 89–90
Vineyard planting, 69–70
Vineyard selection, 64–65
Vineyards, maturing, 80–81
Vineyard workers, 365
Vintage, 202
Viognier, 67, 116
Viral marketing, 228–234
Virginia wine industry, 45
Vision, 44–45, 58
Viticulture, 63–94
Vitis vinfera, 67, 74
Vlogs, 57, 220, 230
Volvo, 166
VP of Marketing, 369
VP of Vineyard Operations, 367
VP of Vineyard Relations, 367

Wadenswill, 410
Wait staff training, 207
Wal-Mart, 150–152, 431
Warehousing, 180
Washington, 25, 200, 390
Washington State Legislature, 390

Water, 83
Weather, 79
Weather monitoring equipment, 82
Weather patterns, 82
Weather stations, 83–84
Web 1.0, 228
Web 2.0, 228
Web 2.0, growth, 233–234
Websites, 230
Wente, 214
West Virginia, 8
Western Management Salary Survey, 364
Wholesaler, role, 183–185
Wildlife, 119
Williamette Valley Vineyards, 330
Wine 2.0, 220, 228–234
Wine 2.0, advent of, 228–229
Wine 2.0, definition, 228–229
Wine and Spirits Magazine, 264, 268
Wine and Spirits Wholesalers Association (WSWA), 187
Wine appreciation, 108
Wine bar, 25, 30
Wine brand position, 166–167
Wine branding, definition, 157
Wine business concept, 24
Wine business strategies, 47–50
Wine business strategy star, 22
Wine business supply chain, 125
Wine Capital of Canada, 250
Wine club, benefits, 225
Wine clubs, 224–225
Wine club, sign-ups, 225
Wine consumers, 167
Wine Country Welcome Centre, 251
Wine dinners, 207–208
Wine distribution, 177–195
Wine distribution, in the US, 181–182
Wine drinkers, basic weekend, 164
Wine drinkers, image oriented, 164

INDEX

Wine drinkers, premium, 163
Wine drinkers, ritual-oriented, 163–164
Wine drinkers, social, 164
Wine economics of finance 323
Wine Enthusiast, 268
Wine/grape source, 128
Wine Group, 6, 10, 53
Wine Institute, 187, 214, 227
Wine journalist, 263
Wine labor issues, 360–362
Wine, library, 202
Wine lists, 194
Winemaker, 367
Winemaker, assistant, 268
Winemaking, 95–120, 130–131
Winemaking equipment, 133–135
Winemaking process, 96–97
Wine Market Council, 162–163
Wine marketing, definition, 157
Wine media, 259–272
Wine media future, 271
Wine of the Year Award, 266
Wine press, 190
Wine price, categories, 97
Wine production, 129–130
Wine public relations, 259–272
Wine rating systems, 264
Winery investments, 321
Winery stocks, 330
Wine sales, 197–217
Wine sales, definition, 198–199
Wine sales, direct, 219–234
Wine sales, language, 213
Wine sales plan, evaluation, 209–210
Wine scores, 264
Wine shipping, 144–145
Wine Spectator, 201, 264, 266, 268
Wine strategy, definition, 19
Wine tourism, 235–257

Wine tourism, benefits, 237–238
Wine tourism, definitions, 236–237
Wine tourism development process, 242
Wine tourism, partners, 243
Wine tourism, programs, 245, 253–257
Wine tourists, motivation, 238–240
Wine trade, 190
Wine trends, 417–434
Wine Village Accord, 251
WineVision, 399
Wine writers, 175
Wine X Magazine, 264
Wolf Blass Winery, 103, 138
Woodbridge, 10
Word of mouth, 229
Wordpress.com, 231
Working with brokers, 188–191
Working with distributors, 188–191
World Commission on Environmental Development, 402–403
World market share, 18
World Trade Organization (WTO), 12, 275
Wyndham Estate, 291

X System, 264

Yeast, 100, 137
Yeast, native, 100
Yellow Tail, 10, 282, 422, 428–429
Yield, 70, 131–132
YouTube, 232

Zinfandel, 8, 12, 66–67, 73, 129, 419
Zoning regulations, 68–69
Zork, 141

Manufactured by Amazon.ca
Bolton, ON